# TOO HOT
# TO HANDLE

## *The Race for Cold Fusion*

FRANK CLOSE

PRINCETON UNIVERSITY PRESS

PRINCETON, NEW JERSEY

Published by Princeton University Press, 41 William Street,
Princeton, New Jersey 08540

First published in Great Britain by W. H. Allen Publishing

This book has been composed in Palatino

Close, F. E.
    Too hot to handle : the race for cold fusion / Frank Close.
    p.   cm.
    Includes bibliographical references and index.
    ISBN 0–691–08591–9
    1. Cold fusion. 2. Cold fusion—Research—Utah. 3. Fleischmann, Martin.
    4. Pons, Stanley. I. Title.
QC791.775.C64C56   1990
539.7′64—dc20                                                        91–11378
                                                                          CIP

Princeton University Press books are printed on acid-free paper, and meet the
guidelines for permanence and durability of the Committee on Production
Guidelines for Book Longevity of the Council on Library Resources

Printed in the United States of America by Princeton University Press,
Princeton, New Jersey

10  9  8  7  6  5  4  3  2

*To the xxxxx from MIT
and the friends and colleagues
who shared the spring of 1989*

# CONTENTS

Preface                                                          1

1. The Greatest Discovery Since Fire                             5

*PART ONE: Genesis*
2. Nothing New Under the Sun                                    17
3. The Sun on Earth                                             31
4. Cold Fusion                                                  52
5. The Chemists                                                 70
6. The Dispute                                                  83
7. Harwell                                                     105

*PART TWO: Deuteronomy*
8. The First Reactions                                         121
9. The Parting of the Waters                                   146
10. Money                                                      172
11. The Caltech Story                                          192
12. From Spring to Fall                                        211
13. International Reactions                                     233
14. Test-Tube Fusion: Science or Non-Science?                  254

*PART THREE: Revelations*
15. The Spy in the Lab                                         275
16. Credibility                                                289
17. 'It's Not Fusion'                                          302
18. The First Anniversary                                      317
19. Assessment                                                 327

Appendices:
  Excess Heat in Calorimetry                                   351
  Fusion Does NOT Give the Earth's Heat!                       354
Notes                                                          357
Index                                                          369

'If an experiment requires statistical analysis to establish a result, then one should do a better experiment.'

*Ernest Rutherford*

'In all the cases reported, the presence of an element has been mistaken for its creation.'

*Chadwick, Ellis and Rutherford (1930)*

# PREFACE

On 23 March 1989 two chemists, Stanley Pons and Martin Fleischmann, stunned the world with their claim to have harnessed fusion—the power of the Sun—in a test tube of water at room temperature. Coming within a day of the *Exxon Valdez* oil disaster in Alaska, and with the nuclear catastrophe of Chernobyl still fresh in people's minds, the sudden possibility of cheap, abundant, and pollution free energy captured everyone's attention.

At the time I was commuting regularly between Oak Ridge National Laboratory in the USA and the Rutherford Appleton Laboratory in England. Thus I saw at first hand the developing excitement within the scientific community on two continents as thousands of us changed research programmes literally overnight and attempted to replicate the phenomenon. As I became professionally involved in trying to understand what, if anything, was happening I realised that here was a rare opportunity for the public to experience science in action, feel the excitement that drives inquisitive minds, and see how discoveries are made, tested, replicated, proven and developed into a new technology. Or if, as was soon clear to many, something was wrong with the claims, it would be possible to show how the standard workings of science can design experiments to answer specific questions, strategically closing in on the truth. It would also show scientists as real people, with urges for power and glory, desirous to be first in the race and to win the prizes that this offered.

What I did not anticipate was that this would prove to be a matter as much of intrigue and scandal as of science. The former is itself a fascinating tale that will surely be told in full elsewhere, of how interested parties helped to orchestrate the episode and stimulate financial investments in ways that have been widely criticised in the scientific community. This led to calls for the resignation of the president of the University of Utah, and suggestions of unethical behaviour.

When Galileo found results that displeased the Pope, his reward was imprisonment. A group of scientists whose data did not agree with the claimed 'cold fusion' received threats of legal action. When

1

I made a television documentary that concluded that there was no evidence for nuclear fusion products, I received a letter from Martin Fleischmann alleging that there was a warrant issued in Utah for the arrest of the film crew, accusing me of having gained 'illegal access to the National Cold Fusion Institute' and of being a media person masquerading as a scientist. My colleagues may offer their opinions on the latter; I can only assert that the first accusation is nonsense and if the allegation is true then I would love to know more about it but, regrettably, my request for further information has been met with silence.

There has been much exposure during 1989–90 of maverick behaviour while the face of serious science has been rather lost. There were instances where some results were presented with a gloss they did not merit, and worse. The editor of *Nature* among others has expressed concern that the young will equate some of the farce of test-tube fusion as typical of science and be discouraged from pursuing this career.

To be concerned is not simply pedanticism. Society relies on science and makes substantial investment in it, not least from you the taxpayer, in the trust that claimed discoveries have been thoroughly and carefully researched. Parts of the test-tube fusion episode failed sadly on this score. If these events became regarded as a norm for science then public confidence would be threatened. It is important that the public see that the test-tube fusion story is *not* typical of normal science and that there is a received body of opinion in the scientific community that the episode was unacceptable.

I have divided the book into three parts. Genesis and Deuteronomy tell the story in a style that will be accessible to any reader. In Revelations I provide additional details which many scientifically-minded readers will demand although its content is still accessible to the general reader. Some references have been added for those who wish to check on detail or proceed further although this work has not been written as a formal review of cold fusion.

The subject matter focuses on the work of Fleischmann and Pons for it was their claims that excited the interest and fuelled the passions. For example, I have not concentrated much on the work of Steven Jones and related experiments that claim to see neutrons at levels near to background; if real, they are of interest to science but have nothing to offer as solutions to the world's energy problems, at least not in the grandiose form claimed on behalf of Fleischmann and Pons. The media often presented Jones' results as supporting those of the two chemists whereas in due course, if not already, his work will be seen as the first *refutation* of some of their claims; his neutrons were orders of magnitude below theirs and he never made any direct measurements of heat.

2

In the final chapter I give a summary. This is a personal perspective which is a considered opinion drawn from eighteen months of researches and over a hundred interviews, and built upon my own experiences over twenty years of what constitutes good science. Any professional, in whatever career, gains an intuitive 'feel' for what their specialty is, how it operates, intuitively knowing when things are wrong. Scientists, especially theoretical physicists, are all awaiting the next revolutionary breakthrough that overthrows the paradigms of theory and practice. When a radical new result arrives, see how they all drop what they were doing and pursue the new promise. That is what happened when the news of cold fusion erupted. Even though their intuitive 'feel' sensed that it was too far fetched to be real, nonetheless it had to be checked. Had there been a verification that nuclear fusion did occur in solids at room temperature, here is one theorist, and I am sure I speak for many colleagues, who would have taken up the new field to see what fundamental implications it had. We crave new discoveries and the attendant excitement and promise—that is what drives many of us into science in the first place. The idea that established science was somehow attempting to censor cold fusion research is utterly out of line with what science and scientists are all about.

An art in science is to judge when an outlandish notion is worth pursuing and when to ignore it. As students we read the textbooks, attend the lectures and are guided into research. But I have never forgotten advice that my research adviser gave me the first time that I was pursuing a dead end: 'It is important to recognise when to quit.'

Given the detail of the account recorded here, we can safely say that it is time to move on, to pursue other routes for harnessing fusion. I hope that this book provides others with insights, thereby enabling a scientific consensus to emerge.

I am grateful to my colleagues at Oak Ridge National Laboratory in the United States and the Rutherford Appleton Laboratory in Britain for the help and opportunities they have given me in the preparation of this book. However, I should emphasise that the conclusions and opinions are my own and not necessarily those of the management of either institution. There has been one undoubted benefit of the cold fusion saga: it has stimulated interactions among scientists from different disciplines and thereby brought together expertise that might otherwise have remained untapped on individual islands. I have experienced this directly in researching this book and am deeply indebted to many new friends, in areas of science that would have remained foreign to me, for their time and patience in explaining things that I never knew and I hope that we all gained from the experience. In particular I would like to thank:

3

W. Appleton, D. Bailey, T. Barnes, C. Batty, G. Baym, D. Beck, A. Boerdel for Canadian news reports and transcripts of press conferences, H. Bergeson, M. Brauer for tapes of the Erice meeting, P. Bond, The British Council; The British Association for the Advancement of Science, R. Bullough; P. Burge and J. Callender of *BBC Horizon*, J. Cameron and Indiana University video, V. Cate,

C. Cookson of *Financial Times*, R. Highfield of *Daily Telegraph* for archive material and general thanks to members of the scientific media in Britain for their help in researching cold fusion,

J. Davies, C. de Tar, G. Smith and J. Dietrich of *Caltech Engineering and Science Magazine*, J. Dismore and *Yorkshire Television*, M. Edwards, V. Eden, S. Fain, K. Farrel, G. Fraser, H. Furth, M. Fleischmann, Sir Charles Frank, M. Gai,

L. Garwin, D. Lindley and J. Maddox at *Nature*

R. Garwin, C. Goodman, G. Garvey, T. Taddeucci and Los Alamos video, T. Greenland, T. Haywood, A. Haymet, R. Hoffman, C. Horowitz, M. Hawkins, D. Hutchinson, J. D. Jackson, S. Jones, M. W. Johnson, P. Kalmus, S. Koonin, N. Lewis, F. Lizzi, E. Loh, T. Londergan, C. Martin, D. R. O. Morrison, Sir Mark Oliphant, P. Palmer, R. Pease, R. S. Pease, R. Petrasso, R. Phaneuf, L. Ponomarev, N. Ramsey, B. Rose, M. Salamon, M. Saltmarsh, J. Schiffer, G. Seaborg, Bikash Sinha, C. Scott, Patrick J. Smith, E. Taylor, P. Vannier, Yogi Viyogi, A. Watson, P. J. S. Watson, Sir Denys Wilkinson, D. Williams, E. Wrenn, Wu Lei. I am indebted to Brian Gee, Gill Gibbins, Jane Gregory and Sheila Watson for their help in bringing the book to completion.

Of course, I will be more than happy to hear from any reader who may wish to point out substantive errors or fill in gaps to the story as I have perceived and written it.

Frank Close
Indonesia, August 1990.

# 1

# THE GREATEST DISCOVERY SINCE FIRE

In a good adventure story the heroes are going to change the world almost single handedly against all the odds, and come through unscathed. The tale has added bite if its characters appear to be real, its action plausible, if it could 'happen to me'. The following story seems to be far-fetched, unreal; yet it happened.

Civilisation is now threatened with possible destruction: the atmosphere is heating up and the seas are rising, trees are dying from acid rain, a hole in the ozone layer appears over the South Pole. Pollution from the dregs of an ever increasing energy consumption threatens to poison everyone in their own waste. Nuclear power, once thought to be a way to buy time while fixing all of the above problems, has become discredited following the Chernobyl disaster. The only real hope for mankind in the long term appears to be fusion—the power of the Sun harnessed on Earth—because it is clean, cheap and available.

On four continents scientists have consumed the equivalent of billions of dollars trying to capture the dream. Teams from the whole of Europe concentrate their work at a single lab in southern England where a huge torus of magnets containing gas at 100 million degrees tries to reproduce the Sun's power. In the USA big name institutions are in the race: Princeton, Los Alamos, MIT. In the USSR and Japan, national teams are also pursuing the dream, but to many observers it seems to be just that—a dream that may not be fulfilled.

Governments are beginning to question whether they can continue to afford this.

Meanwhile in a basement laboratory in Utah two chemists are working on their own and in secret. They announce that they have found the holy grail. It is so simple that anyone can do it, but everyone else has missed the key. There is no need for extreme temperatures of hundreds of millions of degrees nor for expensive magnets: a battery, some water and *palladium metal*, the key ingredient, are all that you need, plus a few hundred dollars to buy the stuff. In addition, all that is required is a chemical room-temperature mechanism which involves nothing more complicated than an experiment in a high school science lab. Fusion in a beaker of water at room temperature: cold fusion or 'test-tube fusion' as it becomes known.

The details remain secret until all the patents are sewn up. Teams of security staff police the laboratory, guarding the precious property, protecting what potentially could lead to the greatest fortune ever; the combined wealth of Rockefeller, Vanderbilt and Onassis is a mere drop in the ocean compared to the promise of test-tube fusion. The two chemists are cover stories in *Time*, *Newsweek*, appear on TV nightly, and their talks are attended by thousands with the fervour of an evangelical meeting. Nobel Prizes beckon.

Scientists all over the world drop what they're doing and try to discover the secret for themselves. From Hungary, Russia and Japan come rumours that they can do it too. The President of the USA and other world leaders demand to know what is going on and government scientists are urged to get to the bottom of this—and fast.

Military strategists in the Pentagon, in the British Ministry of Defence and in India independently realise that table-top test-tube fusion should produce not just energy but also neutrons and tritium, crucial ingredients in making atomic and thermonuclear weapons. National security agents in the USA start watching the two chemists. The number of countries that have H-bombs is limited because the technology for making tritium is so hard to come by. But with test-tube fusion, any tinpot dictator will be able to do it. Middle Eastern states, whose oil based economies could be destroyed by test-tube fusion, will have nothing more to lose and could exploit the weapons potential of test-tube-fusion devices to keep the oil burning. India could develop test-tube fusion cells, using them as a source of neutrons to make weapons grade materials, breaking out of the stranglehold that the USA has held them in since severance of nuclear ties following the Pokran nuclear test in 1974; they set to work at once before the US classifies it as secret and corners the world market on the essential materials. This raises another spectre: as palladium is the key, what power brokerage will it give the countries which have it as a natural mineral resource—USSR and South Africa?

There are many rumours that people in other countries are reproducing test-tube fusion, but in American government labs they are getting nowhere; however, many people do not believe them and assert that the big laboratories are seeing the phenomenon and keeping quiet because the oil companies with so much to lose have bought their silence. In India too people are convinced that the Western nations are seeing test-tube fusion and are keeping it a secret because they have realised that not only does test-tube fusion have great commercial potential, but could revolutionise military strategy. Some US congressmen believe that the Japanese are secretly developing test-tube fusion technology and that the US is already losing the race for the economic bonanza of all time.

Finding out about test-tube fusion quickly becomes a cloak and dagger venture as desperate scientists and others resort to spying to find out the secrets in the Utah lab. The fax machine is bugged and TV videos of their lab are scanned for clues. And gradually the truth becomes clear. In their haste to win a race for patents, mistakes have been made. The two chemists have no nuclear evidence for fusion at all. Governments, military strategists, thousands of scientists, millions of dollars of time and research budgets have all been on a wild goose chase. It is a modern parable of the King and the Invisible Raiment.

If this were an outline for a novel it would probably be rejected as too far-fetched. It would be unreasonable to expect the reader to believe that so many professional scientists and administrators could be taken in to such an extent as the tale required. Surely it is a ludicrous notion to suppose that two scientists on their own in a remote basement lab could claim to solve the world's energy problems so that the scientific community would take up this claim, that world leaders would ask for regular briefings and that there could be nothing in it at the end.

Yet in the spring of 1989 much of this story was enacted in real life. The two chemists were Stanley Pons and Martin Fleischmann who worked in a lab at the University of Utah in Salt Lake City. On 22 March few outside of their immediate circle had heard of them. On 23 March they held a press conference and announced 'We have sustained nuclear fusion . . .' The next day they were celebrities around the world.

Fleischmann and Pons believed that they had produced fusion in their test-tube because when they passed electrical current through it, more heat was produced than they could account for, and in amounts extending over hundreds of hours that seemed to be far in excess of what is possible in a normal chemical reaction. Chemical reactions involve the atomic elements, such as when carbon in coal and oxygen in the air are combined in fire to produce carbon dioxide and warmth. Nuclear reactions by contrast involve changes in the atomic elements themselves, such as the conversion of hydrogen into helium within the Sun, and vast amounts of heat are generated in such processes.

The uncontrolled liberation of fusion energy has existed for 40 years, ever since the detonation of the first thermonuclear or 'hydrogen' bomb. The attempt to harness that power as a controlled source of useful energy has occupied scientists and engineers ever since. Presently available nuclear power comes from the 'fissioning', breaking up, of the atomic nuclei of heavy elements such as uranium and plutonium and has as byproduct both desirable release of energy and unwanted nasty radioactive contaminants. 'Fusion' on the other hand involves the joining together ('fusing') of the atomic nuclei of light elements such as hydrogen. This is a relatively clean source of energy compared to chemical sources such as coal-burning power stations and to presently available nuclear power. Moreover the raw materials are plentiful. For example, there is enough material in a glass of water to run a car for ten years, enough in the top few centimetres of Lake Erie to power New York State for a similar time and enough in the oceans to keep us going indefinitely. The only thing in the way is the fact that all nuclei are electrically charged and 'like charges repel': this repulsion makes it difficult to force those nuclei together.

To force together two protons that are initially far apart, for example those in two separate hydrogen atoms, is like trying to encourage the two north poles of a pair of magnets to join at a distance of less than a billionth of a millimetre. One way to do this is to give them a hefty shove and hurl them together so fast that their mutual electrical repulsion cannot resist, as in the Sun where the temperature at the centre is so great that the protons are highly agitated and occasionally collide at which point the energy-liberating fusion mechanisms take over.

As you get more energy back when two light nuclei fuse than it costs to get them together it would seem that we are bound to win. However a problem is that these nuclei are very small and only a few manage to collide and fuse. You expend energy speeding millions of protons so that they can resist their mutual repulsion, but only a handful actually collide and fuse giving energy back. The net result is that the total energy return is less than the input. Huge pressures and temperatures will be needed to make fusion reactors give out more than they use. Indeed, even at the pressures of the centre of the Sun, where the temperature is some fifteen million degrees, the conditions aren't extreme enough. The Sun only just manages to keep alight and it tends to be those few protons that are moving faster than average that keep it going—which is just as well or it could have used all its fuel and expired before we arrived.

Overcoming this *Catch 22* has cost modern fusion research billions of dollars so far, using arrays of magnets as big as houses in order to contain the fuel at temperatures ten times hotter even than those in the centre of the Sun, and making it dense enough and stable enough that a self-sustaining reaction can occur. There has been great progress in recent years in this quest, but its attainment is still thought to be decades away and the cost effectiveness of this research is being increasingly questioned.

Thus it was that the world took such note of Fleischmann and Pons' claim to have found a novel approach. They knew that palladium soaks up deuterium (a form of hydrogen found in heavy water) like a sponge soaks up water, an electrical current from a battery forcing the deuterium atoms out from a solution of heavy water and into the spaces between the palladium atoms. Deuterium is the most commonly used fuel in attempts to create controlled fusion and the two chemists believed that once the deuterium was crammed inside the palladium within their test tube, the nuclei of the deuterium atoms would undergo nuclear fusion generating heat and either being transformed—as in all nuclear reactions—into new elements like helium or being shattered into their constituent parts, such as protons and neutrons.

If fusion was really taking place, then these products should be found and from the amount of heat measured one could determine the amount of fusion products that should have been seen; the answer was calculated to be a staggering thousand billion neutrons per second. While the media perceived test-tube fusion to be a solution to the world's energy problems, it was the possibility of these cells being an intense source of neutrons, and probably of tritium too, that meant that they could have signifiant military application.

In advance of the press conference the University of Utah had issued a news release announcing that 'A Simple Experiment results in Sustained Nuclear Fusion . . .'. The release included statements that 'the discovery will be relatively easy to make into a useable technology for generating heat and power'; and that 'this generation of heat continues over long periods and is so large that it can only be attributed to a nuclear process'. The announcement came within hours of the *Exxon Valdiz* disaster where the coast of Alaska was devastated by oil spillage. Everyone's environmental awareness was sharpened: 'Pollution free energy from fusion' was the cry; if there was a time ripe for exploiting the story, Easter 1989 could hardly have been bettered.

A premature release of the story in London, picked up by the BBC World Service and the news wire services, led to the University of Utah being inundated with eager news reporters for the press conference. Television cameras, radio microphones and hundreds of fascinated spectators brought to remote Utah the sort of attention usually reserved for an important visiting head of state in Washington or New York.

And if what they were hearing was really true, then this was indeed more important than most of those East Coast press conferences. Pons,

9

the chair of the university chemistry department, told them 'we have sustained a controlled nuclear fusion reaction by means which are considerably simpler than conventional'. He also told them that a thin palladium wire, only ¼ inch in diameter and an inch long, had reached the boiling point of water within a few minutes, that the wire produced about 26 watts of energy per $cm^3$, 'about four and a half times what we put into it' and that in an early stage of the experiments the apparatus suddenly heated up to an estimated 5000 degrees, vaporising a block of palladium, destroying a fume cupboard and damaging the concrete floor.

Martin Fleischmann, Fellow of the Royal Society of London, Britain's most prestigious scientific society through whose halls had walked Isaac Newton and Ernest Rutherford, held up a fist-sized tube and said 'This experiment has to be approached with some caution.' The audience laughed nervously but Fleischmann remained serious: 'If this device worked as effectively as the small scale experiment we have done, it would be generating about 800 watts of heat.'

The reporters were also told that 'evidence that fusion was taking place was the fact that in addition to heat [Fleischmann and Pons had] detected the production of neutrons, tritium and helium—the expected by-products of fusion reactions'.

There was no reason for the reporters to believe other than what they were being told; the credentials of the pair were first rate. However, Fleischmann and Pons were announcing their results at the press conference before they had told any of their scientific colleagues. There was annoyance when it was learned that the claims had first appeared in the *Financial Times* and the *Wall Street Journal*, both prominent business dailies, not refereed scientific journals. Many scientists were astonished at home that evening to see the faces of the two chemists appearing on the television as Dan Rather on the CBS evening news headlined the Utah work as 'a remarkable breakthrough'. This was the first that the community of professional scientists had heard about it, and it caused the pair much trouble during the subsequent months as it became clear that several of the claims were wrong. But that only became known later, following intense scentific detective work; by the evening of 23 March the discovery of the millenium seemed assured.

Within 24 hours there was another astonishing development. News broke that Steven Jones, a physicist at Brigham Young University (BYU)—30 miles from Salt Lake City and rival of the University of Utah—and a team from BYU had been studying cold fusion for years independently of Fleischmann and Pons. The BYU team had seen neutrons, a sign of fusion, and had discussed the experiment with the two chemists.

Careful examination of Jones' experiment showed that it had less in common with that of Fleischmann and Pons than the media advertised: Jones measured no heat and his neutrons were more than a billionfold too few to explain the amounts of heat that the chemists were claiming. Accusations that one group had stolen ideas from the other were soon flying; Jones showed notebooks of his investigations going back to 1986 and cited a paper written in 1985; for their part Fleischmann and Pons insisted that their effort had begun in 1985, all of which added to the perception that the fusion claims were important and the patents worth fighting for. (In the USA priority is given to the first to invent, not the first to file.) The price of palladium rose.

Utah was suddenly the centre of the world's attention. Local news reporters wrote glowing accounts of a world awash with clean power: 'Goodbye $CO_2$ pollution; Goodbye Acid Rain; Goodbye Greenhouse Effect.' The University seized the momentum that the press conference had generated and lobbied the Statehouse to get money for fusion research. The State authorities did not need much convincing; with the world believing that Shangri-La was at hand courtesy of Utah, State politicians seized on this as a once in a lifetime opportunity to revitalise Utah's economy. Governor Norman Bangerter immediately convened an emergency session of the State Legislature to appropriate five million dollars for fusion research.

The debate—if such a one-sided affair can be called a debate—took place on 7 April. Speaker after speaker announced their hopes that the discovery would be the saving of the human race, would do away with environmental pollution and provide energy 'too cheap to meter'. The Governor even went further by paraphrasing the Bible in his urgings for the bill's passage: 'He that doeth nothing is damned, and I don't want to be damned.' There was the point that the scientific community had not yet had time to confirm the phenomenon but this was dealt with summarily with a warning that would, in the following months, become a knee-jerk response of Utah fusion afficionados. It was Eugene Hansen, Chair of the University's Board of Regents, who first articulated the hidden threat: ' "waiting and seeing" could mean that the discovery of the century will be developed by Mitsubishi.'

The State passed the bill 96 votes to 3.

With State money in place, the circus moved on to Washington. On 26 April, only one month after the press conference, up to 25 million dollars was the pot being chased in public, while in private the Utah lobbyists were talking about 125 million.

There were already misgivings in some quarters about the Fleischmann and Pons claims, the lack of details and sloppy descriptions. A leading scientific journal, *Nature* did not publish the paper but none of this seemed to matter before the Utah steamroller. Whether test-tube fusion was right or wrong seemed to be a secondary concern; test-tube fusion equated with cash flow into Utah was dominant. The Office of Naval Research, for whom test-tube fusion could have significant benefits, awarded the Utah team 400 000 dollars for research during the subsequent three years. Bud Scraggs, chief of staff to the State Governor, said that the State's plans wouldn't be dampened even if the Fleischmann and Pons paper failed the peer review process. 'We are not going to allow some English magazine to decide how state money is handled.'

The University had by now hired a professional in the form of one Ira Magaziner, a business strategy consultant who also picked up the 'yellow peril' theme by telling the congressional enquiry: 'As I speak to you now it is midnight in Japan. Thousands of Japanese scientists are trying to verify . . . and develop commercial applications . . . of this new science.'

The keyword was *fusion*; nothing else would have created such expectations. But the foundations for this belief were like the Cheshire Cat in *Alice in Wonderland*; the more one looked at them, the less substance they seemed to have. There were many incorrect statements, false rumours and wrong inferences. In the rush for funds or to exploit the fusion fanaticism, few had the time or inclination to look beyond the surface and see the shaky foundations beneath.

Rarely does such a window on the daily activity of scientific researchers and their ideas open so widely for public view. And although this particular episode in the end solved no major scientific problem, it serves a crucial role in bringing to light the many aspects of what science is all about and what it means to do good science.

When I set out to write this story I thought that it would primarily show how scientists come up with ideas, design a strategy to test them, carry out the experiments, share experiences and attempt to replicate claimed discoveries, thereby establishing new natural phenomena. It would show how the research process occurs, how one group of specialists reaches the end of its expertise and takes advice, how many mistakes are made and truth arrived at circuitously. How there is rivalry, a desire to be first, but played out within established rules of conduct.

Many of these ingredients were indeed present, but as the months passed I became increasingly aware that there was much more than just science at work here. There were accusations of plagiarism, of cheating, of fabricating data; there was name-calling in public as Fleischmann and Pons were called 'incompetent' and 'deluded';[1]

there was deep politics, paranoia and, by the end of the year, threats of writs being issued against the editor of a leading scientific journal and against one scientist whose experiments suggested that there had been errors in Pons' laboratory.[2]

The pressures were money and patents, and how the competition of the market place can interfere with the more detached quest for truth in the laboratory. It is a reminder of Richard Feynman's observation after the enquiry into the Challenger Space Shuttle disaster; 'Nature cannot be fooled.'[3]

All of which beg the following questions.

How did the news media, scientists and governments take on board a claim that made no sense within the laws of physics, was based in part on wrong data and that was shown to be flawed within days of the announcement? With such lack of evidence, how did administrators decide it was worthwhile to pour millions of dollars of taxpayers' money into this fanciful fusion fiasco? How could two reputable scientists make such a claim and, even a year later, still insist that they had evidence for fusion when the credibility of important pieces of their 'evidence' had been destroyed? Was this a delusion, an error or a fraud? What can science and the public learn about scientific procedures from this; what ethical issues are raised; what murky aspects of science does it expose?

And when we have the answers to at least some of these questions, will they help prevent a recurrence? For this is not an isolated incidence of confusion, though rarely are such errors played out in public view with such global excitement; similar things have happened before and some of the signs were already known but went unrecognised.

# I

# GENESIS

The race begins: events to 23 March 1989

# 2

# NOTHING NEW UNDER THE SUN

The announcement of room temperature fusion in 1989 has some of the hallmarks of a detective mystery where appearances deceive. Consider 'the facts'. Room temperature fusion has been discovered independently and simultaneously by two groups in Utah working within a few miles of each other. One team consists of two chemists, Martin Fleischmann and Stanley Pons, in Salt Lake City; the other group involving Paul Palmer and Steven Jones at Brigham Young University consists of physicists. The basic set-up consists of rather basic apparatus out of a school laboratory—battery, heavy water, metal electrodes and meters for heat or neutrons—which is so widely available that the discovery could have been made anywhere at any time in the preceding half century; but somehow it has been overlooked by two generations of scientists until suddenly in 1989 not one but *two* groups in *the same location* simultaneously announce their breakthroughs.

Coincidence in time, coincidence in space; as Sherlock Holmes remarked 'When two strange things happen, they are usually connected.' Understanding how they came about is one of the challenges in this affair.

A first rule in any mystery is to check the facts. This is as true in any detective novel as it is in scientific research or investigative journalism, and in supposing that the test-tube fusion idea had been overlooked for half a century we are already repeating a media 'factoid' that is incorrect.

Good ideas rarely come out of the blue; search carefully and you will often find that someone else has anticipated your inspiration or, at least, thought of something similar to it. Such was the case with the idea that hydrogen fusion might occur at moderate temperatures within solid materials. This was first thought of not by Fleischmann and Pons nor by Palmer and Jones, but had been anticipated in Europe, India and the Soviet Union up to 50 years earlier. An important part of the research process is to search through the

17

literature and find what ideas similar to your own have been investigated in the past, and then by learning what became of the earlier ideas you can avoid their mistakes and build on the successes. The Jones collaboration in part grew out of such a quest, though there were aspects of their work that had been anticipated in India long before, unrecorded and so unknown to them. Fleischmann and Pons set out on a different course initially, but one that had already been charted in Europe in the 1920s and 30s. Let's begin by looking at what happened in the past, as it will give some insights into what took place in Utah in 1989 and help us understand better why people reacted in the ways that they did.

## Helium and airships

Ferdinand Graf von Zeppelin launched his first airship in 1900, and so began the age of the hydrogen-filled craft in which Germany led the world for two decades. In the aftermath of the First World War Germany was forbidden to rearm until restrictions were relaxed in 1926 and airship construction began again. In the meantime an airship had caught fire in the USA, ending the use there of the lethal explosive hydrogen as a lifting gas; it was replaced by the safer helium. However Germany was short of helium and the major industrial producer— the USA, who was extracting it from natural gas—didn't want to supply it so soon after the war. Thus finding new and cheap ways of producing helium came to occupy the minds of two German chemists at the University of Berlin: Fritz Paneth and Kurt Peters. And the recently discovered structure of atoms gave them the idea that they could make helium out of hydrogen.

The 1920s were in the interregnum between the discovery of radioactivity—the spontaneous transmutation of one element into another—and its detailed understanding. Einstein's famous equation $E = mc^2$ already existed but its full import for the release of vast energies in radioactivity and other nuclear processes still lay in the future. The fact that elements could change from one into another was the startling discovery and had renewed interest in an old idea of William Prout, the nineteenth-century British chemist, that hydrogen could be the primordial substance from which all other atomic elements, including helium, were constructed. Ernest Rutherford had discovered the existence of the atomic nucleus in 1911 and this reinforced the vague idea: bring nuclei of light elements together and build the nuclei of heavy ones.

Atoms consist of a densely packed positively charged nucleus encircled by negatively charged electrons all held in place by the electrical attraction of opposite charges. The nucleus contains almost all of the atom's mass, the electrons contributing at most about 1/2000 of the whole. Different atomic elements are distinguished by the amount of positive charge on the central nucleus, which in turn determines how many negative electrons encircle it. The simplest atom, hydrogen, consists of a single electron encircling the nucleus which has a single positive charge; the positively charged particle that comprises the hydrogen nucleus is called the 'proton'. Helium, the next element in the periodic table, contains two electrons encircling a nucleus containing two protons. In the close confines of the nucleus powerful forces take over, holding the protons in a tight grip and overcoming the disruptive electrical forces. Add more protons in the nucleus and in turn more electrons can be entrapped around them. It is the number of protons in the nuclear cluster that determines which element is which.

All elements are made of the same stuff—positively charged nuclei containing protons and encircled by negatively charged electrons. So in the early 1920s some people conjectured that it may be possible that helium—element number two in the periodic table of the elements—could be built from two hydrogen atoms. This was what Paneth and Peters decided to try to do. They set out to create the conditions where two hydrogen nuclei ('protons') could get close to each other, collide and join to make a nucleus of helium.

The key that they thought would enable them to achieve this was *palladium*, a light grey metal similar to platinum in appearance. You may have some if you have any dental fillings; it is also used as a contact in telephone circuits and its malleability makes it a useful substitute for platinum in jewellery. The thing that made it attractive to Paneth and Peters, and years later to Pons, Fleischmann and Jones, was its great affinity for hydrogen—it can absorb up to 900 times its own volume of the gas. The hydrogen pours into 'empty' spaces between the palladium atoms and as it fills up the metal expands by some ten per cent in all directions, which puts it under great internal pressure—many thousands of times greater than atmospheric pressure. Paneth and Peters knew about palladium's ability to absorb lots of hydrogen and they wondered whether packing so much of it in between the palladium atoms might increase the chance of the hydrogen atoms bumping into one another, thereby fusing and producing helium.

The Germans made a thin capillary out of palladium, heated it red-hot and diffused hydrogen through it. The palladium absorbed the hydrogen and, glory be, they found traces of helium when they were finished! Their idea seemed to be on the right track but the quantities of helium were never going to fill an airship. In fact, the amounts were only just measurable.

The news reached Sweden and John Tanberg, who later became the scientific director and manager of Electrolux, the world-famous manufacturer of household appliances. He worked there for nearly 40 years and it was in their Stockholm laboratories that he immediately took up Paneth and Peters' idea.

Tanberg had for some time been interested in the way that hydrogen behaves in the presence of metals such as palladium. The Paneth and Peters work was his speciality and he was enthused by the fact that *some* helium had apparently been found. The palladium seemed to be helping the hydrogen to fuse but not fast enough, so if he could improve this, he might be able to make helium in amounts that would be industrially useful. Thus it came about that Tanberg anticipated the second ingredient in the Fleischmann and Pons story: Tanberg introduced 'electrolysis' as a way of encouraging the hydrogen to come together.

The basic idea of electrolysis is that an electric current passes through some liquid, in Tanberg's case this was water, and the liquid breaks down into its constituent elements (in this case, hydrogen and oxygen). You first attach a battery to two electrodes which are dipped into the water (it is necessary to add traces of acid or alkali to conduct the electric current through the solution—this is known as the 'electrolyte'). The voltage across the electrodes ionises the atoms (ions are atoms that have become electrically charged as a result of having gained or lost an electron from their periphery). The hydrogen atom loses its electron, becomes positively charged and is attracted to the negatively charged electrode, called the *cathode*; the oxygen gains an electron and migrates to the positive electrode, the *anode*, this migration of ions forming the current. Whereas Paneth and Peters used hydrogen gas, Tanberg 'made' the gas from the water, and what's more, he used palladium for the cathode so that the hydrogen was attracted naturally to it.

The electrical forces that were the problem in getting the like-charged hydrogen nuclei together were at least being put to good use in attracting them into the palladium prison. The electrical voltage forced the hydrogen into the palladium. More and more accumulated until it was at several thousand times atmospheric pressure. Under these circumstances he hoped that a really effective way of making helium might result.

He produced helium and also energy, or so he must have thought, for in 1927 he put in a patent application for a 'method to produce helium and useful energy'. The patent office replied with the comment: 'received but not understood.'

By November 1927 the patent application had been rejected on the grounds that the description was so sketchy that 'an expert is unlikely to be able to practise the invention'.

Given what was to transpire with Fleischmann and Pons, 60 years later, there is a bizarre feeling of *déjà-vu* in all of this. Palladium and electrolysis, claims of helium and energy and 'incomplete descriptions such that [even] an expert is unlikely to be able to practise the invention'. In another piece of coincidence even the helium traces were explained; they had nothing to do with hydrogen fusion.[1]

The denouement in the 1920s is dealt with dismissively in the classic book *Radiation from Radioactive Substances*[2] by Chadwick, Ellis and 'Sir Ernest Rutherford, O.M., D.Sc, Ph.D., LL.D., F.R.S., and NOBEL LAUREATE' (the title pages of modern books tend to be more modest). This appeared in 1930 and on page 315 we find what happened to Paneth and Peters (and most probably to Tanberg too).

Several other careful experiments looking for helium had been done but nothing was found. 'The claim of Paneth and Peters was soon withdrawn and the appearance of helium in the experiments received a more prosaic explanation. It was found that a glass surface . . . is able to absorb a detectable amount of helium . . . *from the atmosphere* (my italics) during only a single day's contact with the air.' So all their efforts to create helium from hydrogen had been fruitless; the traces of helium that they saw had been absorbed from the environment and were not a product of fusion. Sixty years later others were to be accused of making this same oversight. And the summary from Chadwick, Ellis and Rutherford in 1930 could equally well be applied to the events of 1989: 'In all the cases reported, the presence of an element has been mistaken for its creation.'

## Neutrons and fusion

One of the authors of that famous book was James Chadwick. His moment of real glory came in 1932 when he discovered the neutron—a constituent of atomic nuclei that is the missing link in this story and will be both star and villain in later chapters.

The positive charge in a nucleus is due to the several protons it contains, but the protons are not alone; there is a second type of particle within nuclei—the neutron. A neutron and proton are almost identical but that where the proton has positive charge, the *neut*ron has none—it is *neut*ral. With the sole exception of the simplest nucleus, namely the single proton at the centre of hydrogen, all nuclei contain both protons and neutrons. Neutrons help to stabilise nuclei because neutrons when closely packed also feel strong attractions, both to other neutrons and to protons, but being electrically neutral neutrons do not suffer from the electrical disruption, so the strong attraction effectively works better for them.

None of this was known in 1926 when Paneth and Peters attempted to join two hydrogen protons to make helium, as the discovery of the neutron only occurred in 1932. We now know that two protons will not directly make helium; two protons cannot survive the electrical disruption on their own and the stable helium nucleus contains neutrons as well. The hydrogen nucleus consists of a single proton; helium contains two protons and one or two neutrons, these two 'isotopes' being known as helium-3 and helium-4 (written $^3He$, $^4He$ —He being the chemical symbol for helium and the 3 or 4 referring to the total number of constituents within the nucleus).

To make helium requires both protons and neutrons, so instead of hydrogen you use *deuterium* or 'heavy hydrogen' whose nucleus consists of a proton and a neutron tightly gripped without any electrical disruption. It is 'hydrogen' because there is only one proton and 'heavy' because it has twice the bulk of the usual variety, and was discovered by Harold Urey in 1931, a year before Chadwick proved the existence of the isolated neutron. In the shorthand this would be written $^2H$, but being so fundamental it has a special name and symbol of its own, $d$ for the deuteron, the nucleus of the deuterium atom (symbol $D$). When deuterium replaces hydrogen in water we have what is known as 'heavy water', which occurs naturally in small amounts in sea water. (There is a third and rarer form of hydrogen, known as tritium, where the proton is joined by two neutrons. The nucleus of tritium—the 'triton'—has the symbol $^3H$ or simply $t$.) Paneth and Peters would never have been able to make helium from hydrogen as they were missing the neutrons; replace the hydrogen with deuterium and you may have a chance of success.

Immediately following these discoveries John Tanberg returned to his electrolysis measurements but with *heavy* water, $D_2O$ instead of $H_2O$. These experiments were essentially the same as those that Fleischmann and Pons performed 60 years later.

Tanberg told Torsten Wilner, his long-time collaborator, about his plans. Tanberg had loaded a thin palladium wire with deuterium which that night he intended to connect to a high voltage and then discharge the electricity through it leading to a sudden increase in pressure and temperature. He hoped that the tightly packed deuterium atoms would be set in motion, their nuclei bumping into one another and fusing.

By this stage the quest was no longer simply one of producing helium but also of liberating energy. Understanding of the atomic nucleus was progressing rapidly and awareness was dawning of the awesome energies latent within. The rearrangement of *electrons* among neighbouring atoms involves the transfer of 'chemical' energy, taking in or releasing heat (as when carbon and oxygen combine to produce heat in a fire). The rearrangement of *protons* and *neutrons* also involves changes in energy, 'nuclear energy', the only essential difference from chemical being one of scale; the energies involved in nuclear reactions are some millions of times greater than those in chemical processes. When the nuclei of light elements join together or 'fuse', energy is released; what is more, the amount is more than the energy needed to push these nuclei into one another overcoming the electrical repulsion, so if we could accomplish this efficiently we would have a plentiful and clean supply of energy. This is what is behind modern schemes to produce 'fusion' power. That is what Tanberg was probably the first to try and do.

Tanberg had already realised that vast amounts of energy might be extracted from this process. He informed Wilner that he had calculated the energy that would be released and it was over a million kilojoules, equivalent to a thousand grams of dynamite. After thinking for a moment he added that Wilner should go home so that if anything untoward happened he could 'tell people what it was'!

No disaster struck then, nor ever.

Tanberg's notes record that he produced electrical discharges and occasional deafening bangs, but saw no radiation or radioactive residues. And no helium! Either there was no helium or it was too faint to detect with the equipment then available. Nor did he ever measure more energy in the bangs and heat than he put in from the electricity supply.

Fleischmann and Pons's ideas had been anticipated though for all the wrong reasons, primarily a misguided attempt to make helium to fill airships, and all this because the USA would not sell the gas to postwar Germany. Coincidentally restrictions on helium sales from the USA also enter the Jones story too, though this time it is India rather than Germany that is involved.

The story here also goes back some 50 years to S. K. Bose, the famous Indian physicist whose name is associated with 'boson' particles and with Bose condensation, a phenomenon connected with the behaviour of liquefied helium. When liquid helium was first made it soon became an essential laboratory tool as a coolant in low temperature experiments. Supplies of liquid helium are an absolute necessity in many modern science laboratories. You aren't looking for quantities that will fill air balloons but, nonetheless, you need amounts that call on industrial preparation and, as was the case for

Paneth and Peters in Germany, the question was where to get it. Any source would be useful.

In the 1920s a team of British geologists discovered hot springs while surveying Northern India and the seismic faults around the Himalayas. They noticed that there seemed to be a lot of helium gas in the vicinity. Bose became interested in this and set his colleague Dr Shyamadas Chatterjee to work to see if there was enough to use as a source for liquid helium for their labs. There are two springs in particular that interested them, 200 kilometres north-east of Calcutta and some 30 kilometres apart near Shantiniketan where the Indian poet Tagore lived. They are about 72 degrees celsius and Chatterjee found that one, slightly hotter, had 2 per cent helium in its bubbles, the other having about half this amount. He also looked for signs of abnormal radioactivity in the vicinity while seeking for the source of the heat and gas but nothing significant turned up.

That was in the 1930s; today there is a helium collection plant at the springs. Each spring is about the size of a swimming pool and produces enough helium for use, after liquefaction, at the Cyclotron (small nuclear accelerator) in Calcutta. This helium source is very useful to them, both for its proximity to Calcutta and because most commercial helium comes from petroleum wells and the USA, the major producer, has restricted trade with India following disputes in the 1970s over India's development of atomic weapons from fissile materials that the USA had supplied for peaceful uses.

But in addition there remained the puzzle of how the helium came to be in the springs. There is a jungle region in northern India where there are several more hot springs and here again there seems to be an abundance of helium. Chatterjee wondered about this as it began to seem more than simply coincidence. Then in 1978 he saw a paper by three Soviet scientists which reported that there were anomalous amounts of helium-3 in some metals; the Soviets suggested that it was being produced by fusion—fusion induced by cosmic rays. Chatterjee then had the idea that maybe fusion could be acting within the Earth in the vicinity of the hot springs; that 50 kilometres down beneath the ground the pressure and temperature caused the Earth to be a fusion engine. This would produce helium-3 and tritium; the tritium is unstable and decays producing helium-3 which would add to that directly produced. The helium would seep up through fissures, and hence its natural occurrence near the hot springs.

Chatterjee told this idea to Bikash Sinha, Director of the Calcutta Cyclotron, during a car ride back from the springs in 1984. At the time Sinha thought it wasn't very likely and forgot about it.

Halfway around the world, in Utah, Paul Palmer of Brigham Young University was fascinated by helium in the atmosphere.

Palmer originally specialised in acoustics but became interested in the physics of the Earth and in measuring ages by radiocarbon dating. He developed a detector for decays of carbon-14, a radioactive form of carbon, that can be used to this end. Though working a Brigham Young University he was a graduate of the University of Utah in Salt Lake City and knew Martin Cook, one of the Utah faculty. Cook believed that there might be evidence for a 'young Earth' in the fundamentalist sense of literal biblical interpretation, due to the presence of helium-3 in the atmosphere in quantities whose origins are hard to understand unless they are from fusion which, according to the conventional wisdom, required great heat and hence could only have happened when the Earth was being formed). If the Earth is five billion years old, all the helium would by now have risen away, yet it was there in amounts suggesting that it had appeared in the atmosphere only recently. One possibility therefore was that the Earth is only a few thousand years old, a second theory is that ancient helium-3 trapped inside the Earth is being ejected by volcanic eruptions and the third idea, Palmer's, was that the Earth is indeed billions of years old, as carbon dating assured him, and that fusion is taking place even now. The question was where and how.

This remained only a vague idea until he read the same Soviet paper from the 1970s as Chatterjee had seen and which dealt with the abundance of helium-3 and helium-4 in minerals. The paper claimed that there was no helium-4, the common form of helium, only helium-3. The problem here, as in the case of the atmosphere, was how the helium-3 got there.

The Russians had suggested that muons—particles that arrive in cosmic rays and are in effect heavy versions of electrons—are the seed that catalyses the fusion.[3] Palmer did not discover the paper until 1986, but he was lucky because in his department at Brigham Young University in Utah was Steven Jones who had spent the previous four years doing custom-built experiments with muons, attempting to induce fusion in the lab. Thus Palmer and Jones came together and started the collaboration which attempted to replicate fusion under pressure, cold fusion in metals, to see if the Earth's heat and the helium-3 could indeed be due to natural fusion in the rocks.

They did not know of the Indian hot springs nor of Chatterjee's ideas. Nor did they know of the old Swedish experiments attempting to make helium from hydrogen in the laboratory. They were also unaware that less than 50 miles away from them Fleischmann and Pons—who were also ignorant of these historical works and of the parallel developments at BYU—were beginning to muse about creating hydrogen fusion with the aid of palladium in a test tube.

In India Bikash Sinha had all but forgotten Chatterjee's remarks, when on 24 March 1989 he first heard the reports from Utah claiming that Fleischmann and Pons had performed fusion in a test tube. Like many other physicists, Sinha's first thought was that 'this must be nonsense'. Then somewhere in the excitement Sinha remembered Chatterjee and it all linked together for him: hot springs, helium, fusion. At this stage no-one in India yet knew about Jones and Palmer, nor what had motivated them. So, quite independently, in India, ideas similar to Palmer's in Utah had emerged but the circumstances and the responses to the ideas varied widely.

Sinha flew to Bombay, home of the Bhabha Atomic Research Centre, and returned to Calcutta with some heavy water. He bought some palladium and set up an experiment. It was the association of hot springs and helium that had made him think that there could be something to Chatterjee's idea after all, and maybe to the claims by Fleischmann and Pons. It was after the Indian team had begun their experiments that they learned about Jones and how his experiments had been motivated by the same ideas as Chatterjee had had. However, whereas Chatterjee had left it as an interesting hypothesis, Jones and Palmer had designed an experiment right away to try and simulate the effect in the lab.

So helium had, by a series of random paths, been the common ingredient: in Germany, Sweden and India, to Fleischmann and Pons and also to Palmer and thus to Jones. The irony is that in all of the experiments on test-tube fusion that began in 1989 following the March announcements, and which variously claimed fusion evidence from heat or from radiant neutrons, no measurable production of helium ever took place.

## Modern fusion

When Paneth and Peters tried to turn hydrogen into helium they were following in the traditions of the ancient alchemists who wanted to convert base metals into gold. With the discovery of atomic structure and nuclear transmutation it is now possible to do that, but the cost is greater than for digging gold out of the ground, so nuclear physics has not undermined the gold standard. The payoff is not in the end products so much as in the energy that can be tapped. If this energy is converted into heat as the radiant particles pass through water, for instance, it can produce steam to drive a turbine and generate electricity.

So far the emphasis has been on fission, where the nuclei of heavy elements such as uranium fragment into smaller pieces. This produces energy and also unwanted lethal products whose disposal is the source of much political concern. The materials for fusion are plentiful and deuterium, available in water, is relatively clean by comparison. (Tritium, however, is radioactive.) The interest in nuclear processes has occupied tens of thousands of scientists and engineers for over a generation. They have devised sophisticated instruments to detect and measure the energies of even the smallest traces of the products. Power stations have been designed and successfully operated on the basis of that knowledge. The occasional and unfortunate accidents have been due to inadequate safety procedures, not to lack of understanding in the basic nuclear processes; had there been some previously unknown nuclear process taking place at room temperature in solid materials and that had been somehow overlooked, then it is likely that we would have discovered this long ago in some disaster. It is worth bearing this in mind when assessing claims to have found 'overlooked' nuclear pathways.

The basic players in the attempts to produce commercial fusion on Earth are deuterium ($D$), tritium ($T$) and helium ($He$); atoms whose nuclei have two three or four constituents. All nuclei consist of neutrons and protons, and it is the rearrangement of those constituents when nuclei come together and break apart that releases energy. The figure on page 28 shows the make-up of these simplest nuclei and illustrates what happens when two deuterium nuclei collide. They could simply bounce off one another but this does not liberate any energy. Energy is released when either a neutron or a proton is shaken loose, the reactions being written ($n$ referring to neutron and $p$ proton),

$$d(np) + d(np) = {}^3He(npp) + n$$
$$d(np) + d(np) = {}^3H(pnn) + p$$

The combined mass of the final products is smaller than the combined mass of the initial nuclei. The mass that 'disappeared' has been converted into energy as described by Einstein's equation $E = mc^2$, which implies that a mass $m$ is equivalent to an amount of energy $E$ (where $c$ is the velocity of light) and so mass can be converted into useful energy. In these examples the energy is manifested as kinetic energy: the energy of motion as the products fly apart from each other.

These energies in nuclear reactions are enormous compared to the amounts involved in chemical reactions. Atomic energies are written in units called eV, short for electron-volts, one electron-volt being the energy an electron gains when accelerated by a one volt potential. Energies released and absorbed in chemical processes are about 1 eV per atom. The nuclear processes liberate a million times more energy, which is measured in MeV for mega (*million*) eV.

27

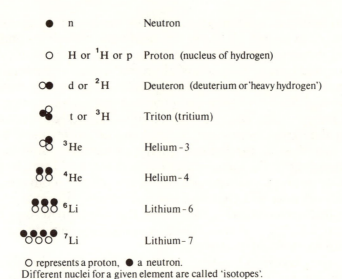

| | | |
|---|---|---|
| ● | n | Neutron |
| O | H or $^1$H or p | Proton (nucleus of hydrogen) |
| O● | d or $^2$H | Deuteron (deuterium or 'heavy hydrogen') |
| t or $^3$H | | Triton (tritium) |
| $^3$He | | Helium - 3 |
| $^4$He | | Helium - 4 |
| $^6$Li | | Lithium - 6 |
| $^7$Li | | Lithium - 7 |

O represents a proton, ● a neutron.
Different nuclei for a given element are called 'isotopes'.

The names, symbols and make-up of the nuclei of the light atoms.

d + d →
n + $^3$He
p + t
$^4$He + γ

d + t ⟶ $^4$He + n

d + p ⟶ $^3$He + γ

The products of nuclear fusion involving deuterons with hydrogen isotopes.

D + $^6$Li ⟶ $^4$He + $^4$He

D + $^7$Li ⟶ $^4$He + $^4$He + n

Some reactions between deuterons and lithium nuclei.

Figure 1

28

In the first reaction above the neutron carries away 2.45 MeV of energy, while in the second the proton has 3 MeV.

Another possibility is that the two deuterium nuclei combine to form helium-4; here again the mass of helium-4 is less than the combined masses of two deuterium nuclei and the 'spare' mass is manifested as electromagnetic radiation far beyond the visible spectrum and known as a gamma ray (denoted by $\gamma$). This gamma ray carries away 24 MeV of energy but this process is very rare, occurring some ten million times less frequently than the neutron or tritium production channels.

These neutron and tritium production processes occur about 50:50 and modern attempts to generate useful energy in fusion experiments have tended to use beams of deuterons for which these are the fusion products. If you can get hold of the rarer tritium you may liberate nearly 18 MeV through the reaction:

$$d(np) + {}^3H(nnp) = {}^4He(nnpp) + n$$

A problem—*the* problem in the attempts to fuse nuclei together and release their internal energy—is that all nuclei carry positive electrical charge. It is the attraction of *opposite* charges that holds the negatively charged electrons in the atomic periphery where they encircle the positively charged nucleus, but the corollary is that like charges repel; two protons, each one positively charged, *repel* one another. You want to force those nuclei together while nature is designed to prevent it.

At this point it must seem paradoxical that atomic nuclei containing several closely packed protons exist at all. This is even more astonishing when one realises just how compact the nucleus is; for example, scale a typical atom up to the size of a football stadium and the nucleus will be smaller than the football. The key is that when protons are touching they feel a strong attractive force, more powerful than the electrical forces that are trying to force them apart. It is this 'strong nuclear force' that holds the nucleus together but protons only feel it when they are in close proximity; once apart they feel the electrical repulsion.

It takes only a few thousands of eV (keV) to bring two nuclei together and if they fuse you get over a million eV (MeV) in return. In practice it is very difficult to get two such small nuclei to collide and fuse, in beams of billions most of them simply miss one another. To make a useful fusion reactor needs high densities of the deuterium fuel and, it has been traditionally assumed, temperatures greater than those in the centre of the Sun so that fusions occur frequently enough that more energy is liberated than consumed. This is the approach on which most of the effort—and the money—has been spent in recent decades. Recently some people have been trying a different approach, seeing if they can change the nature of the atoms such that fusion can occur at a useful rate even at room temperature. The former, the world of the tokamaks—particle beams—and mega dollar budgets, may be termed 'hot fusion'; the latter attempts to generate fusion at room temperature are known as 'cold fusion'.[4]

# 3
# THE SUN ON EARTH

When I was a child my father would tease me with two conundrums: 'What happens when an irresistible force meets the immovable object?' and 'If you discovered a substance that destroyed every solid, how would you contain it?' I suspected that he was cheating, that the questions were based on false premises. There is nothing that is infinitely irresistible nor that is truly immovable, and as for the idea of a 'substance impossible to contain'—*that*, I decided, was a fiendish invention designed to get me into a muddle as I was sure that such a substance would be a manifest impossibility in practice. However I was wrong: the quest for fusion is perhaps an example of these problems.

The holy grail of limitless energy, 'harnessing the power of the Sun on Earth', is metaphorically an irresistible force and today, 40 years after the quest began and with over twenty billion dollars already spent trying to achieve it, the barriers to be overcome seem to be immovable. But the truly irresistible and the immovable are abstractions, great prizes warrant intense efforts and so in the case of fusion the quest continues. We know that fusion occurs: thermonuclear explosions or 'H-bombs' are uncontrolled examples of fusion, we have the Sun as evidence of a self-sustaining controlled fusion machine and in the laboratory we can replicate the nuclear processes that fuel it. So the question is 'merely' one of engineering: can we replicate these extreme conditions and create controlled nuclear fusion reactions while using less energy than at present?

Energy is released when the nuclei of the light atomic elements join, so bringing them together is the primary goal. However, incredibly strong electrical repulsions keep the nuclei at atom's length. Tremendous pressures and heat, millions of degrees, are needed to overcome this. The centre of the Sun where fusion occurs is at a temperature of around fifteen million degrees but practical fusion needs a much higher temperature than this—hundreds of millions of degrees. At such a temperature the deuterium and tritium fuels are so hot that they would vaporise every material known, stripping the atoms apart into 'plasma' of electrons and nuclei roaming free of one another. It is this that bears on my father's second question—how to contain the uncontainable: producing these extremes and then controlling and containing them are the problems.

To contain the hot fusion fuel scientists in the USA, Europe, Japan and the Soviet Union have built doughnut shaped machines of magnets (called 'tokamaks' from the Russian acronym for toroidal magnetic chamber) which bottle the deuterium in powerful magnetic fields.

One of the basic principles behind these attempts to achieve thermonuclear fusion in tokamaks was discovered inadvertently and in an unrelated context as long ago as 1905.[1] Two physicists at the University of Sydney in Australia had a piece of copper tubing, used as a lightning conductor, which had been crushed as the electrical current passed through it. What had happened was that the passage of electrical current created huge magnetic forces which constricted the tube. A similar effect occurs when electrical current passes through a column of plasma; the plasma contracts radially—known as the 'pinch effect'—locally increasing its density to the point where ultimately thermonuclear reactions might occur. In 1988 Princeton University attained a record 300 million degrees in their tokamak, but still used orders of magnitude more energy than they produced.

In the USA, Lawrence Livermore Laboratory and Rochester University are following an alternative approach called 'inertial fusion' where the deuterium and tritium fuel is in a small glass bubble which is momentarily blasted by powerful lasers, but here too practical economic fusion seems to be years in the future, and some 700 million dollars may be needed to upgrade the facility and keep the research moving forward.

In recent years, however, some scientists have been trying a new approach, 'cold fusion', which involves 'changing' the behaviours of atoms at room temperatures.

The main idea here has been to exploit the existence of particles known as 'muons' which in some ways behave like electrons, but are some 207 times heavier and are unstable. Muons can replace electrons in atoms but the muon's greater mass causes the atoms to shrink (in the case of hydrogen by a factor of 207 in all directions) and the effects of the repulsive forces drop enough that the atoms can crowd together more easily and fusion can occur at low temperatures. Among the several groups pursuing this route is that of Steven Jones of Brigham Young University, Provo, Utah.

Fleischmann and Pons thought that they could achieve cold fusion by another route. They believed that packing atoms together inside crystalline solids could produce so much pressure that the atoms

32

would be squeezed together and hence fusion would occur in a test-tube.

Deciding which of these approaches to back is like entering a lottery, but informed prejudice gives some guide to the odds.

The cold fusion with muons, 'muon catalysed fusion', *can* occur and has been seen to happen; however, it does not seem to happen fast enough to be useful as an energy source. In the 1970s it was only a curiosity but around 1980 it had a renaissance when people discovered that in certain complicated molecules, involving both deuterium and tritium, the presence of a muon caused fusion to occur much faster than had been previously thought possible. Jones entered this reborn field in 1980. Today muon catalysed fusion is still tantalisingly short of the 'breakeven' point where the energy given back equals the cost to run the reactor. Muon catalysed fusion may be only one miracle short of practical realisation, but as yet we do not know where that miracle will come from, nor even whether it exists.

Since 1986 Jones has put most of his effort into solid state cold fusion, a similar approach to that also pursued by Fleischmann and Pons. The betting odds here are extraordinarily long; theory suggests that it is not possible to liberate useful energy this way nor is there any natural phenomenon that gives us confidence that this can occur. There is controversy over whether this sort of fusion occurs under pressure inside the Earth, or inside the heavier Jupiter, but even if it does it is at levels that are useless for commercial fusion. Fleischmann and Pons claimed that they had produced net power output from this approach at levels that would potentially solve the world's energy needs. Investigating their claims and describing how governments reacted to them form the theme of most of this book.

So at present the 'hot fusion' route, even though expensive, seems to be the best bet. If controlled practical fusion is ever achieved, the investment will pay off.

The media love to compare controlled hot fusion with 'harnessing the power of the Sun on Earth', and I have used that idiom in this chapter heading; yet it is not really true. Not only must the temperature be more extreme than the Sun to be practical, but the fuel is different.

The Sun's fuel is hydrogen nuclei—protons—which alone are useless for commercial fusion. There is no nucleus consisting of two protons alone and so no energy can be liberated by a pair of protons making such an object. Were this the whole story there would be no Sun. However, the chance of two protons fusing isn't exactly zero. There is a radioactive process—beta decay—which enables one of the protons to shed its charge, in effect becoming a neutron at the instant of fusing (the electrical charge being carried off by a positively charged form of the electron, known as a positron), the proton and neutron fusing to make a deuteron and liberating energy. Even so this sequence of events is very unlikely to happen and occurs some million times slower than would have been the case if there were neutrons there to start with. It is this slim chance of two protons going through this sequence that limits the speed at which the Sun burns up its abundant fuel. Were there not so much, so dense and so hot, the Sun would not light at all. So for practical controlled fusion on Earth, protons alone are useless; for all practical purposes, two protons 'cannot fuse'.

Once the deuteron has formed, the fusion in the Sun can race ahead. Its copious protons fuse with the new deuterons forming $^3He$ and a radiation. The most likely fusion processes are then $^3He + ^3He = ^4He + 2p$ (the two protons being returned to the Sun's supply to help keep the cycle going), the $^4He$ combining with $^3He$ to make beryllium and lithium which in turn fuse with protons to produce more $^4He$. The net result is that once the protons have overcome the initial fusion hurdle they are converted to $^4He$, radiating energy in the process. At no stage is tritium produced, nor do $dd$ or $dt$ reactions play any significant role in the solar cycle.

Contrast this with the search for practical fusion on Earth. This is a hard enough task even without the added difficulty of overcoming the proton–proton fusion barrier, and so $dd$ and $dt$ fusion is where the effort is being put. Fusion of light nuclei is the common principle in the Sun and in tokamaks but the fusion fuels are rather different. Even so, in the public's mind fusion and the Sun go together and the fascination with fusion is sometimes akin to ancient sun worship.

The promise of utopia has attracted the attentions of both charlatans and serious scientists. On occasion even the latter have lost their usual reticence and blundered. An extreme case of charlatanism occurred in Argentina during the Peronist era of the 1950s. Later in that decade the British media announced, with some pride, that the eminent scientist Sir John Cockroft was '90 per-cent certain' that controlled thermonuclear fusion had been achieved at the Harwell Laboratory. Unfortunately some of these failings pervaded part of the test-tube fusion story of 1989.

### Peron's fusion reactor

In science as in manufacturing, quality control filters the good from the bad. In science it involves peer review—other scientists examine, criticise, and question your work and only when you pass this rigorous vetting can you be truly confident that the work is valid. In the funding and support of science and technology it is important that careful examination should take place and the advice of experts be listened to. But if central government acts arrogantly, ignoring the advice of the experts, listening to the advice of favoured confidants to the exclusion of others and deciding that it 'knows best', the outcome can be devastating.

In Argentina in the 1950s the Peronist regime provided fertile ground for various nonsenses to flourish and this provides a parallel and a contrast to what happened in 1989 with test-tube fusion in the USA and Europe.

The saga began with a desire for Argentina to prove that it was as good as the major technological powers. The regime decided that the man who would lead the miracle was Kurt Tank, an aircraft designer who left Germany after the war for a new life in South America. Tank had what appeared at first sight to be good credentials, having designed the Focke Wulf 190 aircraft which had given Germany air superiority for eighteen months during the Second World War. As a result of his past reputation, he was chosen to lead the construction of a new Argentine Air Force.

What the authorities failed to realise was that in the few years since the war had ended, aircraft design had moved forward a long way, and there had been a rapid development of jet aircraft of which Tank had little or no real experience—he had not been involved in this critical new phase. Moreover, he seemed to have no awareness of his limitations. He talked glowingly about nuclear powered aircraft but this was beyond Argentina's ability and so they settled for more conventional designs. At least they chose the lesser of two evils, but even so Tank managed to create havoc. He brought in the Pulqui Mark 1, which lost three test pilots in crashes before being scrapped as an idea to be replaced by the Pulqui Mark 2, which also crashed in trials killing more. We do not know how much this was due to poor manufacturing standards or poor pilots, but in any event the authorities dispensed with Tank's services and he went off to Egypt and then India, where he designed the Hal-Marut, a plane suitable for twice the speed of sound but with engines capable only of less than half that speed. He was still dreaming of nuclear powered aircraft.

For a brief period Peron was besotted with Tank until he started decimating the Argentine's supply of test pilots. In this golden period Tank also sold Peron on nuclear ideas and brought even greater financial ruin as a result.

In pursuing his dream of nuclear powered aircraft Tank came to know an Austrian refugee named Richter who had been a nuclear physicist in Hitler's Germany and was another of the new generation of South Americans. Richter told Tank of his own revolutionary ideas—that three years after the explosion of the first hydrogen bomb, he knew how to make a controlled fusion reactor. All he needed was the money and support and he could produce it for Argentina using local materials.

Richter's story was unearthed by Mario Mariscotti who wrote a book in Spanish called *The Secret of Huemmel Island* in which he tells how Argentina, during the Peronist regime, had mistakenly thought that fusion was its for the asking. Mariscotti is a professor of physics at the University of Buenos Aires where he had become intrigued by the fact that the university had an old cyclotron[2], a primitive atom smasher built in the early 1950s, which would make it among the first half-dozen to have been built in the world. The cyclotron was a research facility through into the 1980s, but was an expensive and remarkable piece of equipment to have appeared in the 1950s in a country where, neither before nor since, has there been much cash for anything scientific. It turned out to have been bought by Richter, and as such was at least one good thing that came from his connections with Peron. Mariscotti was intrigued and as there are no government archives he had to piece the story together.

He learned that Tank had introduced Richter to Peron and the physicist managed to persuade the dictator that he could project Argentina into the 21st century ahead of its competitors and, in turn, make Peron appear one of the most enlightened leaders of the century. Needless to say, this argument impressed Peron greatly.

Peron gave him Huemmel Island, in a lake in the foothills of the Andes, to build the reactor. How much money went with this is not known but the scale of the ruins—the shell of the reactor is still visible—suggest that it must have been several million dollars even in 1950s prices, perhaps 1 per cent of the gross national product of Argentina, or roughly the equivalent of a whole year's research budget for a developed country.

His experiment consisted of a huge spark-gap into which he passed a mixture of hydrogen and lithium gases. The spark was supposed to ignite the fusion reaction or miniature thermonuclear explosion. Contrast what Richter was doing with what happens in the Sun. The centre of the Sun is at a temperature of fifteen million degrees and at a density 150 times that of water; in Richter's experiment, the density of the gases would be a mere fraction of water density and the spark that was supposed to ignite the thermonuclear explosion would have a temperature of only a few thousand degrees.

After only three weeks Richter announced his experiment to be a success. No-one appears to have questioned this claim, examined the data, or checked whether any experiment had even been done. Almost before this 'success' was announced, Richter was building on Huemmel Island 'The Reactor' that would power the nation. He decided it should be ten metres high with four-metre thick reinforced concrete walls but the workmen were sloppy and the concrete cracked. This first reactor was demolished and a second one begun, even larger than the first, and built in a hole in the ground—the same scale of construction as a U-boat shelter.

Before the reactor had a chance to work (or more accurately, fail) Peron fell from power. The new military leaders realised that Richter and the entire enterprise was flawed, but didn't know what to do with him. Then they discovered that Richter had tried to enhance the energy at the centre of his reactor by supplementing the sparks with acoustic energy. To do this he had bought some loudspeakers from the government and the sparks had damaged them. The regime thus charged him with damage of government property and jailed him for a few months until he was released under a general amnesty. In the 1980s he was living in comfortable retirement in a suburb of Buenos Aires, still convinced that his experiment worked.

This happened because one man, Peron, had a dream. Richter was taken on board because Tank was in favour and recommended him. There was no consultation, no 'peer review', no controls built into the system. We may compare and contrast this with reactions in the USA in 1989 to requests for funding of research into fusion following the announcements from Utah. Achieving fusion as an energy source is a dream today made more urgent by the fear of impending environmental disaster.


## 'The mighty ZETA'

Fifty miles west of London and fifteen miles south of Oxford in the English countryside is the United Kingdom Atomic Energy Authority's laboratory at Harwell. During the 1930s and throughout the Second World War the site had been an isolated airfield and it was here, immediately following the end of the war and a clamp-down on the interchange of information with the USA, that it was decided to set up Britain's civil atomic energy programme. The red-brick government-issue buildings stand isolated within a security fence, surrounded even today only by a few houses, fields, orchards and small villages housing some of Britain's finest racehorse stables.

In 1958 Britain's Sir John Cockroft, Nobel Laureate in Physics and Director of Harwell Laboratory, under pressure from the media hesitatingly announced that he was '90 per cent certain' that the laboratory's ZETA (Zero Energy Thermonuclear Assembly) machine had produced a controlled thermonuclear fusion reaction. Both Sir John and the Harwell Lab had the highest reputations, and so the confident media were unequivocal.

The media wanted a story of national scientific success, to restore public pride following the launch of the Soviet Sputnik. The media pressures led Sir John to speculate more than the scientific data allowed and the national reaction was out of all proportion to the real facts. Here is a cautionary tale with some parallels repeated in the 1989 episode, and also some differences.


The early work on controlled fusion had begun in the 1940s independently in the USA, USSR and Britain. Among the people involved in these conceptual developments were Andrei Sakharov in the USSR, Edward Teller in the USA and in the British universities G. P. Thomson at Imperial College and Peter Thonemann in Oxford. It was realised very early that the fusion of deuterium nuclei (deuterons) could prove to be a copious source of neutrons with possible military applications. The work was classified secret and in Britain the university workers joined closed laboratories, Thomson's group moving to AEI (Associated Electrical Industries) Laboratories at Aldermaston and Thonemann moving to the Atomic Energy Research Establishment (AERE) at Harwell.

News of Richter's work in Argentina, claiming to produce energy from nuclear reactions involving light elements, appeared in the media in 1951. Although these turned out to be insubstantial, for a while they created concern from a security point of view.

The 1950s were the height of the Cold War, and research into thermonuclear energy release was hardly discussed among different institutions in a single country let alone between nations. The USA exploded its first thermonuclear device in 1952, and within a year the USSR had done likewise. Work on controlled thermonuclear fusion proceeded secretly in both nations and in Britain. But in April 1956 an unexpected development took place: Kruschev and Bulganin, the Soviet leaders, visited Britain accompanied by the physicist Igor V. Kurchatov, Director of the Institute of Atomic Energy of the Soviet Academy of Sciences, who looked like a bearded patriarch. He had led the team that developed the Soviet Union's atomic bomb and had begun seeking peaceful applications of nuclear power. These included the USSR's first nuclear power station and the nuclear-powered ice-breaker *Lenin*. Kurchatov visited Harwell and suggested that he might give a lecture on Soviet work into thermonuclear reactions in gas discharges.

This created a minor panic. Work was going on in this specific area at Harwell, where it was classified secret, and now a leading Soviet was offering to come and tell all! The offer was accepted and the talk given on 25 April 1956 in the main Harwell lecture theatre (now known as Cockroft Hall) was attended by laboratory scientists and G. P. Thomson's AEI team from Aldermaston. However, Sir John Cockroft was in the USA at the time, did not attend the lecture and so was unable to get at first hand insights into what the Soviets had, and had not, done.

Among the products of the fusion between two deuterons (*dd* fusion) are neutrons. None had been seen at that time in the Harwell experiments that preceded ZETA but Kurchatov announced that the Soviets were seeing some neutrons which were due to fusion occurring, but were unlikely to be from thermonuclear fusion because the rate of their production did not vary with current in the expected way.

Once this was in the open there was less reason to maintain such extreme secrecy about the fusion research programmes and some relaxation occurred.

Since 1955 at Harwell construction had been taking place of ZETA—a 3 metre diameter doughnut shaped vacuum chamber containing deuterium plasma. Experiments with ZETA were underway in 1957 and Thonemann's team successfully demonstrated that they could produce quasi-stable highly ionised plasmas with densities of $10^{13}$ to $10^{14}$ (ten to one hundred thousand billion) particles per $cm^3$, at temperatures of about 5 million degrees for times of milliseconds. These times were long enough to enable them to study the behaviour of the plasma and its response to the magnetic forces, necessary first steps along the way to controlled thermonuclear fusion.

The team was systematically searching for neutrons, a sign of fusion, and it was in August 1957 that they first appeared. That night the team changed the plasma, replacing the deuterium with hydrogen, and the neutrons disappeared. This control experiment[3] showed that deuterium was important in producing the neutrons and was suggestive that fusion between two deuterons (*dd* fusion) was taking place. A more definitive test involved measuring their energies.

When two deuterons fuse together at low energy, the neutron is liberated with an energy of 2.45 MeV. The ZETA team detected these neutrons using scintillation counters and boron counters, and by activating silver (neutrons have a large 'cross-section'—propensity to interact—with boron and silver, hence the use of these elements to detect them) and showed that fusion was taking place in ZETA on a substantial scale. This was a significant discovery in that fusion had not been seen in toroidal discharges before.[4] The question that then needed to be answered was whether the *dd* fusion was being induced by collisions within the hot plasma—'thermonuclear' fusion—or from some other non-thermonuclear processes such as collisions between deuterons that had been accelerated to high energies during the initial pulse of current and other deuterons of comparatively low energies. Only in the former case, where the plasma is in equilibrium, its particles randomly moving around with a distribution of energies, does the concept of 'temperature' have a true meaning and the possibility of sustained *thermo*-nuclear fusion exist.

In a thermonuclear fusion reaction, owing to the completely random directions in which the particles of the hot plasma move, the mean energy of the neutrons emitted will be independent of their direction. However, in the non-thermonuclear fusion process above, a neutron emitted in the direction of the moving deuteron will have a higher energy than one emitted in the opposite direction, much as an object thrown from a moving vehicle travels faster when thrown forwards. To tell which of these was the case required measuring the neutron energies precisely enough to get a sense of their distribution.

The energy of the neutrons had been measured by means of scintillation counters where neutrons knock on (electrically charged) protons, akin to H. G. Wells' invisible man being detected in a crowd by his bumping into the visible people. This technique is sufficient to measure energies and identify *dd* fusion, but does not have the resolving power to give the mean energy and its distribution.

This was stated clearly in the paper that Thonemann and his team published in *Nature* on 25 January 1958: 'To identify a thermonuclear process it is necessary to show that random collisions in the gas between deuterium ions are responsible for the nuclear reactions. In principle, this can be done by calculating the velocity distribution of the reacting deuterium ions from an exact determination of both the energy and direction of emission of the neutrons. The neutron flux so far obtained is insufficient to attain the desired accuracy of measurement.'

Sir John Cockroft, the Director of Harwell, had won a Nobel Prize for his 1932 work where, with the prototype high energy nuclear accelerator, he and Ernest Walton had made the first artificial disintegration, or *fission*, of atomic nuclei. Here with ZETA he was overseeing the first example of the reverse process—the controlled *fusion* of light elements and detection of the products, the neutrons. His 1932 work had been greeted with great interest and in 1958 he rightly decided that the news from ZETA merited a press conference. The publication of the paper in *Nature* was thus accompanied by a press conference at Harwell, which was held in the hall where Kurchatov

40

had spoken nearly two years earlier. Although the paper stated that the measurements of the neutrons were not accurate enough to determine if the fusion was *thermo*-nuclear, there was a great desire by the media that Britain's prestige should be boosted by the ZETA 'triumph' following the nation's 'backwardness in Sputnik research' (the Soviet satellite having been launched in October 1957).

Cockroft clearly believed that the source of the neutrons was thermonuclear fusion. When the media sought assurances that they were, and received no response from the ZETA team, they continued to pressure and asked Cockroft to quantify his expectations. In one of the few misjudgements of his brilliant career Cockroft said that he was '90 per cent certain' that thermonuclear fusion was occurring.

That was the 'quotable quote' which the media picked up and put into the headlines. The *Daily Mail* proudly stated 'The Mighty ZETA: Limitless Fuel for Millions of Years', the *Daily Telegraph* was only slightly more reticent: 'Hydrogen power 90 per cent certain; Energy from sea in twenty years; Machine will surpass heat of Sun', and on television one scientist said 'This discovery is greater than the Russian Sputnik.'

The British public gained an expectation that was out of all proportion to the facts. None of the ZETA team had claimed this at the press conference or in the *Nature* article, though, hedging bets, the paper also said this is 'not inconsistent with . . . a thermonuclear process'. Yet the editorial in *Nature* said that a '*thermonuclear* (my italics) reaction has now been made' and the article was headlined 'Controlled Release of Thermonuclear Energy'.

At this stage there was still the possibility that the neutrons were indeed from thermonuclear fusion and the proof awaited precision measurements of the ZETA team's energy spectrum.

ZETA had been behind closed doors in Hangar 7, a huge aircraft hangar, 100 m long, which is now converted into a lab. At the other end of the building some very different experiments had been going on involving nuclear physicists who knew of the existence but few of the details of ZETA. Among these were Eric Taylor and Ted Wood who entered into a collaboration with Basil Rose, who specialised in rather different areas of low energy nuclear physics and was a member of the cyclotron[2] group at the lab.

Rose had been present in the audience at the press conference and afterwards commented to Taylor that 'What is needed is a detector of neutrons that has good energy resolution and is very efficient' (so as to be able to determine a distribution of energies of the neutrons and determine whether they were being produced uniformly in all directions—as in thermonuclear fusion from a static plasma—or instead were on the average moving in some particular direction). By chance

Rose has spoken to the right person as Taylor had been working with a hydrogen diffusion cloud chamber which is potentially well suited for such a task. So their collaboration was born and, as Rose remarked to me, shows 'the importance of having people from different disciplines talking with one another and pooling their expertise.'

A cloud chamber is good for detecting low energy, electrically charged particles as they leave behind a trail of liquid drops as they pass through the mist in the chamber. The resulting trail is similar to the vapour trail of a high flying aircraft, showing where the plane, or atomic particle, has been. Neutrons, being electrically neutral, do not leave a direct trail, but when they hit protons in the hydrogen-filled chamber the protons recoil, and it is their trails that reveal the impact of the neutrons. And as protons have essentially the same mass as neutrons the range of the proton trail and its direction with respect to the incident neutron beam[5] accurately determines the energy of the incident neutron. So this is what the trio took over to ZETA to make an accurate measure of the neutron energies.

A small hole in the shielding around ZETA allowed neutrons from a discharge to escape and to enter the cloud chamber. Under the normal conditions of operation, the current in ZETA flowed away from the hole and the cloud chamber. In the experiments of Rose's group, ZETA was also operated in 'reverse', where the current flowed towards the chamber. The trails in the cloud chamber enabled Rose and his colleagues to determine the mean energies of the neutrons in the two configurations. If the neutrons had been produced by thermonuclear fusion in a stationary plasma the neutrons' energy spectra would have been the same in both cases, peaked at 2.45 MeV. However, what they found was that the peak was at around 2.3 MeV in the normal mode (current flowing away from the detector) and 2.6 MeV in the reverse case (current flowing towards the detector)—like the object thrown from the moving vehicle against or with the motion.

It was still possible that this was thermonuclear fusion taking place in a mass of gas moving in the same direction as the current, but other measurements made this seem very unlikely. The explanation offered by Rose and colleagues was that the neutrons were produced by collisions between slow deuterons and deuterons moving with an energy of nearly 20 keV in the direction of the current. ZETA was 'the biggest low energy accelerator in existence'.

This result was an embarrassment following the media stories of January, but there was no hesitation about publishing it. The media again dealt with this in mixed ways. The *Daily Express* announced 'Shock for Britain, ZETA is not the success claimed'. It might more accurately have added '. . . claimed, by the media'. In fact, the scientists were making considerable progress. The first controlled fusion reactions had been seen and their nature was now being clarified. Thus Harwell scientists were, in effect, now able to 'look inside' ZETA and were learning more about the behaviour of the plasma. *The Times* of 16 June was nearer the mark when it said '. . . comments have gone farther in the direction of pessimism than appears to be justified' and reminded readers that 'the objects of the ZETA experiments were to produce and maintain high temperatures and to study the conditions in which these temperatures are produced as an aid to later stages in design.'

Morals from this episode that may have some parallels with the test-tube fusion story of 1989 include the delicate balance between the world of science and the media. When there is a great wish for something to be true but the evidence for it is not yet complete, it can require great strength to resist the pressure to say more than the data warrant. Apparently innocent remarks, made parenthetically, can end up as lead headlines and the caveats can be lost, and 'quotable quotes' get quoted, distorting the real message. The media so built up the expectations for ZETA after the first conference that the subsequent news was perceived as 'failure'.

However, there are also significant differences with both Richter in Argentina and the 1989 episode in Utah. For ZETA none of the published experimental results was wrong, the limits of accuracy of the experiment were clearly spelled out and the main results were soon confirmed by other groups—Aldermaston in Britain and Los Alamos in the USA. By contrast, in 1989 not one of Fleischmann and Pons' published claims of significant nuclear products stood up to scientific scrutiny.

Apart from the embarrassment, the exposure of ZETA was beneficial. It became an essential research tool in the early quest for controlled thermonuclear fusion, helped to clarify the scope of the problem of heating, containing and controlling the deuterium fuel, and set the parameters for the development of the subsequent fusion research programme. If practical hot fusion is ever achieved, ZETA will have helped it happen.

It also had some immediate benefits for the fusion programme in Britain. Several of the people involved in the ZETA story, directly or as spectators, are still working or living in retirement in and around Harwell: many of them feel that the publicity and interest generated in ZETA made many politicians and administrators aware of fusion, and helped to gain support for the funding of Culham Fusion Laboratory which was built during the early 1960s. Culham has been doing forefront work in the fusion field for over two decades, and today is the site of the Joint European Torus (JET). The scale of the enterprise is so large that fourteen nations combine their efforts at this single lab. It has achieved temperatures of 250 million degrees in its fuel and plans a series of experiments into the early 1990s. Its goal is getting nearer but is still tantalisingly out of reach. Similar efforts are taking place in Japan, the USSR and the USA.

Culham is only six miles from Harwell; for many years a model of ZETA was on display in its entrance hall as a reminder of Britain's heritage in this field.

## JET, the Joint European Torus

The European Fusion Programme is centred on Culham in Britain, home of JET, the Joint European Torus. This tokamak consists of D-shaped magnets forming a doughnut 4 metres high and 2.5 metres across. The fusion research programme at JET began in 1983 and will run until 1996 according to present plans.

There are three separate criteria that must be met before a commercial fusion reactor can be built: temperature, density and 'confinement time' must all exceed certain minimum values.

The temperature will have to exceed 100 million degrees (or 10 keV in the units the fusion scientists use; 1 keV corresponds to 11.6 million degrees). Below this temperature any fusion is too slow to sustain. Second, the plasma must be dense enough—enough particles in a given volume—that individual particles have a good chance to find each other; this critical density is of the order of $10^{20}$ ions per cubic metre. Third, the plasma has to retain its energy for at least a second in order that enough reactions take place to deliver more energy than was consumed in setting the system up.

By 1989 JET had been able to exceed each of these values individually but has not achieved them simultaneously. Its highest temperature is 28 keV, density $1.2 \times 10^{20}$ and energy retention time 1.2 seconds. The problem is to achieve them together in such a way that the 'fusion product', the multiple of all three quantities, exceeds a certain critical value. The fusion product has unusual units: keV sec per $m^3$; the keV is the temperature, sec comes from the confinement time and per $m^3$ gives the density. The goal is for its value to exceed $50 \times 10^{20}$ keV sec per $m^3$.

Progress at JET can be measured by the improving value achieved for the fusion product. In the early 1970s when the idea of JET was first conceived, the best recorded value fell short by a factor of 25 000; by 1990 JET was within a factor of 20.[6]

JET heats the plasma with RF (radiofrequency) power, like a

microwave oven. The ions in the plasma gyrate back and forth in JET's magnetic containment field, and by tuning the RF power to this frequency, the radio waves resonate with the ions, passing on energy and raising the plasma's temperature. The technological catch though is that the confinement time drops as the RF power pushes the temperature up, and so the critical value of the fusion product is still out of reach. In future machines the aim will be to raise the temperature without need for so much extra heating.

There are two stages to a working reactor. First one must achieve 'breakeven' where fusion reactions produce as much energy as go into the plasma. JET is very near to this and its present programme may achieve this. However, this will still be a factor of five away from the full goal: 'ignition'. Ignition involves fusion which produces enough energy both to heat the fuel and to maintain the temperature of the plasma as new fuel is added.

So far JET has used only deuterium as fuel. In the final phase of the project, in 1991, tritium may go into the machine.[7] This will yield information on the behaviour of a genuine thermonuclear plasma, but has hazards in that tritium is radioactive and the reactions will produce so many neutrons that the machine will become radioactive. Workers will be unable to have unprotected access to the device and remote handling will be needed for the innards.

## European problems

Even though JET is moving the search for economic fusion forward, not all are convinced that it is worthwhile. The European Panel on Science and Technology Options Assessment (STOA) commissioned a report on fusion in 1988. Written by Colin Sweet of the London-based Centre for Energy Studies, it appeared at a sensitive time when budget wrangling threatened to delay fusion. Sweet's report cast doubts on the eventual feasibility of fusion as a viable power source and implied that the large sums of money could be better used elsewhere. 'Forecasts about fusion have become couched in an aura of over-optimism' he wrote. This appeared while the European Commission had yet to approve the 600 million pounds fusion budget. 'A full and proper economic evaluation is overdue and should be done as a precondition to the allocation of resources beyond 1991.'

This uncertainty hung over the European Fusion Programme at the start of 1989 and in Britain, home of JET, there were further questions being asked about the future of fusion and nuclear energy in general.[8]

Electricity in Britain was then produced by the CEGB (Central Electricity Generating Board): a state-run organisation which the Thatcher government was preparing to 'privatise'—turn into a public company. In 1988 the CEGB spent 150 million pounds on nuclear energy research, of which twenty million went into studying environmental effects. The UKAEA (United Kingdom Atomic Energy Authority) is the biggest outside contractor to the CEGB, and the research cost the CEGB 65 million pounds a year. Privatisation has now split the CEGB into three and the environmental research and development work is to be cut back.[9] The place of nuclear research in the relationship between the CEGB and the UKAEA was unclear at the turn of the year, and then in February 1989 the House of Lords Select Committee on Science and Technology published a report on Research and Development in nuclear power.[10] They saw little point in continuing home-grown fusion research and advised that future work should have a European dimension. This reinforced the government's decision of October 1988 to trim five million pounds from the UK fusion budget; this did not affect JET but threatened 150 jobs at the adjacent Culham Laboratory.[11] The House of Lords Committee report was even more cause for pessimism, suggesting that the UKAEA should brace itself for major cutbacks including site closures. The Committee noted that the UKAEA has a 'magnificent record in scientific and technical achievements' but stressed that future research and development in nuclear energy will attract considerably less resources than in the past. Restructuring of the Atomic Energy Authority would have to be done, but there was concern whether it would be able to find enough non-nuclear work to take up the slack. The promise of test-tube fusion could open up new avenues for them.

## Hot fusion in the USA

The magnetic confinement fusion programme cost the US taxpayers 350 million dollars in 1984 and funding from Congress has remained at this annual figure ever since, even though inflation has continued to eat into it. Salaries of the many scientists, support staff and engineers and the overheads all inflate, using up increasing amounts of the money even before any experiments begin, so real research dollars in 1989 were probably less than one half of what they were five years earlier. By 1989 many fusion scientists in the USA were becoming increasingly nervous about their future. Research programmes were being stretched out in time and slowed down, and experiments were having to be cancelled as priorities were reassessed.[12]

One example is that of Princeton University, home of the Tokamak Fusion Test Reactor, or TFTR. Experiments have been using deuterium alone, and the temperature of the plasma has been pushed up to 300 million degrees. This is hot enough for fusion to occur but so far they have been unable to confine the hot plasma long enough and at high enough densities to generate more energy from fusion than is used to keep the machine working in the first place. What they want to do next is to use deuterium and tritium together in the machine, since the *dt* reaction releases more energy per fusion than does the *dd*, and so Princeton hope that the faster energy release might enable them to cross the breakeven threshold. But tritium is not easy to come by, nor is it cheap. Due to the shortage of money, Princeton have had to postpone this experiment time after time.

There were further problems for hot fusion in 1989 with the arrival of the new administration of President Bush. This brought in new people such as Robert Hunter as head of the Department of Energy Office of Basic Research, and one of his first acts was to reorder some of the research funds.[13] This caused some research activities to be trimmed at Oak Ridge, Los Alamos and Princeton, where the *dt* experiment was postponed for at least another three years. It also led to some staff cuts and by the spring of 1989 the leaders of the fusion programmes were anticipating that after five years of no increases to match inflation, Congress would make a twenty million dollar *cut* in the fusion budget. The labs were already trying to plan how to deal with that.

The *Catch 22* problem was that for several years there had been few big headlines about fusion as no dramatic breakthroughs had been occurring. So public and congressional interest was low and this in turn kept budgets tight, preventing the building of the new facilities or running the new experiments such as the *dt* one at Princeton that could produce the breakthrough. Only then would there be the excitement that attracts new funds. Even if fusion becomes an energy producer, realistic commercial production will not be until the mid-21st century, which is beyond the interest span of the average member of Congress.

One project that is in the balance is the Compact Ignition Tokamak, or CIT, that would produce the conditions for the plasma to ignite. Its supporters claim that the CIT would be the first step towards a test reactor that would burn plasma for long periods. There is general agreement in the hot fusion community that CIT will bring significant advances to the field, but the problem is that it will cost 700 million dollars which requires new funds and imposes new pressures both within and outside the community. Within the community the question is which site will win the CIT, and both inside and outside there is the question of 'how much of *my* money will be sacrificed to support the CIT'—areas of science that have no immediate connection to fusion nevertheless get funded from the same sources, such as the Department of Energy. There is little doubt that some of the 700 million dollars would be at the expense of others and some practitioners of 'small' science, whose projects can sink or swim on fluctuations of tens of thousands of dollars, see calls for 700 million with less than total enthusiasm.

Within the fusion community there is argument over the siting of the CIT. One argument is that if building the CIT involves cutting into present research programmes then it may be preferable to run the present experiments for longer. Princeton, which is the dominant fusion lab with funding of 96 million dollars a year and which first proposed the idea of the CIT, is the natural site for it. However, not everyone agrees with this. As the overall budget shrinks, Princeton's star status causes tensions which will increase further if CIT exacerbates the financial disparities. Ronald Parker of the Massachusetts Institute of Technology has proposed that the CIT be built on a remote site rather than in Princeton town since it will use radioactive tritium, and while the amounts pose no threat, nonetheless people will protest and this could delay approval: best be realistic and avoid potential trouble at the outset. Arguments against this include: 'MIT want to keep power away from Princeton', and 'putting it elsewhere costs another 250 million and there will be protests anyway'.

So in early 1989 the US hot fusion community faced some internal dissension and considerable financial pressure. Self-protection was a high priority for these Goliaths when along comes David in the form of two chemists who announce that they can perform fusion in a glass of water! Naturally the DOE take notice since if this is true it will solve their financial problems. Naturally the hot fusion teams sit up. Within days of the Utah press conference Representative Walker convinced some members of the US Congressional Energy Research and Development Committee to shift five million dollars away *from* hot fusion *into* test-tube fusion. So already the hot fusion community was under threat and facing possible redirection of its projects.

The full committee refused to redirect the funds and, paradoxically, the very fact that test-tube *fusion* is news has grabbed public attention and ironically could be the headline that revitalises interest and in the longer term attracts money from Congress—for *hot* fusion. This was acknowledged in the US Congressional Hearings on test-tube fusion held on 26 April, 1989. Mike Saltmarsh, Director of Fusion Energy at Oak Ridge National Laboratory said 'Whatever the final outcome, the renewed discussion of the potential promise of controlled fusion power has been very healthy'. And Harold Furth, Director of the Princeton University Plasma Physics Laboratory made similar remarks: 'Whatever the reality and utility of "test-tube fusion energy", some progress may have been made . . . in focusing the world's attention on the potential value of a realistic long-term strategy for the achievement of the fusion energy goal.' So as with the birth of Culham in Britain, there was again the possibility that fusion could profit.

## Tritium and the military connection

Fusion takes place in the Sun—protons fusing to deuterons eventually producing helium and liberating heat. Although the detailed processes and fuels differ from those in the hot fusion programme, the end result—self-sustained power—is the same.

Uncontrolled fusion has been achieved since 1952 with the detonation of thermonuclear weapons. The end result, destructive explosions, differs from magnetic fusion but the fuel, tritium, is the same. Tritium is an essential fuel in thermonuclear weapons; it is also a product of *dd* fusion—the very process that the Utah chemists claimed to be able to make happen inexpensively in a test tube. The US military were already spending vast sums on making tritium for warheads and the reactors that were used for this process had been closed, pending repairs, in 1988 as a result of nervousness about reactor safety following the Chernobyl accident. The repair and building new reactors would cost billions of dollars, so when test-tube fusion entered the scene the military took note at once, recognising the potential of test-tube fusion as a source of the much-needed tritium. This sort of application of test-tube fusion also impressed Indian Government scientists who decided that the western nations would soon classify test-tube fusion as a secret; thus India mounted an immediate test-tube fusion research effort so as to 'get in on the ground floor'.

The US owns about 100 kilogrammes of tritium, including what is stored in bombs. Tritium has a 12.3 year half-life, which is the same as saying that 5.5 per cent of its atoms decay each year. So to keep its 100 kg stock of tritium fresh, it has to manufacture 5 to 6 kg annually.

To make tritium you can fuse a neutron onto a deuteron, or alternatively add a neutron to $^6Li$ (lithium-6) which splits to yield one atom of tritium and one of $^4He$. The $^6Li$ reacts more easily and so this is the approach used the most. The USDOE (Department of

49

Energy) has done this since the late 1940s at the Savannah River Reactor on the border of Georgia and South Carolina. There are four reactors which burn enriched uranium (containing 80 per cent $^{235}U$) as fuel. This produces the intense flux of neutrons; heavy water flows through the reactor to slow down the neutrons so that they react better with the lithium.

The lithium is stored in aluminium, stays in the reactor for up to a year where it is bombarded with neutrons. The tritium produced is held within the aluminium and released as gas after heating and then stored in metal canisters. The military reactors at Savannah River are getting old, one having been closed permanently and the other three being closed pending extensive repairs. Therefore the DOE may have to build new reactors at a cost of 8 billion dollars, and it will be ten years before these are ready.

The US military is worried about the long delay while the new facilities are being built and also that the US is selling tritium to civilian users in the US, Britain, Japan and Canada. As a result sales have become limited, and in particular this affects the fusion programmes which want tritium for *dt* fusion.

In Britain too there are impending problems. British Nuclear Fuels (BNF) makes tritium at Chapel Cross in South-West Scotland, using neutrons made in an ageing Magnox reactor. This is due to close by the mid-1990s and BNF is unlikely to have replacement reactors ready until several years after this.

The military and political pressures to find new ways of making tritium and the vast amounts of money involved were stimulating early in 1989 new ideas on how to make fusion—for tritium production. Bogdan Maglic, Director of the Advanced Physics Corporation in Princeton New Jersey proposed accelerating heavy ion beams in closed collision orbits. This could make one gram of tritium in half a day and as a byproduct give back some of the electrical power—one megawatt—from the fusion processes involved. This is a long way short of breakeven since much energy is needed to accelerate the ions in the beams, but breakeven is not the aim here; the goal is tritium for the military. Other ideas were around and one cynic remarked 'Everyone and their grandmother is trying to get money based on the US tritium shortage.'

Only a few months earlier the USDOE had decided that the prospect of nuclear fusion by the 'conventional' route was still too many years away to meet the Pentagon's urgent needs for tritium. So when Fleischmann and Pons announced test-tube fusion as a source of energy—which was the 'angle' that the media took up and portrayed it as a clean source—the news that they apparently saw tritium as a fusion product was lost on most media, but it made many scientists concerned and others excited.

Hot fusion budgets on the line; tritium production for weapons; India believing that test-tube fusion would become a classified secret in the West. So test-tube fusion was more than just a media event with dollars on the line for the University of Utah, there were many private agendas riding along with it. As a result it was easy for protagonists to claim that opinions on test-tube fusion, particularly its validity or otherwise, weren't always made on purely scientific grounds, but that self interests were the driving forces.

# 4
## COLD FUSION

Cold fusion is the brainchild of Charles Frank, who first came up with the idea in 1947.[1] He was at Bristol University in England where the physics department was leading the world in the study of cosmic rays, and whose head, Cecil Powell, was in process of winning the Nobel Prize for his role in this.

The Bristol team was in the process of discovering that in addition to the light electron and heavy proton and neutron there are particles of intermediate masses which they named 'mesotrons'. In particular they identified two mesotrons with nearly identical masses but different properties: the slightly heavier one they named 'pion' and the lighter one 'muon'. The pion is unstable and decays, producing the lighter muon in the process. The idea of two particles so near in mass was troubling for some people and made them wonder if some misinterpretations had occurred; in particular Patrick Blackett, another Nobel Laureate, raised a question as to whether there might be other explanations of the pion decay. Charles Frank looked into this and showed that the pion decay interpretation was correct, but it was in answering that question that Frank had his brainwave and stimulated a whole new field of research.

We are continuously bombarded by cosmic rays, atomic particles ejected by distant stellar catastrophes. Whirled through the magnetic fields of space until entrapped by the Earth's own magnetic field, they crash into the upper atmosphere shattering the nuclei of atoms and projecting fragments and other particles down to Earth. At ground level we are protected from their full force by the intervening umbrella of air but, nonetheless, some of these particles manage to penetrate. Among them are muons, negatively charged particles similar to electrons in all respects but for the fact that they are unstable and are 207 times heavier; they are in effect 'heavy electrons' and can replace electrons in atoms to make 'muonic atoms'.

The simplest example is muonic hydrogen, where a single muon encircles a proton.

In ordinary hydrogen the bulky proton is almost at rest relative to the flighty lightweight electron which encircles it. The mean distance of the electron from the central proton defines the atom's size. Now suppose that the electron's mass is increased 207 times (so that it is like a muon) while it remains subject to the same attractive force holding it around the proton. Quantum theory implies that it would have to move 207 times closer to the central proton to maintain the stability of the atom. Now we cannot change an electron's mass but we can replace an electron by a muon which, being 207 times heavier, is almost equivalent. Thus muonic hydrogen is similar to ordinary hydrogen but for two important features. First, the muon being 207 times heavier than the electron causes it to orbit the proton 207 times closer in, so muonic hydrogen is 207 times smaller than ordinary hydrogen. The second difference is that muons are unstable, having a mean life of only 2.2 millionths of a second, so the muonic hydrogen atom doesn't live long.

Frank came up with his idea of muonic atoms and the consequent possibility of cold fusion as the result of a picture of the cosmic particles that the Bristol physicists had obtained.

Cecil Powell's contribution had been to take images of the particles by sending emulsions to the upper atmosphere in balloons, where cosmic rays hit them. Then the balloons return to ground, and the pictures are developed. Atomic particles in the cosmic rays show up as dark tracks, and then the hard work comes in interpreting them, deciphering their code, and learning about the basic entities of matter. It was from this that Powell helped to establish the identity of the muon as a heavy sibling of the electron and distinct from another particle, the 'pi-meson' or 'pion'.

In 1947 tracks of mesotrons (pions, muons and possibly other new particles) were very much the centre of attention in Bristol. The physicists lived for the emulsion trails and double checked to see if they were interpreting them correctly. One of the images showed the trail of a particle more massive than an electron which suddenly turned through a sharp angle making an L-shape, and whose thickness and character changed at the kink. This suggested that a heavier mesotron (the pion) had died and that the second branch of the L was due to its progeny—a muon. This image of a pion transmuting was as important to particle physics as Rutherford's discovery of the transmutation of the elements had been 40 years earlier.

But was this the right interpretation? There was another possibility for this L-shaped trail: rather than showing the death of a new mesotron it might instead show a known muon that somehow had been captured by the matter around it, extracted energy from it and then shot off again: this was the suggestion that Blackett had made, and Charles Frank began to think about it. He concluded that the picture indeed showed the death of a pi-meson, which was important in the history of particle physics, but the idea of capture and energy extraction—not by the pi but by the *muon*—then began to develop in his mind.[2]

He realised first that a muon in the vicinity of a hydrogen atom would orbit 207 times nearer to the central nucleus than would the electron, and so the muon would get inside the electron's orbit and take over the electrical attraction of the proton; it is an atomic 'excuse me' dance routine. The electron now finds itself *outside* an electrically neutral system, namely the positive proton and the close-in negative muon; the electron escapes and where once was a hydrogen atom is now muonic hydrogen.

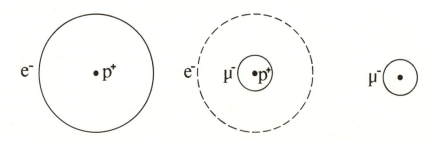

Figure 2. Muonic hydrogen atom. A hydrogen atom (a) is invaded by a muon, μ (b) which orbits the proton more closely than does the electron. The electron's attraction for the positively charged proton is shielded and the electron is released, leaving a muonic hydrogen atom (c).

So it appeared that making muonic hydrogen would be straightforward. What's more, this would be so small—207 times smaller than ordinary hydrogen atoms—that it would behave much like a neutralised proton, or 'neutron'. And here was the bonus: the positive charge of the proton is so effectively shielded that it will now be able to encroach much closer to the nucleus of a neighbouring atom without being repelled; the chance of bumping into it and undergoing nuclear fusion, 'cold fusion', thereby became a real possibility.

The fusion would be between the proton of the muonic hydrogen atom and the nucleus of one of the atoms in the surrounding material. In the presence of deuterium, for example, the proton fuses with the deuterium nucleus making helium-3 and releasing energy—but no neutron signal this time. This happens very quickly and the energy genrated by the fusion can kick out the muon, releasing it to play its role as matchmaker again, acting as a 'catalyst': hence 'muon catalysed fusion' was born.

In 1947 when Frank had his insight the existence of muons had already been known for ten years and the existence of muonic atoms had also been predicted. However, nobody had made the conclusive step of considering the consequences of this, namely the chain of muon catalysis reactions. Frank had come up with his idea due to the accident of circumstances: the work surrounding him in Bristol, and the association of ideas—Blackett's idea for the pi-meson being applied successfully for the muon.

The possibility of fusion is the good news. The bad news is the second difference from ordinary hydrogen that I alluded to, namely that whereas hydrogen is stable, muonic hydrogen is not. Living only for a couple of microseconds on average, the muon cannot go from atom to atom catalysing fusion and liberating energy indefinitely; the dream of the ages isn't so easily fulfilled. And further hindering the fusion is the fact that the produced helium nucleus has positive charge and attracts the muon to it, away from the deuterium where it is wanted. The longer that the muon sticks to the helium, the less time is left for it to catalyse further fusions. Nonetheless, two microseconds is a long time compared with many things that happen in and around the atomic nucleus, and the muon has a chance to initiate several fusions before it dies. The crucial question is whether it manages to liberate at least as much energy from fusions as was expended in running a particle accelerator to produce that muon in the first place.

Today this ideal is tantalisingly just outside of reach. It is trying to understand the details and hopefully finding the key to economic success for muon catalysed fusion that has occupied Steve Jones, and several others, for the last ten years. His inspiration goes back to the early idea of Charles Frank, plus the fact that muon catalysed fusion was observed in 1956 by accident and the interest has grown in fits and starts ever since.

Frank's idea was taken up by Andrei Sakharov in the USSR. In 1948 he wrote an internal report of the Lebedev Physical Institute in Moscow, a paper[3] in which he considered the expected efficiency of muon catalysed fusion, and whether it could be practically useful for producing energy and neutrons.

The first theoretical calculations on various characteristics of muon catalysed fusion appeared during the 1950s, by Yu. B. Zeldovich, Sakharov and S. S. Gershtein in the Soviet Union; the ideas were not much pursued elsewhere.[4] The idea had arisen for Frank through his location and questions that had come up about muons; Sakharov had been developing fusion in secret and picked up the notion because he would have been on the lookout for any papers about fusion, of whatever sort.

But the western world took note when, late in 1956, Luis Alvarez in Berkeley California saw a weird track in a bubble chamber photograph of atomic particles. At first he thought that he had discovered a new particle, but then an astrophysicist friend of one of Alvarez's colleagues[5] recognised that it had the characteristics of proton–deuteron fusion—*pd* fusion—catalysed by a muon. Edward Teller then worked through the implications with Alvarez and, unaware of Frank and Sakharov's works, they reported 'a short but exhilarating experience when we thought we had solved all of the fuel problems of mankind for the rest of time. While everyone else had been trying to solve this problem by heating . . . to millions of degrees, we had apparently stumbled on the solution involving low temperatures instead.'[6]

The announcement of the discovery was made by the Atomic Energy Commission in Washington and the same day, 28 December, the scientists gave a public presentation to their professional colleagues at the American Physical Society (APS) meeting in Monterey California. The *New York Times* headlined 'Atomic energy produced by New, Simpler Method'.[7] Atomic fission had been known to the general public for a decade; the awareness of fusion bombs was only four years old, so the possibility of a 'third and revolutionary way to produce a nuclear reaction without uranium, as in the fusion reaction, or million degree heat as in the fusion reaction' generated a lot of excitement.

The scientists stressed that it was little more than a laboratory curiosity, the energy coming from the fusion of only a few hydrogen atoms and 'was scarcely enough to register on highly sensitive measuring instruments'; although the process had no immediate commercial value it suggested 'possible industrial uses of immeasurable importance'. Alvarez was fully aware of the 'bad news' part of the equation—the muon's brief life which only allowed it to catalyse a handful of fusions before it perished. He commented that if the process was to have practical importance 'we would have to find a different catalysing particle which has properties similar to the muon but has a lifetime of at least ten or twenty minutes.'

J. David Jackson, a young Canadian theoretical physicist working at Princeton New Jersey, read about this discovery in his *New York Times*[8] (plus ça change!) and within a month had made a pioneering and thorough evaluation of muon catalysed fusion, including studies of the 'sticking probability'—namely the chance that the muon sticks to the emitted fusion fragments rather than being released to catalyse more fusion.[9]

The emerging calculations implied that muon catalysed fusion would not release as much energy as was needed to create the muons; even if the muon were as stable as Alvarez's 'ten or twenty minutes' the sticking makes the useful production of energy out of reach. This seemed to kill the idea, and with no obvious practical use it faded from mainstream interest.

The calculations by Frank and the Soviet theoreticians had dealt with the fusion of proton and deuterium or of two deuterium nuclei. The deuterium nucleus, 'deuteron', contains a proton and a neutron; the heavier form of hydrogen called tritium has a nucleus ('triton') consisting of a single proton and two neutrons. Tritium is nasty stuff, and ingesting it can be fatal. Its main use is as the fuel in thermonuclear devices. Recently people have begun to contemplate it as possible fuel in muon catalysed fusion and it is from this that current interest derives. Jackson's 1957 paper had made the first study of muon catalysed fusion in mixtures of deuterium and tritium but had concluded that there too the energy return did not offer much practical hope.

The renaissance of this subject goes back to some Soviet experiments in the mid-60s, where V. P. Dzhelepov and colleagues noticed that muon catalysed fusion of two deuterium nuclei depended on the temperature.[10] This was surprising because temperatures that we are used to in the room involve the motions of atoms and their electrons, but the nuclear processes occur on energy scales utterly unlike these; it is the muon that provides the seed, not the temperature of the room. Some experimentalists publicly stated that, in their opinion, the measurements were wrong!

S. S. Gershtein and his student E. A. Vesman then identified the missing ingredient in the existing theories.[11]

The implantation of muons into materials forms not simply muonic atoms, but also complicated molecules whose nuclei are surrounded partly by muons and partly by electrons. The electrons and muons in these molecules can take up several different configurations, each with their own characteristic energy. Gershtein conjectured that there could be a 'resonance'—that is, if the energy released during muon capture coincided with an energy level in one of the complex molecular states, this molecule could store the energy and the overall chance of interactions occurring would shoot up dramatically. Vesman developed this and showed that he could explain the temperature dependence of muon + DD fusion if there is a rather loosely bound state in a complex molecular system involving three deuterons where the nuclei were encircled partly by electrons and partly by muons. It took ten years of experiments in the Soviet Union to find the definitive evidence and so confirm Vesman's resonance theory.

Leonid I. Ponomarev[12] then applied the resonance theory to the case of deuterium–*tritium* ('DT') fusion and discovered that in a complicated molecule where two deuterons and one triton were encircled by two electrons and one muon the theory predicted that the fusion rate leapt enormously relative to what had been expected from the earlier simple theories. Ponomarev's calculation implied that there were 100 million to a billion fusions per second, that is 100 to 1000 in a muon's lifetime, if nothing else (such as sticking) intervened. This was so important that he and some colleagues spent the next two years computing and checking their calculations until they were totally satisfied with the correctness of the result. Ponomarev then announced it to the world at an International Conference on Nuclear Physics in Vancouver during August 1977.

In 1977 when Ponomarev announced this possibility, no experiments on muon catalysed DT fusion had been done. Each DT fusion released 17 MeV of energy, and so as few as *eight* sequential DT fusions are sufficient to release more energy than is contained in a muon at rest (its 'rest energy' or mass). The DT route suddenly appeared to be promising, though there is still a long way to go as it takes 20 to 40 times this to make a muon in a particle accelerator. V. Dzheplov, who had earlier discovered the temperature dependence of muon catalysed fusion and stimulated first Gershtein and Vesman and now Ponomarev, organised an experimental investigation at Dubna (a nuclear accelerator near Moscow). A crucial test was on 4 June 1979, when a decisive experiment involving muon catalysed fusion in a pure deuterium–tritium mixture started. Ponomarev was at home and awaited news all night long. At last, at 5 a.m., he received a phone-call which said 'Go to bed; we see the neutrons'[13] (the evidence for fusion). Soon thereafter the machine was shut down for renovation and the experiments ended.

But now interest in cold fusion was beginning to grow and experiments began in 1982 at Los Alamos New Mexico. Among the participants was Steven Jones of Brigham Young University.

### Steven Jones

In 1839 Brigham Young, the US religious leader, visited Britain and established a mission that contributed many converts to the Mormon church in America; among the many who emigrated at that time were Thomas and Mary Jones from Llanaber in Wales. The Joneses became Mormon pioneers who, at the request of Brigham Young, founded the town of Mink Creek Idaho just over the border from Utah. Four generations later, in March 1949, Steven Jones was born nearby in Portacello Idaho. Today he is a professor of physics at Brigham Young University (BYU) in Provo Utah, is a world authority on the experimental investigations of muon catalysed fusion, and is a quietly spoken contemplative man with a dry sense of humour who describes his fair slightly greying hair as 'palladium blond'.

His interest in physics developed at high school in Washington State where his family moved in 1953. For an exhibition he built a small electron accelerator with a cloud chamber as detector for the particles produced by the electron collisions. 'It did not work,' he told me with a laugh. In 1967 he entered BYU as an undergraduate student of science. At the end of the first year he went to France and Belgium on a two year mission for the Mormon Church, returning and graduating at BYU in 1973. Having decided that he wanted to make a career in physics he enrolled at Vanderbilt University in Nashville Tennessee to study for his doctorate in experimental high energy particle physics. His thesis project involved experiments that had been anticipated by his exhibition model from school: he spent many months during his studies taking part in experiments at the world's largest (and working!) electron accelerator in Stanford California receiving his PhD in 1978.

His background and adherence to his religious code of conduct has moulded Jones' personality such that he presents a calm exterior yet can fight very strongly for what he believes to be right. He found the fast-paced cut and thrust of particle physics, with its aggressive competitiveness, not to his taste. As a result, after graduation he set out into an area of research which was more leisurely paced, where there was time to reflect and find the answers without the pressures from perpetual competition. He found his niche in Idaho at the National Engineering Laboratory and, soon after he arrived there, news of the Soviet interest in muon catalysed fusion reached the West. Thus Jones became involved in its renaissance.

The scientists of Idaho were designing a gold-lined stainless steel container that could hold a mixture of deuterium and tritium gases at up to 3000 times atmospheric pressure and at temperatures ranging from hundreds of degrees centigrade to freezing and below. The 'bomb' was sent 700 miles south to Los Alamos National Laboratory in New Mexico where it was exposed to a beam of muons produced in the laboratory's particle accelerator.

They discovered that the rate of muon catalysed fusions grew as the temperature and pressure increased. Near the end of 1982 they achieved the first interesting number of eight fusions per muon, where the energy released exceeds the rest energy of a muon. After the Fleischmann and Pons press conference in 1989 Jones recalled this early achievement of 'breakeven' in muon catalysed fusion and joked that they had wondered whether to call a press conference in 1982 and announce that energy output from fusion had been accomplished. That they did not was because they had achieved only conservation of energy in the sense that the amount released exceeded that contained in the mass of a muon. Self-sustaining fusion requires much more than that. One must take account of the inefficiencies such as the fact that many muons fail to initiate fusions at all, they are produced in an accelerator that has to be kept running and only a fraction of the power input ends up in beams of muons and energy is expended in the motion of the neutrons and alpha particles produced in the fusion reactions. To produce practical breakeven some 300–1000 fusions per muon will be needed (estimates of the number vary) in contrast to the eight.

By 1985 the fusion rate achieved in their experiments was improving, but still remained tantalisingly short of practical breakeven. The way forward needs better detailed understanding of the 'sticking'—the unwanted occasions where the muon attaches to the produced helium instead of initiating further fusions. The experiments show that this seems to depend both on temperature and pressure, but other experiments show different results; when these matters are fully resolved then it may be possible to design experimental circumstances where optimal fusion occurs.

In 1985 Jones moved back to Provo to take up an appointment in the physics department at his Alma Mater, BYU. He began in the autumn term and just before starting this new career he and Clinton van Sieclen wrote a paper on the theory of cold fusion, which concentrated on the muon catalysed fusion but had some prescient remarks about the possibility that fusion might be influenced by pressure and materials. In this paper they coined the phrase 'piezonuclear fusion' from 'piezo' meaning 'pressure'.[14] It was published in March 1986, five months after Jones' arrival at BYU, and he decided to give a talk about it in the departmental weekly colloquium. In the audience was Paul Palmer, a geophysicist, and it was from this that Jones became involved with solid state fusion and set out on the road that led to his interactions with Fleischmann and Pons. Ironically, having entered muon catalysed fusion initially to get away from the frenzy of high energy particle physics, Jones was about to be catapulted into greater media attention and controversy than a career in particle physics is ever likely to have given him!

## The Brigham Young experiments

Jones has given talks about muon catalysed fusion all over the USA and in Europe, so it is ironic that it was after speaking at his new home institution, Brigham Young University, on 12 March 1986 that the seminal interaction occurred with Paul Palmer.[15]

Palmer had become interested in the anomalous presence of helium-3 in water near thermal vents where superheated water gushes up from the Earth's mantle into the ocean depths. He had already begun to wonder if fusion could be responsible for the helium production when he discovered the Soviet paper on helium in metals—the same paper that the Indians had read.

L. Khabarin, B. Mamyrin and V. Yudenich had discovered in 1978 that when they heated several metals to drive out helium trapped within, the gases were almost entirely helium-3 rather than helium-4; normally helium-4 is much more abundant, the helium-3 being the rare form. Another surprise was that the helium-3 was not distributed uniformly throughout the metal, but appeared to be highly concentrated in small clusters as if it had been formed there. They suggested that it had been created by fusion within the metal and that muons from cosmic radiation might have been the catalyst.

As a result of this, Palmer had in the back of his mind been thinking about muon catalysed fusion, and when the newly arrived Jones announced his talk, Palmer naturally took notice and went along to listen.

Palmer was inspired by what he heard, as his notebooks[16] record:

> Colloquium yesterday by Steve Jones set me thinking. He talked of spontaneous fusion under pressure e.g. $pd \rightarrow {}^3He$. Well, when sedimentary material at a continental margin gets pulled down in a subduction zone at a plate, fusion could take place as the pressure increased. I looked in a table of isotopes and found that the concentration of deuterium is 1.5 parts of 10 000 (parts of ordinary water). If all this fused . . . it would give 2 million callories (of heat) per gram[17].

One thousand calories is enough to heat one gram of rock to 3000 degrees, so the deuterium in one gram of water could make two kilograms of rock red hot! There is always some moisture around, many rocks containing more than a gram of water either crystallised or trapped in the spaces between the rock's atoms. So if all the deuterium fused it could generate substantial heat in the Earth.[18]

But does all of the deuterium fuse, or even any of it? Palmer now thought about testing the idea, and this brought him back to his original thoughts about the helium.

Palmer believed that the helium-3 in the atmosphere comes from the fusion of two deuterium nuclei or 'deuterons'. Each deuteron comprises a single proton bound to a single neutron, so a pair of deuterons is a total of two protons and two neutrons. These fuse together and then break apart, releasing energy in the motion of the products. One of the products is helium-3 which consists of the two protons bound to a single neutron; the spare neutron from the original gang of four particles speeds off. When a *proton* and deuterium fuse they also give helium-3, accompanied not by a neutron this time but by radiant energy, a gamma ray. So the basic fusion mechanisms of deuterium fusing either with another deuterium or with a proton give helium-3. (See the figure on page 28.)

Palmer then notes 'Measure the $^3He$ that outgasses from the lava. Its simple! These data must be available.' The ratio of hydrogen to helium-3 that outgasses would allow computation of the fraction of spontaneous fusion per average water molecule, since hydrogen is in the pristine water whereas helium-3 is produced by any fusion. His last remark that day mused on how it might happen: 'Maybe the rock catalyses the reaction!' Here was the germ of the idea that fusion may be aided by deuterium being compressed within solid materials.

He was excited enough to pursue it further. His notes the next day were headlined 'Catalysis of DT (deuterium–tritium) fusion in the Earth'. He had talked with two colleagues about the possibility of fusion occurring when water is drawn down into the Earth's mantle at boundaries between tectonic plates. They had found a paper in the literature describing the ratio of helium-3 and helium-4 in volcanic regions and were unknowingly getting close to the origins of Chatterjee's ideas in India. Palmer's notes read: 'It is exciting. When gas comes from volcanic regions the ratio of $^3He$ to $^4He$ is eight to fifteen times greater than found in the atmosphere and crustal rock.'

This looked promising: $^3He$ is produced by fusion, there is an excessive abundance of the substance in volcanic gases so maybe here was the evidence staring him in the face that fusion is taking place among the volcanic rocks. However, the paper in which Palmer had found the news had an explanation of its own, namely that helium formed in the early Universe, and trapped inside a new born Earth four billion years ago, is being released in volcanic plumes. Palmer recorded this fact:

[They] attribute this to primordial helium . . . being released . . . I attribute it to the **generation** of helium-3 by the fusion of deuterium and tritium by something or other. I would guess that the 'something or other' are negatively charged ions of fluorine, chlorine and maybe oxygen. Maybe under pressure free electrons do it.'

62

This was a guess. There seemed no doubt that the helium came from deep in the earth but 'at this time [the data] do not indicate the cause . . . *Is it primordial or fusion generated*'? There was the crucial question. How could he test it? 'Is there a mantle probe that is not contaminated with possible fusion-generated helium-3? How about meteorites? How about the Soviet deep wells?' Palmer continued:

> The best test might be a lab experiment, squeezing hydrogen and deuterium with catalysts in a diamond cell. The trouble with this is that if it didn't work it might be because the conditions and catalysts are wrong.
>     How to proceed? Search all leads and write a proposal? Or do the work and write a patent and proposal?

Palmer discussed his ideas with Steve Jones and the pair wrote up memoranda. Palmer's paper entitled 'Experiments in Cold Fusion' was dated 28 March 1986, and Jones wrote 'Comments on Catalysed Fusion' on 1 April.[19]

To plan the strategy Jones and Palmer got together on 7 April with Bart Czirr, a physicist from the University, and Johann Rafelski, a theorist who had been interested in Jones' work on piezonuclear fusion since 1985, and who was about to move from Cape Town to the University of Arizona. (Rafelski later made theoretical input into the project and was a co-author on the paper.)

Palmer's guess, recorded in his notebooks of 13 March, was that the rocks had catalysed the fusion, and so at the 7 April meeting the group discussed various metals that could be prime candidates for the process. These elements are common in minerals and Jones' notes of that day mention aluminium, nickel, lithium, platinum and, significantly, *palladium 'because it dissolves much hydrogen'*. They discussed ways of simulating the turmoil or 'non-equilibrium' conditions of the Earth and among these ideas they considered electric discharges. It was this that later led them to build an electrochemical cell.

Already they were aware that if this turned out to be true, it would be revolutionary for geophysics and possibly for fusion research. What had started with Jones as a search for muon catalysed fusion, then theory on piezonuclear fusion in his paper with van Sieclen and with Palmer as a puzzle about the gases emitting from volcanoes, was now turning into a search for entirely new routes to fusion. Enthused by the tantalising clues that Palmer had unearthed they gradually realised that they could be onto something big here.

Now it happened that a few days earlier Jones had called up Lee R. Phillips, an attorney associated with Brigham Young University, in order to file a patent arising from his muon-fusion work. Jones and Rafelski went to see Phillips following the discussion. The patent for the muon work came to nothing, but then Rafelski suggested that, seeing as the attorney was present, they might as well get him to notarise the logbook in case the discussion should one day prove seminal. Thus it was that Jones notebooks are signed by Phillips with the statement

*The catalysed fusion process outlined above was explained and formulated on or prior to 4/7/86 (April)*

The US convention is to put the month before the day in dates. In view of the subsequent disputes over who thought of what and when, this notarisation turned out to be most prescient.

Up to this time Jones had been concentrating on his muon work for which the DOE had been providing the funding. The new ideas were an outgrowth from this, but were sufficiently novel that they couldn't be pursued within the terms of his research contract with the DOE. The agency provided funds for a specifically approved line of research only, so Jones now decided to tell them of the new ideas. Ryszard Gajewski, the administrator in charge of basic energy research, was enthusiastic: 'The work was promising in the sense that it identified a new way of effecting nuclear fusion even though there was no strong indication that it might be practical.'[20]

Jones and Palmer's BYU team were focused on geophysical questions and, in particular, the abundance of helium-3. In the course of their literature search, on 16 April Palmer found a 1978 paper by the three Soviets—B. Mamyrin, L. Khabarin and V. Yudenich—in which they reported finding excessive amounts of helium-3 in various metals. This encouraged the BYU group as it suggested that metals, in minerals such as rocks, could indeed play an essential role in catalysing the formation of helium, presumably through fusion. What excited them even more was that, surprisingly, no one seemed to have followed this idea up (and indeed no one had, as B. Mamyrin confirmed in 1989).

On 22 May 1986 Palmer built their first electrochemical cell to try and force hydrogen into metals. The idea here is similar to that used by Tanberg many years before, namely to fill a beaker with water and pass an electric current through it by connecting a battery to two terminals that are dipped into the liquid. The negative terminal is called the cathode and the positive is the anode. They added a small amount of hydrochloric acid as 'electrolyte' to carry the current, the oxygen from the water bubbles off at the anode and the hydrogen appears at the cathode. Palmer made the cathode of copper and hoped that migration of hydrogen to the cathode would result in some of the gas entering the metal simulating the passage of ocean water through minerals of rock. As all water contains some 'heavy hydrogen' (deuterium), where the single proton of the conventional hydrogen nucleus is accompanied by a neutron, there should be some deuterium produced at the cathode too. Deuterium and hydrogen can fuse and so there was the chance that fusion might take place within the cathode of the cell.

If a proton and deuteron fuse, the products are helium-3 and a gamma ray, a burst of high energy light (no neutrons are produced here). The helium-3 is what Palmer was trying to produce as it was the presence of this gas in the atmosphere that had stimulated his idea in the first place, but detecting it would not be easy. So instead, Palmer looked for the gamma ray which would be produced in association and as such would be proof that the fusion process was occurring.

When gamma rays hit sodium iodide crystals, they deposit their energy, exciting the atoms which re-emit the energy as visible light. This is then amplified by photomultiplier tubes and converted into electrical impulses. So by turning gamma rays into electrical current a recorder can then measure the current and hence in effect the number of gammas. This number in turn is a measure of the amount of fusions that have taken place.

They ran the cell all day, after which time a green crust had formed on the cathode. Something was happening certainly, as the discoloration proved, but was there any fusion? To answer this they brought the sodium iodide detector over to look for the tell-tale gamma rays. First they measured the background—the number of gamma rays in the natural environment such as those produced by the concrete in the laboratory and compared it with the number of gamma rays when the electrochemical cell was present. The rate appeared to go up very slightly when the cell was there, but by such a small amount that Palmer wrote in his logbook 'You wouldn't bet on cold fusion on the basis of these results; but neither would you bet against it.'

The next day they tried again but this time they added ten per cent of heavy water to the liquid, thereby increasing the amount of deuterium—the fusion fuel. However, they still had no luck as here again, there was no clear signal.

But Palmer's log that day contains a significant remark: 'Bart [Czirr] suggests that we use *palladium* as the metal cathode as it has the ability to let hydrogen diffuse through it at will. It should work fine.' With this new development, the Brigham Young University team were replaying Paneth and Peters, and Tanberg's tactics of the 1920s and were moving towards the same road that, unknown to them, Fleischmann and Pons were following.

By increasing the amount of deuterium they were raising the chance that two deuterons fused together. Half of the time this gives rise to tritium and a proton, and (almost) the rest of the time gives helium-3 and a neutron. The neutron seemed to be the easiest of these to detect as proof that fusion had occurred and so Jones' colleagues Bart Czirr and Gary Jensen set about building a neutron detector, working on the project throughout the summer of 1986. Meanwhile Palmer was experimenting with different electrolytes— acids and alkalis that carry the current through the cell—including sodium hydroxide, sulphuric acid and traces of other impurities in attempts to mimic the elemental make-up of seawater in the presence of minerals. If fusion was at work, then something catalysed it, and until they knew precisely what the 'something' was they would have to try all combinations.

By September 1986 a crude neutron counter was working and recorded a slight increase in the amount of neutrons relative to the background when the cell was present. But was it significant? The amount being recorded was at the limit of the detector's sensitivity and the scientists felt that it didn't (yet) prove anything one way or the other. The detector was recording neutrons all right, but they couldn't be sure if there were neutrons coming from the cell over and above those coming from natural sources, such as cosmic rays and general radioactive 'dirt' around the lab. To settle this, they decided that they would have to improve the detector still more. Bart Czirr and Gary Jensen worked on this for the next several months in between teaching duties as the new university year had now begun. Jones meanwhile returned to his major field of muon catalysed fusion.

Building and testing the neutron spectrometer—capable of both counting neutrons and measuring their energies—took all of 1987 and much of 1988. During the early part of 1988 the focus of research gradually moved back towards the piezonuclear fusion. In January Jones, Palmer and a third colleague, Prof. Larry Rees, organised a student research topic around this and in March one student, Paul Dahl, made a presentation at the University about his work. Around this time they sent away some samples from the metal cathodes for examination. Although they had been unable so far to be sure whether or not either neutrons or gamma rays were being emitted from the cell, if fusion was indeed occurring then there should be evidence within the cathode—helium or tritium should be present, though in very small quantities. These specimens were sent to Harmon Craig of University of California, San Diego, whose paper on the helium abundance in volcanic atmospheres had stimulated some of Palmer's early thoughts. The samples were later sent to Al Nier of the University of Minnesota, a leader in the field of helium isotope studies, but any traces of helium-3 or tritium were too small to detect.

Energy applications for muon catalysed fusion, Jones' main research line up to that time, were appearing to be very remote and in August 1988 Jones decided to mount a vigorous effort to pursue the piezonuclear fusion. The pace of events now quickened. Rafelski visited Jones on 16 August and 'urged more study of piezonuclear fusion'; on 24 August the various members of the fusion team met at the University and decided to make this their primary aim with detection of gamma rays (by a sodium iodide detector) and neutrons both on the menu.

It was on 20 September that Jones received Pons and Fleischmann's proposal from the Department of Energy. Rafelski was also consulted about their proposal.

In view of the subsequent interaction between the Brigham Young and Utah University groups, with accusations from some Utah administrators and from Pons that Jones had stolen ideas from them, it is important to ask where the BYU programme had reached by the fateful 20 September 1988. It is clear that by 24 August they had already formulated their ideas and been moving towards confirming them, but the *definite* proof still awaited the energy sensitive neutron spectrometer.

Jones' notes, which he has circulated to show the BYU version of events, do not directly address this issue.

First he notes that as the gamma ray detector was easiest to set up they started with that but, as before, obtained only inconclusive results. The improved neutron detector was finally ready 'later in the year' and 'within a few weeks' the results were positive enough that they formed the basis of the paper that was sent for publication on 23 March 1989, the paper that should have been sibling to that of Fleischmann and Pons.

The sequence of the research that led to the final publication is as follows. The dates are relevant in that they merge in with meetings and other interactions between the BYU and the University of Utah team and with the developing programme of the latter group.

The definitive experiments began on New Year's Eve 1988. These consisted of a series of individual runs lasting about a day apiece, with and without cells forming 'foreground' and 'background' data, respectively. They then plotted the ratio of the number of neutrons seen in the two cases whose energies were consistent with that expected for neutrons produced by *dd* fusion. A ratio significantly larger than one would imply that an excess amount of neutrons was being detected when the cells were being operated, and hence be evidence for fusion taking place.

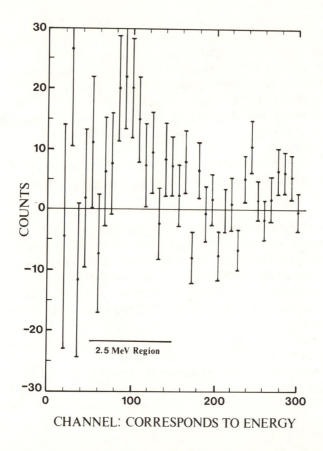

Figure 3. The BYU spectrum showing the difference between the signal and the scaled background. Note the tendency for an excess in the region indicated (channels 50 to 150), which is the energy region where neutrons coming from *dd* fusion should appear.

Three of the first five runs showed ratios between 1.5 and 2, implying excess neutrons of between 50 and 100 per cent. Then in run number six, taken towards the end of January, the ratio hit 3.5; a very dramatic outburst of neutrons had been detected from the cell. The neutron production rate appeared to rise after about an hour of electrolysis and then dropped dramatically after eight hours. Jones showed these data to Fleischmann and Pons on 23 February. He then

68

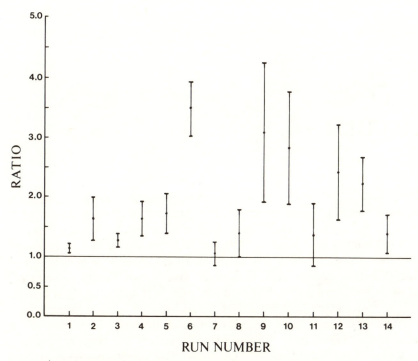

Figure 4. Ratio of foreground rate to background rate for each run, in the 2.5 MeV-energy region of the spectrum. Reproduced by courtesy of *Nature*.

made several shorter runs, which thereby accumulated fewer neutrons and had correspondingly larger uncertainties. These data were all to hand by early February and he prepared to speak about them at the APS (American Physical Society) meeting set for 1 May, sending in an abstract for the announcement of the talk. In the remaining weeks he refined the calibration of the detector by exposing it to beams of neutrons of known energy, produced by a van der Graaf accelerator. However, the publication of the abstract for his talk was claimed by some in the University of Utah as having brought the cold fusion work into the *public* domain and excusing their own subsequent announcement.[21]

# 5
## THE CHEMISTS

In the phone directory there are the barest summaries of thousands of lives—a name, place of abode and the occasional glimpse of intrigue. You may find a rare name whose five or six entries must be related, so giving insights into a family structure advertised by their addresses. A number reveal more of themselves in the pages of *Who's Who*.

Martin Fleischmann is a Fellow of the Royal Society, Britain's premier scientific society. His entry in *Who's Who* shows that he has won The Palladium (*sic*) Medal of The Electrochemical Society and has been professor and head of the chemistry department at Southampton University since 1983. These are the bare bones of a long and distinguished scientific career. The family portrait reveals more; father of three children and approaching his ruby wedding with his wife Sheila, whom he married at the age of 23. Out of four billion lives on the planet, the shape of one begins to emerge.

These are the surroundings of a man. Insights into the actual personality are restricted to a few words under the heading 'recreations'. What these are and how they are written can tell much or little. On the same page of *Who's Who* as Martin Fleischmann's entry there is a barrister whose recreations include 'coping with multiple sclerosis', a surgeon who enjoys 'history, literature, art and indulging the senses' and an energetic biochemist keen on 'tennis, squash, sailing'. Characters begin to appear. Fleischmann's entry reports 'music, cooking, skiing, walking'—nothing very remarkable here perhaps, but they helped to bring him to the most intense period of his scientific career.

Fleischmann was born in Karlsbad, Czechoslovakia on 29 March, 1927. He attended primary school in Czechoslovakia, had a year off school in 1938–39 due to the troubles with the rise of fascism and the impending threat from Nazi Germany, and as life in Central Europe became increasingly precarious the Fleischmanns joined the growing band of refugees who settled in England. Whatever effects these traumatic events may have had in forming his world view and concerns about military adventurism, they can hardly have limited his intellectual growth. He went to high school in Worthing, a pleasant town in the south-east of the country, where he was equally interested by science and arts. 'I seriously thought of studying English literature—or becoming a cook!' he told me with a laugh, 'but then Imperial College London offered me a place in mathematics.'[1]

Imperial College had a style that appealed to him. At school the students were never given standard equipment and had to make things in the lab for themselves. At Imperial College there was a similar emphasis. 'We were taught to design an experiment of a size that was best suited to what you wanted to measure.' In scientific experiments 'scaling' the apparatus optimally can sometimes affect one's ability to see an effect accurately or at all.[2]

Fleischmann graduated in 1948 and began to study for his doctorate in chemistry, which he obtained in 1951. With his wife Sheila, whom he had married during his doctoral studies, the new Dr Fleischmann took up his first post at King's College, University of Durham in north-east England. In 1952 Durham University had two campuses, one in the ancient cathedral city of Durham nearly encircled by the River Wear and the other twelve miles away in the industrial city of Newcastle on the River Tyne. It was at the latter that Fleischmann worked, initially as a fellow sponsored by ICI, the major chemical company, and then as a lecturer. In 1957 the University of Durham split in two; the college at Newcastle became a university in its own right and today Durham and Newcastle are two separate institutions. Fleischmann stayed at the new University of Newcastle and was promoted to 'reader', one rung below the top post of professor. (In England the title Professor is usually reserved for one or two members of a department. 'Reader' corresponds to Full Professor in North American usage while Lecturer corresponds roughly to Associate Professor.)

He stayed at Newcastle for fifteen years. By this time his international reputation was well established. He had a gift for coming up with innovative ideas, suggesting links that others would never dream of, and testing them in a rough way to see if they were flawed or might instead lead somewhere. If they appeared likely to do so he would then map out a strategy and advise students or co-workers on how to refine them and carry them through to fulfilment. Thus it was no surprise that in 1967 he was chosen to become the Faraday Professor of Electrochemistry at the University of Southampton on the English south coast.

A colleague at Southampton, Dr Alan Bewick, described Fleischmann as 'more innovative than any other electrochemist in the world'.[3] As can be the case with such creative minds, many of the ideas are regarded as crazy, but occasionally one is crazy enough to lead to real breakthroughs. Fleischmann has had some excellent ideas including methods to measure the surface diffraction of X-rays, and he has made a detailed study of electrolysis in solutions that do not conduct electricity. One of his discoveries was originally thought by some to be wrong as no one was able to replicate it for nearly two years; however, Fleischmann was eventually proved to be correct.[4]

His stature increased and with such international renown it was natural for him to serve for two years, 1970–2, as the President of the International Society of Electrochemistry. It was this reputation that attracted Stanley Pons to Southampton in 1975, and it was Pons' passion for skiing and his interest in cooking which overlapped so well with Fleischmann's and brought the two men close.

## Stanley Pons

Bobby Stanley Pons was born on 8 February 1943 in Valdese North Carolina. In the late nineteenth century his family helped to found this small town in the foothills of the Blue Ridge Mountains and by the time that Stanley Pons was growing up it had a thriving textile industry, his father owning several mills and employing over 1000 workers.

Stanley Pons was the eldest of three sons and had a scientific urge very early, experimenting with a chemistry set and doing minor mechanical work on his father's car. He graduated from Valdese High School, studied chemistry at Wake-Forest University in North Carolina and then had to face a hard choice between his parents' wishes that he return to work in the family businesses and his own strong interest in chemistry. Paul Gross, a chemistry professor at Wake-Forest, recognised Pons' ability and recommended that he go to graduate school. Chemistry won and Pons enrolled at the University of Michigan in Ann Arbor.

He was totally committed to his research, his hard work and enthusiasm moving him rapidly on. Within two years he and his supervisor, Harry Mark, had completed a controversial piece of work, pioneering a way to measure the spectra of chemical reactions on the surface of an electrode. They submitted a paper for publication and the editor sent it for reviews before deciding whether to publish it, ask for revisions or reject it. As Prof. Mark recalled, 'the reviewer said we were out of our minds but within two years people conceded that we were correct.' The paper[5] was published in 1966 but Pons left Ann Arbor, his PhD work uncompleted, lured back to the family businesses.

His father owned several mills, not just in North Carolina, and Stanley Pons worked in one of these in Louisiana where he met his present wife, Sheila, and then he became manager of a restaurant that the family owned in Palm Beach Florida. Here he developed his interest in cooking, a passion later shared with Fleischmann.

Although he had terminated his PhD course and left active research he nonetheless kept abreast of developments in chemistry. From 1967 to 1975 he worked in the family businesses and then decided to return and finish his PhD. To start again from scratch in the USA would mean that it would be five years or more before he had any hope of receiving his doctorate, but in England, with his early research experience behind him, he could complete the degree sooner.

Friends of his knew of Fleischmann as an eminent electrochemist. Moreover by then (1975) Fleischmann had become chair of the chemistry department at Southampton University. Through this introduction Pons enrolled at Southampton and though formally supervised in his studies by Alan Bewick, Pons and Fleischmann soon became good friends, sharing their common interests in cooking and skiing.

By 1978 Pons had completed his thesis and received his PhD from Southampton. He returned to North America, first lecturing at Oakland University in Rochester Michigan and then from 1980 to 1983 at the University of Alberta in Canada. He was eager to make up for lost time and published prolifically. There is no simple guide as to how many publications impresses committees the most. Do too little and no one takes any notice; do a vast number of papers and you may be accused of superficiality or lack of care. It was in this period that Fleischmann and he began to discuss projects, but they did not start working on any until Pons joined the University of Utah, moving to Salt Lake City in 1983.

The year 1983 was seminal for the English universities, under financial pressure from the Thatcher government to economise, and also for Martin Fleischmann. In order to meet the financial cutbacks the universities were forced to encourage staff to retire early. This was a popular option and, at 56, Fleischmann was in an ideal position for it. He retired from the University but retained an honorary professorship.

It is ironical to contemplate what this economy measure could have meant to the University had the test-tube fusion work succeeded: of the billions of dollars in royalties, Southampton—where Pons had got his start in the field and where Fleischmann had spent over twenty years—would have had claim to none.

Now that Fleischmann was retired he had no teaching commitments and in consequence had much free time for research. He started visiting Utah regularly—'Stan's attitude to cooking is similar to mine and, of course, the snow is good!'—staying in Pons' home sometimes for several weeks at a time.

Pons' list of publications[6] shows no direct overlap with, or preparation for, the test-tube fusion programme. Fleischmann by contrast has studied hydrogen and precious metals for several years. They co-authored three papers together in 1985—their first work together though not related to test-tube fusion. In the following years the number of papers rose, culminating in sixteen in 1988 (during which Pons wrote 36 papers but still none relating to test-tube fusion). It is interesting that, according to media reports, their idea to pursue test-tube fusion goes back to 1985, the year that they first did published research together. To produce so many papers jointly, plus scores more separately, during 1985 through 1988 can have left little time for intensive work on the test-tube fusion project.[7]

## An idea is born

For most of us the mental picture of an experiment goes back to school laboratories. Clean the test tube, add the right amount of chemical according to the textbook instructions, heat it up, watch the colour change on the litmus paper, a straightforward sequence of steps that has been performed by thousands before us and will be repeated by as many after. Occasionally the unexpected happens.

I still remember the dirty liquid in my school test tube suddenly turning a brilliant turquoise, then shocking pink and then back to grey all in the space of a few seconds. This wasn't 'supposed' to happen; it was much more exciting than what the textbook had promised! Unsuccessfully I tried to repeat it and asked the teacher for an explanation. It was clear the he didn't know, nor did he seem much to care; 'Oh it's probably something or other. Now stop wasting time and get on with what you're supposed to be doing.'

It was definitely 'something or other', but what? The excitement of delving into the unknown was mine for the first time that day. I suppose that some trace of a previously used chemical was still in the tube which had not been properly cleaned. It was a sensation to me, a novice, but the experienced experimentalist learns the tricks of the trade, knows what mistakes to avoid and when weird things happen tries to investigate or to file them away in the memory—just in case.

Every professional experimental researcher, in whatever field, has their own private list of fluky data, things that didn't happen quite in accord with expectations and are likely to be due to uncontrolled circumstances—such as the dirty test tube—rather than signalling a major discovery. The real world is more quirky that the idealised laboratory of the school textbooks. Charged particles from outer space (cosmic rays) are bombarding us all of the time and occasionally cause the electronics in the sophisticated detector circuits to misfire. Everything is radioactive. We are all emitting gamma rays as a result of radioactive decays of potassium nuclei within our bodies. Invisible to the eye, these rays can be picked up by sensitive custom-built detectors. If not accounted for they could contaminate a specimen under study or be mistaken as an important signal of something happening within the experimental investigation, whereas they are actually part of the 'noise'—the 'background'.

'Background' is the bane of experimentalists' lives. If not correctly understood it can be the Achilles' heel of a complicated chain of investigations. Unexpected responses of the detectors may be background—not every burglar alarm that sounds signals a break-in; 'something else' triggered it. However it may be a genuine 'signal' as when an intruder trips the alarm. Experience teaches a lot as to whether or not a phenomenon is likely to be a genuine signal or background noise, but there is still a twilight zone, the tantalising occurrences which are probably noise but might possibly be signal. Such things deserve a second look. The genuine effects are reproducible; the quirky—such as UFO sightings—are filed away but rarely prove significant for science. Unless, that is, two groups happen to tell each other and it turns out that both have had the same or similar experiences. That in essence is what started Pons and Fleischmann on their quest for test-tube fusion. But misunderstanding the 'background' plagued them and many other of the experimenters trying to replicate their work.

Both Fleischmann and Pons are electrochemists, experts at studying the effects of electric currents passing through liquid conductors (an 'electrolyte'), the most familiar example being electroplating. In electrolysis you attach a battery to two metal *electrodes* which are dipped into a sodium chloride solution. Four volts of electricity will split the sodium chloride into its constituent sodium and chlorine atoms, whereas a temperature of 40 000 degrees would be required to do it by heat alone. Fleischmann and Pons in their lectures liked to use this example to present electrochemistry as 'high energy chemistry'. The effect of undirected heat energy of 40 000 degrees being achieved by the directed use of a mere 4 volts illustrates the 'efficiency' of the technique; they also realised that electrochemistry can induce the effects of high pressures—high temperature and pressure being two of the conditions for achieving fusion.

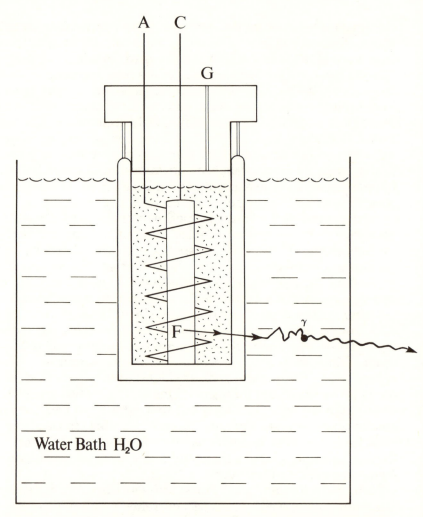

Figure 5. Electrolysis cell and water bath. A battery is connected to the anode and cathode via leads A and C respectively. The anode is a coil of platinum wire surrounding the cathode rod at the base of C. G is a hole in the seal to let out gases. The cell is filled with heavy water (shown as dots) and is surrounded by a bath of ordinary water to maintain constant temperature. If fusion occurs inside the cathode, at F, a neutron may be emitted which enters the water bath, slows and is captured at $\gamma$, emitting a gamma ray of energy 2.2 MeV.

When you electrolyse water it splits into its constituent hydrogen and oxygen: positively charged hydrogen ions are attracted to the negatively charged terminal, the *cathode*, where they react to form hydrogen gas, and the oxygen forms at the positive terminal, the *anode*. They knew that palladium has a natural affinity for hydrogen and that if palladium is used as the cathode the hydrogen ions initially migrate into it; as more and more hydrogen goes in so it gets under ever greater pressures.[8] Martin Fleischmann told me that he had always been puzzled by the behaviour of hydrogen in palladium. Some earlier research in 1972 had led him to believe that very high pressures could be attained, and also that the hydrogen ions behaved as free in the palladium crystal lattice, moving around and probably bumping into one another. He also knew about work in the USA and USSR from the mid 1970s on metallic hydrogen. Some theorists believe that hydrogen becomes a metallic solid where its electrons freely flow when compressed by two or three million atmospheres pressure; there have been speculations that metallic hydrogen could be useful as a fusion fuel. The gravity of Jupiter crushes its hydrogen so much that pressures of this magnitude might occur at the centre of the planet, forming metallic hydrogen which undergoes fusion thereby generating heat within it.

Here we see some similarity, though remote, with the motivations of Palmer and Jones. There are other, less well publicised reasons for interest in attempts to make metallic hydrogen and in the hydrogen–palladium experiments which bear on some subsequent responses to Fleischmann and Pons' work. Metallic hydrogen could have applications as a room temperature superconductor, and could be a powerful fuel useful in nuclear devices and as an explosive. Fleischmann has been particularly worried about possible strategic implications of their research, believing that much of the negative reactions to it are prejudiced by this and that national security interests have orchestrated attempts to suppress or trash their work.

Fleischmann told me more about the background to his work with Pons. 'There was this idea about metallic hydrogen, but also the idea that hydrogen or deuterium might cluster and fuse within the palladium had been in the back of my mind for some time. One has many ideas all one's life and there are many points where they can come together cleanly enough to design an experiment. Then you need an area of laboratory in which to do the work. I had always intended to do it at some stage.'

One of the problems facing him was that one cannot simply set up any experiment in a public building; health and safety regulations have to be obeyed. The exact wording of these regulations varies from place to place and, having retired from Southampton, he was freer to do 'way out research'. Setting up the experiment in Utah and keeping it secret turned out to be easier than in Britain.

'I could show you exactly where we were when we first discussed it,' he continued. 'We were hiking in Milcreek Canyon in Utah. If I had been on my own I probably would not have done it.'

In forcing deuterium into the palladium host, the energy of the system changes. Fleischmann and Pons estimated that the chemical potential[9] of the deuterium could be raised by 0.8 eV through application of electrical potentials to the palladium cathode. They estimated the sort of pressures that would be required to achieve such conditions by compression and the answer astonished them. Their calculations came up with the number of a thousand trillion trillion times atmospheric, $10^{27}$, a number larger than the number of atoms in your body. What had started out as a quest for metallic hydrogen now became a serious hunt for fusion. Temperature—an intense degree of heat—was the route that the big machines, the hot fusion tokamaks, had been following in their attempt to induce the nuclei to meet; Fleischmann and Pons believed that they had stumbled on another way—intense pressures provided by the natural make-up of solid palladium. If nature has indeed provided a miracle key to solve every problem, this certainly looked promising. Martin Fleischmann told me how over a whiskey in Pons' kitchen he remarked, 'It's a billion to one chance, shall we do it?' and Pons replied, 'Let's have a go.'

### First attempts

Fleischmann and Pons started their experiments with the hope that fusion of deuterium—'*dd* fusion'— would occur when deuterium ions were forced into palladium by electrolysis. They were in effect about to repeat experiments that had been done, unknown to them, by Tanberg in Sweden 50 years earlier, passing electrical current through a solution of heavy water, splitting it into its constituent oxygen and deuterium. But to make the current flow you first add a little acid or alkali (electrolyte). Some people who later tried to replicate the Fleischmann and Pons experiment before full details were circulated thought that the specific form of electrolyte was not important and guessed its form incorrectly. We do not know even today whether or not this is the case, but Pons certainly felt very strongly that it should be *lithium* deuteroxide, *LiOD*. Having deuterium in the electrolyte clearly was a useful strategy; as for the lithium, Fleischmann said to me 'There are a lot of reasons for choosing this. The surface of the palladium is better preserved, allowing the deuterium to infuse better; it is easy to prepare clean solutions; and if I look you straight in the eye and say ''If you think of *any* nuclear phenomena what would *you* choose''—well, it is obvious isn't it?'

This was said to me a year after the announcement at the press conference when Fleischmann and Pons still did not know what caused the heat. Fleischmann was alluding to the strategic importance of lithium-6, which is present as 7.5 per cent of all naturally occurring lithium and is an important ingredient in thermonuclear weapons as it can be cleft by neutrons to make tritium, the essential fuel, and can fuse with deuterium to make helium-4, releasing energy in the form of gamma rays *without* accompanying neutrons.[10]

Wanting to keep their plans secret, and convinced that 'no one would fund such a crazy idea' they paid for the initial experiments out of their own pockets.

They began with some exploratory tests to see what effects if any were there to be measured and to help design an optimal experiment—to scale it properly, in the idom that Fleischmann uses. The project suddenly became interesting when they were forcing deuterium in to a 1 cm$^3$ block of palladium ('charging the palladium') by means of a 1.5 amp electrical current. It took considerable time to charge the palladium fully and Fleischmann was in Britain when, one night, Pons' son was sent to the laboratory to reduce the current. Whether he carried out the instructions correctly or not we do not know, but the next morning the block had disappeared, vaporised, and part of the fume cupboard housing the experiment and the floor nearby had been destroyed. This excited them—maybe the idea was right after all—and also made them nervous, both for their own safety and for what might happen if the news leaked out. Pons called Fleischmann who said 'We had better not talk about this on the phone,' already acutely aware of where this might lead.

Was this a nuclear exposion or a mundane one caused by a build-up of pressure as the deuterium flowed into the palladium and filled some cavity within; a 1 cm$^3$ block of cast palladium will almost invariably have some 'voids' inside. Another suggestion was that the electrolysis caused the liquid level to drop as more of the liquid turned into gas; this exposed the top of the palladium above the liquid surfae and gas began to leak out, releasing in a rush the considerable energy that had been stored with the trapped deuterium, sending the temperature up above the melting point (1554 degrees Celsius) or even boiling at 3000 degrees.

I asked Fleischmann what was known about this and he laughed as he said 'Not enough!' Early after the press conference Fleischmann and Pons advertised the explosion of their palladium block as prima facie evidence for fusion, but a year later Martin Fleischmann presented it to me in a more detached way. 'That experiment was an indication that we should continue. If you had that sort of phenomenon occur you have to take note of it. A nuclear explosion is what it *could* have been, but you do not know that that is what it *was*.'[11]

Naturally this event made a big impression on them. In their paper on test-tube fusion they wrote of it 'WARNING. IGNITION!' while in talks they stressed 'Under no circumstances use large current densities in big samples and subject to large sudden thermal shocks.'

Whatever the real cause for the explosion of the block in Pons' laboratory may have been, the event focused their expectations and also urged increased caution and security. The dangers of creating a miniature nuclear device worried them and also increased their concerns for safety. They decided to scale down the experiment and worked with a thinner piece of palladium, an $8 \text{ cm} \times 8 \text{ cm} \times 0.2 \text{ cm}$ sheet rolled into a cylindrical shape, and also lowered the current to less than 0.2 amps instead of the 1.5 amp current that blew up the block.[12]

The basic idea was to keep a careful accounting of the power input and output, including all the known chemical reactions that will transfer electrical energy into heat, and then see if the sums balance or whether, as they hoped, 'excess' power was being produced by 'fusion' reactions in the cell. At any particular moment the books might not appear to balance (for example electrical energy input could be stored to be released as heat later, or energy taken up while forcing the deuterium into the palladium may be returned later if the deuterium leaks out and recombines in the atmosphere), so the relevant question was whether there was a net excess output of energy over a long period of time.

The essential principles are as follows (a more detailed discussion of some of their results follows in the appendix 'Excess Heat in Calorimetry' on page 351). You put in a certain amount of power into the cell. If the cell was sealed off and no gases escaped then all of the power would go into heating the electrolyte. You measure the temperature rise and from this determine how much power has entered (to make this conversion it is necessary first to have calibrated the cell by, for example, placing a small resistance heater in the cell, putting in *known* amounts of power, measuring the temperature rise and determining the relationship between temperature and power).

Fleischmann and Pons spent many days pre-saturating the palladium with deuterium. They then began to take measurements to see if any heat was being generated, suggestive of fusion occurring.

For nearly two weeks they used very low currents, about 0.05 amps, and discovered that the cell *cooled down* more than expected. This was because the deuterium was continuing to be dissolved into the palladium and this takes energy out of the surroundings, acting like a refrigerator. After this time they doubled the amount of current and found that the energy books almost balanced. After checking this balancing they increased the current further to three times the starting value, and discovered an excess heat output of ten per cent over and above what they could account for going in. Finally they upped the current to 0.2 amps for another two weeks and found that the heat excess rose to 25 per cent.

If this heat excess was due to fusion, then there should also be some proof in the form of electromagnetic or nuclear radiation, such as the production of neutrons. They noticed that background radiation in the laboratory appeared to rise by up to 50 per cent while the experiment was running but they had no means of knowing whether or not this had anything to do with the experiment as the natural background radiation from cosmic rays can vary substantially during long time periods. But already one thing was clear: the heat could not simply be due to deuterium *dd* fusion as this process should produce neutrons at a rate a billion times larger than they were recording. This would have triggered their detectors dramatically, yet all they saw was a small increase in the background level of radiation. Had they seen billions of neutrons as well as the heat then they would have known for sure that fusion was occurring; and may already have begun to worry about their health from prolonged exposure. So they were left with a tantalising hint: heat, yes; neutrons, no. This decided them that more work was needed to verify whether there was indeed fusion, and so they began to plan out a detailed strategy and designed an experiment—'scaling it' in the sense that Fleischmann had learned in his days at Imperial College. At this stage they were in a similar position to Jones in the sense that both groups had suggestive hints of phenomena whose confirmation would require more sophisticated detailed experiments.

Unfortunately at this point Martin Fleischmann entered hospital for a serious operation which interrupted the project for almost a year. It was not until 1988 that they set out a detailed strategy for carrying out a series of experiments to confirm their preliminary observations that nuclear fusion may be possible at room temperature inside metals. This involved varying the experimental conditions in all possible sensible ways, trying out different amounts of electrical currents, and changing materials, the shapes and sizes of electrodes, and the electrolytes, studying how the heat production varied with the mean temperature of the cell. In effect, they wanted to vary the experimental conditions enough to clarify the source of the heat and confirm if it was fusion. If this proved successful then they would, in the longer term, extend the experiments to find out what conditions were best for building controlled fusion reactors.

They decided to use palladium rods, a few centimetres long, with diameters ranging from 1 mm up to 20 mm. They estimated that it would take them four months to build the cells, the measuring equipment and constant temperature baths. It takes longer to saturate large rods with deuterium than to saturate the smaller ones; the latter took up to a fortnight and they estimated that the largest rods could take up to a year. As a result of all this they realised that to complete the measurements on the range of electrodes and to see how they depend on the other experimental conditions would take nearly three years.

This had now become a major programme that would need external funding and so they drew up a detailed proposal in the summer of 1988 and submitted it to the Basic Energy Research Programme of the DOE (Department of Energy) in Washington DC. The DOE sent it out to referees for opinions, one of whom was Steven Jones, who received it on 20 September. The days of innocence were nearly over.

Referees of DOE proposals usually remain anonymous, and Jones could have done so had he wished. However, as his work was so similar and geographically near by, he suggested to the DOE that there should be contact and openly advertised his presence.[13]

The work had already begun when, late in the year, Pons first learned about Jones. It may help us to understand some of the subsequent developments better if first we have an impression of what was going on in the University of Utah chemistry department in 1988, up to and around the time of the interaction with Jones.

Pons was chairman of the chemistry department, and his name appeared as co-author on over 30 papers in the first eight months of 1988 alone,[14] which is an astonishing output from a single individual if detailed concentrated effort was put into all of them. His name also appeared on over 100 papers during the five years when the fusion experiments were reportedly in process.[15] This suggests that the fusion experiment was far from an all-consuming enterprise during that period; indeed the amount of detailed progress had been rather limited by 1988. The main experimental fusion programme began in the latter part of that year.

In October Pons assigned this as a project to his research student, Marvin Hawkins. By November Hawkins was working on it seriously, operating the cells, taking readings with Pons showing supervisory interest. Martin Fleischmann came to Utah at the turn of the year and, during January in particular, put a lot of effort into the experiments. These experiments involved thin rods of palladium and formed the bulk of the data in the paper of March 1989. Pons was much occupied as chair of the chemistry department and the detailed day to day work with the fusion cells was carried out by Hawkins and Fleischmann.[16]

# 6
# THE DISPUTE

How far had Jones and also Fleischmann and Pons reached in their researches by late 1988? Jones had been doing research funded by the US Department of Energy (DOE) since May 1986, had hints of neutrons and by September 1988 was in the process of starting to use a neutron detector sensitive enough to measure their energies and determine whether indeed they came from fusion. So his results were preliminary but it would only take a few weeks' work, using the neutron spectrometer, to know for sure. It was on the last day of 1988 that he began the experiments whose data were to appear in the published paper. By the beginning of February 1989 he had enough to convince him that it was real, and he agreed to go public by talking about it at the American Physical Society meeting in Baltimore the following May.

Fleischmann and Pons had done very different things. They had a vaporised block that led them to an *idée fixe* about test-tube fusion being real. Their observations that nuclear fusion may be possible inside metals at room temperature were also only preliminary, as were those of Jones: they had 25 per cent heat imbalance at best, were at the *start* of their real research and estimated that a full three years would be needed to do what was necessary. This included having to study rod sizes, materials and changes in the amounts of electrical current to see how the effect depended on these, which was all for the future. Jones had already used a variety of materials, one of which was palladium, and was preparing to obtain the final proof with his detector that he had just spent two years building.

So both groups had come a similar distance in the sense of having the indications that there was something interesting to pursue further. The essential difference is that Jones was nearer to the finishing line, as his route was different to that of the chemists. Jones had tried various materials and what he now sought was rather clearly defined—are there neutrons of a specific energy or not?—and his apparatus was almost completed and ready to go; the chemists by contrast had to do a broader range of experiments as their problem was more subtle—is there fusion and if so by what process?—and the necessary materials and apparatus had yet to be assembled.

That was the situation when the DOE sent Fleischmann and Pons' application to referees, including Jones, for opinions.

The theoretical underpinnings of the proposal were weak. Practical fusion needs a combination of high temperature—to hurl the nuclei together against their electrical repulsion—and/or pressure, so that there is a greater density of them and so the chances of bumping rather than missing are increased. Traditionally it has been believed that both are needed, as in the Sun for example where fifteen million degrees and a central density more than 100 times that of water are only just enough, or high pressures and temperatures exceeding 100 million degrees aimed for in the hot fusion programmes. The chemists believed it would be possible at room temperatures because the electrolysis forced the hydrogen and deuterium in the palladium creating conditions that Fleischmann and Pons regarded as equivalent to $10^{27}$ times atmospheric pressure. But this figure can mislead, as was pointed out by many people as soon as the Fleischmann and Pons paper saw the light of day, and by some, at least, of those who had sight of their DOE application. The true pressures are a few thousand times atmospheric, which may sound a lot but is quite normal in solids and will not give you any fusion benefits on its own. Secondly, and this was not so widely appreciated, the deuterium nuclei are farther apart in the palladium than in the heavy water, and being further apart are even *less* likely to fuse when in the palladium than in the heavy water! As heavy water does not spontaneously give off heat from its deuterium fusing, so is it even less likely that it would when inside palladium. The pressure might help but not enough to make commercial fusion, not at the level to create measurable heat.

The fusion rate in $D_2O$ is calculated to be at an unmeasurably slow rate of $10^{-65}$ per pair per second at room temperature. However matter itself is predicted to be intrinsically unstable on such a scale: the quantum theory implies that evaporation of matter through gravitational interaction with black holes would occur at a rate of around $10^{-50}$ per second, a thousand trillion times faster than spontaneous fusion. It was from comparing numbers such as these that the majority of scientists suspected that watts of power from test-tube fusion was an impossibility.

Nonetheless, it is experiment that decides, not theory, and Jones recommended to the DOE that the chemists' proposal be funded despite his unresolved reservations. One of the referees was sufficiently negative about the proposal that Fleischmann, at least, was nervous about the idea.

Jones also suggested to Ryszard Gajewski, the administrator of the Advanced Energy Program at the DOE, that he inform Fleischmann and Pons that the BYU group had been doing similar work on piezonuclear fusion since 1986 and that cooperation between the two universities, which are only some 50 miles apart, would be a great benefit. As Jones pointed out, his neutron detection was the missing ingredient in the Fleischmann and Pons proposal which was almost exclusively on their particular expertise, namely, measuring the heat output. Gajewski passed this on to Pons who phoned Jones in December to discuss the matter. There were several subsequent contacts, with Pons seeking written information about the Brigham Young University neutron spectrometer.

Jones mailed all the details to Pons as the detector was now working and beginning to gather the data that would be the basis of the Brigham Young University–Rafelski paper. Jones even offered to let Pons use the neutron spectrometer for himself and remarked 'Pons seemed pleased with the offer.'[1]

At least, that is how Jones remembers it. Pons' version is somewhat different in colour. Pons disagreed about the neutron detector. Whereas Jones believed that Pons was pleased with the offer, Pons reportedly said 'We never needed Jones' spectrometer, never wanted it.'[2] Pons agreed that Jones originally argued with the theoretical aspects of their proposal but later became convinced by them and only then offered cooperation. 'Jones called up. Said "tell me more", then revealed himself as the reviewer of the DOE proposal and only later suggested that we collaborate. . . In all my scientific life I have never seen [such] a situation.'[3]

Fleischmann told me that in his opinion Jones should have declared his interest to Gajewski more explicitly, not merely that he was interested in the problem, and should have informed Gajewski of his results at that time so that there would be no ambiguity about claims for priority.[4]

This is where some of the conflict began. Jones' best data did not exist until 1989; by the time that he met with the two chemists he had measured 'run number 6' (*see* Figure 4, page 69) which is shown in his paper as the most dramatic signal and which proved to the BYU team's satisfaction that they were right. However, in September when he first saw the chemists' proposal the 'good' data did not yet exist, indeed, did not exist until early in the new year. It is in this sense that the criticism in the above paragraph has its substance.

I will assess this more in the conclusions, by which time the evolving events will have enabled us to understand the pressures on the participants better. However, although one has some sympathy with Fleischmann's point, and with hindsight might wish that more explicit details had been known, in reality this was not so clear cut. First, at the time Jones had no idea what a storm was going to erupt about test-tube fusion and in notifying the DOE of his possible conflict of interest he was already being exceedingly conscientious and could, for example, have chosen instead to have remained an anonymous reviewer as far as the chemists were concerned. Second, Jones' experiment was at the stage where its results were moving foward the fastest. Could Jones have been urged into 'fast forward' mode in a desire to beat the chemists? I asked Jones this and he replied 'Hardly. We were going as fast as we could.' Indeed, Jones took time out in October 1988 to visit Europe, speaking at a meeting on muon catalysed fusion and having a brief vacation with his wife, hardly the behaviour expected of someone who is frenziedly responding to the sudden emergence on the scene of competition.[5]

Initially the communications between Jones and Pons were by telephone and mail; a formal meeting did not take place until February 1989. In the meantime a significant development took place. In December 1988 the American Physical Society (APS) had invited Jones to talk about muon catalysed fusion at their May 1989 meeting. On 2 February 1989 Jones sent in to the APS an abstract of his intended talk which stated in part 'We have accumulated considerable evidence for a *new form of cold nuclear fusion* which occurs when hydrogen isotopes are loaded into various materials, notably crystalline solids (*without muons*)' (my italics). These sudden developments in a period of five weeks when Pons and Jones had been having discussions, culminating in the above *public* claim that could have influenced patent priorities, fuelled the suspicion in the University of Utah camp that they had been upstaged.

How had Fleischmann and Pons' programme been progressing since September?

In the autumn of 1988 Pons had set his graduate student, Marvin Hawkins, the task of measuring heat produced in cells containing small cylindrical rods of palladium, 1 mm, 2 mm and 4 mm in diameter and 10 cm in length. Hawkins collaborated with Fleischmann during the latter's visits to Utah and they planned to extend the data samples to include rods of 8 mm up to 2 cm diameter. At this stage Pons was involved primarily with his job as departmental chairman and only followed progress relatively remotely; Fleischmann and Hawkins were making the detailed investigations.[6]

Charging up the smallest rod with deuterium took only two days, the 2 mm rod took a week and the 4 mm one took a month; an 8 mm rod took several months to charge before any detailed experimental measurements could begin.[7] In each case, after charging, measurements of heat had to be made for several hundreds of hours and repeated at several different levels of current, between 0.05 and 1.5 amps corresponding to a current entering each square centimetre of surface of the rods of 8 and 64 mA. (mA means a milliamp or *thousandth* of an amp, not a *million* amps as incorrectly stated in one newspaper.) They also wanted some data at higher current densities, 250 mA/cm$^2$, but were nervous that increasing the amount of current could lead to another disaster – the vaporised palladium block still haunted Fleischmann—so they decided to achieve this by keeping the total currents approximately unchanged but using rods that were only 1.25 cm long instead of 10 cm, the effect being to increase the current density eightfold.

Given the long times required to charge the larger rods, you can see that after getting a first qualitative result with the smallest rods within a few weeks, a proper quantitative analysis to test the variation with rod diameter and current, and so establish the nature of the effect would take a long time. When the news of Jones' work first came to Pons' attention late in 1988, the 4 mm diameter rod was not yet charged and ready for test, nor were all rods giving heat—the chemists referred to these rods as being 'dead'. In particular no heat came from 8 mm rods, one of which they started charging at Christmas and which was still being charged in March.

What is more, at this stage they were concentrating on what they knew best, namely the heat measurements, and were giving little attention to nuclear effects, though some search for tritium—which can be done by a chemical process—was made on a 1 mm rod that produced heat. This was interesting; there is some tritium occurring naturally in the heavy water and when they measured the tritium levels after the experiment had been running for some hours they found that the tritium level had *dropped*. They decided that this was because tritium from the heavy water had fused with deuterium in the rods.

No measurements of neutrons or other signs of nuclear processes had been started. Indeed, even by February 1989 when the first meetings with Jones took place, where Jones was claiming to see neutrons—a sure sign of fusion—Fleischmann and Pons were seeing only heat.[8] *Fusion* was the word that would excite the world following the press conference, and although they were convinced from their heat data and also from their vaporised palladium block that fusion was occurring, they would have to see neutrons to convince the scientific community of it. The problem was that as late as February 1989 they had no suitable neutron detector to use.

## SUMMARY OF EVENTS UNTIL JONES SENDS NEWS
## TO AMERICAN PHYSICAL SOCIETY
February 1989

| | |
|---|---|
| 1985–6:<br>Palmer idea at Jones seminar<br>Jones talks with Rafelski<br>Crude detector → hints of<br>   neutrons | Fleischmann–Pons idea<br>Sometime 1985–8: Experiments<br>   with –<br>   palladium sheet<br>   block vapourises |
| 1986–8:<br>Building neutron detector<br>Jones concentrates on muon<br>   catalysis | 1985–8:<br>FP co-author 29 papers on other<br>   topics |
| August 1988:<br>Decision to concentrate on solid<br>   state fusion<br><br>20 September:<br>Jones receives FP DOE application | Summer:<br>Decision to pursue test-tube<br>   fusion<br>DOE grant application drawn up |
| Jones announces his presence to Pons ||
| | October–November:<br>Hawkins begins experiments with<br>   rods |
| December:<br>APS invites Jones to speak on<br>   muon catalysis at their May 1989<br>   meeting<br><br>31 December:<br>Jones begins experiment that will<br>   appear in his paper | Experiments with small diameter<br>   rods show heat. Charging of<br>   large rods begins |
| January 1989:<br>Hints of neutrons at limits of<br>   measurability<br><br>Late January:<br>'Run number 6' gives possibly<br>   significant effect | Fleischmann in Utah |
| 2 February:<br>Sends abstracts to APS for talk in<br>   May. Abstract refers to new<br>   'solid state' fusion | February:<br>First accusation that Jones pirated<br>   ideas from FP proposal |

BYU left hand column; University of Utah right hand column

On 14 February, Fleischmann had visited Harwell laboratory in England where he has a consultancy. Harwell is one of the world's leading nuclear laboratories and, as such, owned state of the art neutron equipment combined with high level expertise in electrochemistry. He asked to speak confidentially with Ron Bullough, the Chief Scientist of the United Kingdom Atomic Energy Authority (UKAEA) whose office is at Harwell, on the grounds that he (Fleischmann) had something 'very important that might require a major decision'.[9]

Fleischmann told him that they were seeing heat coming from their cell in amounts that could only be understood if nuclear fusion was happening. His belief that it was fusion was due primarily to the heat, but he also told Bullough that since October there was some evidence in Utah that neutrons were being produced too. (It is not clear if this refers to his own experiment or to the fact that Jones had by then made his work known to them.) Fleischmann then showed a proposal that he and Pons had submitted to the Department of Energy (DOE) in Washington requesting financial support for them to carry through a definitive series of experiments over a period of three years. The DOE had sent this out to five referees for opinion. Fleischmann said that one of the referees had said that it was nonsense and that the reaction of this referee had made him nervous about the validity of their experiment. Nonetheless he believed that they were seeing genuine effects that were inexplicable unless fusion were taking place.

What Fleischmann and Pons needed was their own proof of the neutrons and Harwell was well equipped with the necessary measuring instruments. Fleischmann discussed radiation with them, aware that if fusion was indeed happening in the cell there could be significant health risks although he did seem to be feeling all right. He then bought a health physics monitor to help his neutron detection and returned to Utah with it the next day, 15 February.

So in the last half of February Fleischmann and Pons had a detector with which they could count neutrons. By this stage they had a 4 mm rod that was running with a 0.8 amp current. As the efficiency of the detector was very low (it was small and, they estimated,[10] only recorded less than one in a million of any neutrons emitted by the cell), they counted neutrons at the cell for 50 hours. Then they counted neutrons 50 metres away from the cell to get an estimate of the background level due to neutrons coming from cosmic rays or the natural environment of the laboratory. They compared the readings taken remotely with those at the cell and found that the rate at the cell was three times that of the background. This they decided was evidence for neutrons coming from the cell.

When this became general knowledge later, there was general unease at this way of measuring the background; the neutron flux 50 metres away from the cell measured for 50 hours could be very different than that at the cell site, even if the cell had been absent, measured for a different 50 hours. The cosmic ray flux varies and the different materials around the laboratory can produce very different amounts of background neutrons from one place to another. To get a realistic background measurement required, at least, taking readings at the cell site when the cell was not operating. A problem was that they did not have the time to do this; cells were being charged up and time was running out, every spare moment being needed for preparing and measuring active cells. There is a natural time scale to the experiment, charging the cells, inducing the phenomenon for hundreds of hours and monitoring at different currents; you cannot hurry science. To do everything properly would take time—time that they no longer had.

They did this over two weeks, but there was no means of knowing the energies of the neutrons, the essential information that would tell them whether they were from the fusion of deuterium—that would be the final proof. Fleischmann and Pons worried that the numbers of neutrons that they appeared to be seeing were billions less than should have been the case if their heat data were correct and due to fusion. Each watt of power from fusion should release a thousand billion neutrons.

There was already conflict behind the scenes. Earlier in February, while Fleischmann was away in Britain getting the health monitor from Harwell, the University of Utah attorney, Norm Brown, had spoken to the Brigham Young University attorney Lee Phillips and suggested that Jones had pirated ideas from his sight of the proposal that Fleischmann and Pons had sent to the DOE. This accusation (which followed after the submission of Jones' abstract to the APS on 2 February) was reported to Jones who in turn passed it on to the DOE funding agent Ryszard Gajewski. Gajewski questioned Pons about this and Pons apologised for the lawyer's insinuation, both to Gajewski and to Jones on 21 February.[11]

Immediately after this unfortunate incident Fleischmann and Pons paid a visit to Jones in the morning of Thursday 23 February to see what his experiment was producing. For their part Fleischmann and Pons took along one of their palladium cells (but one that never produced heat—their heat producers were at that moment fully occupied in gathering data). Fleischmann and Pons by now believed that they were seeing neutrons, though they were puzzled by the numbers. Then they saw Jones' underground laboratory, the neutron spectrometer and its data, which were much more extensive than theirs. In particular Jones had a complete count of neutrons and also of their energies, so he could tell which neutrons came from cosmic rays and which were possibly the result of fusion. Fleischmann and Pons couldn't do this and Jones invited them to bring their (working!) cells over to Brigham Young University so that his neutron spectrometer could measure a detailed spectrum. They agreed and set the following Monday, 27 February, as the test date.

It was clear that Jones could do much more detailed neutron measurements than could Fleischmann and Pons, but his data weren't particularly impressive. His evidence for fusion neutrons consisted of what to the eye looked to be only a moderate increase of neutrons in a small range of energy, though admittedly in the right region to be from fusion. Fleischmann and Pons, who did not realise that they had measured their neutron signal incorrectly, thought that their cell was producing 10 000 neutrons per $cm^3$ of palladium each second (in fact the totality of neutrons from cosmic rays, from the concrete in the laboratory, from radon and other gases seeping through the air vents may have been responsible for almost all of these and the fusion neutrons, if there were any, were far fewer), and so Jones' data didn't appear very impressive. So it was a shock to them when, during lunch that Thursday, Jones announced that his Brigham Young University team was preparing to publish its data and offered to let Fleischmann and Pons publish simultaneously. When Jones made the offer he said he was 'attempting to establish an open and cooperative relationship' and to help mend the broken fences of the previous day's rancour.

Fleischmann recalled this period to me. 'We were given a long account of tritium generation at Mount Mauna Loa and all that and then "by the way here is the neutron spectrum". Jones gave the impression that he intended to write a paper which was going to be on the Mount Mauno Loa business with a bit on the neutron spectrum but it did not turn out that way at all. The Mount Mauna Loa thing got completely obliterated. I thought their data were interesting but marginal. I said, and it is engraved on my mind, "you should get an inorganic chemist to make some interesting compounds that might possibly show the *dd* fusion *and wait*. And we should wait too. It is premature this thing." But no, they had to go ahead with this rather inadequate data.'

I asked him 'if Jones had gone ahead and you had not published at that time, would that have killed your priority as Jones was claiming only to see neutrons and few of those in a rather different set-up to you, while your results were totally concerned with heat?'

Fleischmann replied 'There was this problem. We had not got any neutrons to speak of—well, we had a few from one experiment, we had some tritium. What were we to do? We had to tell the university. You cannot keep it secret. The university felt obliged to go for patent cover.'

On Monday 27 February Fleischmann and Pons were due to bring their working cell to Brigham Young University for the neutron spectrum to be measured, but a graduate student had to go to a funeral and so they suggested it would be better to come at the end of the week instead. However, by the end of the week they still didn't show. Instead a meeting was proposed by the Utah University President Chase Petersen in a long phone call with the Provost of Brigham Young University during Friday 3 March.

---

### SUMMARY OF FLEISCHMANN AND PONS INTERACTIONS WITH HARWELL AND JONES
#### February to March 1989

| | |
|---|---|
| Early February | Jones sends abstract for talks to APS |
| | Accusations from Univ. of Utah that Jones pirated ideas |
| 14 | Fleischmann at Harwell; obtains neutron dosimeter |
| 15+ | Neutron measurements |
| 21 | Pons apologises for accusations against Jones |
| 23 | FP visit Jones. Jones announces that he will publish |
| 1 March | F notifies Harwell that neutrons are being seen |
| 6 | Heads of BYU and UU scientists and authorities at BYU |

---

What had happened in the intervening five days?

Fleischmann and Pons believed that they were seeing ten thousand neutrons per $cm^3$ per second and on 1 March Fleischmann phoned Ron Bullough, the Harwell Chief Scientist, to tell him the news. Harwell had much more sophisticated neutron detectors on site, able to measure energies with better precision even than Jones could. Bullough offered Fleischmann their facilities as Fleischmann and Pons would be able to develop their work much faster and still retain control and priority over it. Fleischmann decided to send over all the equipment, cell, fluid and palladium, by air to Harwell for them to do the neutron measurements.

So by 1 March Fleischmann and Pons believed that they were on course to getting immediate confirmation of the neutron energy spectrum from Harwell independent of Jones, whom they were increasingly viewing as an unwanted competitor. Thus they had 'no need of Jones' neutron detector'. And on 2 March their confidence received a further boost with the news that the DOE had approved funding of their research programme.

92

On 3 March the heads of the two universities spoke by phone, and discussed the importance and complexities of the test-tube fusion research at the two institutions. Brigham Young University and the University of Utah are great rivals and the presidents saw this as an opportunity to 'fuse' the two institutions to their greater mutual glory. In particular Petersen of Utah reportedly claimed that 'there are billions of dollars at stake and Nobels in the offing'. Thus it was that on Monday 6 March the summit took place at Brigham Young University involving administrators and principal scientists from the two universities. Before the meeting began, at 9 a.m., Chase Petersen, the Utah University president, met with his opposite number at Brigham Young University and with Joe Baliff, the Brigham Young University Provost, to discuss the agenda. Petersen already had an agenda of his own drawn up and this was agreed to with the proviso that it included a brief historical summary of research done at Brigham Young University. The signifcance of all this is that Utah had already convinced themselves that test-tube fusion could be an economic bonanza of unimaginable magnitude, and the first shots in the skirmish for priority had already been fired in February with the accusations that Jones had pirated the Utah chemists' work.

Petersen began by pointing out how wonderful an invention practical test-tube fusion could be and that the potentially vast proceeds from the invention could be extremely valuable to the University of Utah.

Already the hyperbole was out of all proportion compared to the evidence. Fleischmann and Pons were seeing 25 per cent unaccounted-for heat *sometimes* and had not yet more than the hope of proving that there were neutrons proving fusion; Jones claimed to have seen a few neutrons at levels which, while interesting for science and understanding aspects of the Earth, bore no large scale practical benefits—what use is a billionth of a watt? Jones made this point by taking a small flashlight out from his pocket and making a strong cautionary comment that he would be 'extremely surprised if enough power could be generated by the process to power even a flashlight.'

Jones then remarked that, even if there were some commercial benefits, he failed to see how they could be vouchsafed for Utah, and reviewed the history of his own group's work at Brigham Young University. He showed the notarised notebooks of April 1986 demonstrating the range of materials, including not just palladium but several other metals, that was being used two and a half years before they learned of the work by Fleischmann and Pons. Jones' record of that meeting notes that 'None of these dates were questioned or challenged by Fleischmann and Pons or anyone present, nor did anyone raise any questions about the proposal-review process. They did not allege that the Brigham Young University work had pirated any ideas from their own work.'

The reception was positive and Utah President Petersen commented to Brigham Young University President Holland, generously, on the 'remarkable coincidence that such similar research had sprung up independently at the two universities.'

The focus of the meeting now turned to the question of how these results should be released. The University of Utah team made it clear that they still had a long way to go and would like another year to eighteen months to continue their research before announcing it. Jones responded by reminding everyone that the DOE had been funding his work for nearly three years already, that he had positive results ready to publish, that the DOE funding agent had encouraged him to go ahead and that he was due to speak about them at the Spring Meeting of the American Physical Society in Baltimore during 1–4 May.

Fleischmann and Pons knew that they had no hopes of completing their work by then and that they could lose the race if Jones went public there. Utah President Petersen suggested strongly his hope that Jones would not give the talk, a suggestion which shocked Jones considerably, insisting that the American Physical Society had invited him to give the talk, he had completed a three year research programme, had done all the necessary cross checks and controls, and had definitive results to present.

Petersen backed off and said he would not ask Jones to cancel his talk, and instead it was agreed that both groups would submit papers *simultaneously* to a journal, and within three to four weeks in order to have them accepted and hopefully published before Jones' scheduled talk in May.

At this point there is once again dispute about the nature of any agreements. Jones believes that 'it was also agreed by all that no public disclosure of the research would be made by either group prior to the simultaneous submission of the papers.' Pons, by contrast, asserts that there was no agreement not to publicise and it would be hard to forge such an agreement given the tricky detail that Jones had already submitted an abstract of his talk to the APS, the wording of which appeared rather explicit to knowing eyes. However, believing this to be the agreement, Jones cancelled a previously scheduled talk in the Brigham Young University physics department set for 8 March, only two days hence.

The two groups prepared to write their papers and got ready for joint submission, which was set for Friday 24 March. The arrangement was that they would all meet at the Federal Express Office in Salt Lake City Airport and send the papers off together to *Nature*. But during the week immediately following the meeting two fortuitous things happened to the chemists that at the time looked too good opportunities to miss.

Fleischmann's hope of 1 March that he would fly out his cells to Harwell had not worked. What he had not bargained for was that the 'fusion' cell would be deemed a radioactive hazard and so be prevented from getting the urgent clearance for transport across international borders. They were therefore still needing the measurements of the energies of the neutrons, which were necessary as a proof that they were indeed neutrons produced by *dd* fusion and not somehow spurious. Their first piece of fortune in mid-March was apparently finding a way to do this.

Any neutrons produced within the palladium will shoot out of the cell into a surrounding water bath. As they slow down they are captured by protons in the water and in the process emit a gamma ray, which can be detected by crystals of sodium iodide, *NaI*—a standard procedure for detecting and measuring the energies of gamma radiation. Pons called in Bob Hoffman, a radiologist, to take a spectrum in the laboratory supposedly for health and safety reasons, and to identify the prominent signals. There are a series of peaks that show up in the detector like a range of mountains as there are lots of gamma rays at different energies coming from the many naturally radioactive materials that are all around and within us. In particular there are prominent naturally occurring peaks at 1461 keV caused by gammas from potassium-40 ($^{40}K$) and at 2615 keV from thallium-208 ($^{208}Tl$) which provide important landmarks when identifying the various peaks appearing in the detector. Any gamma coming from the capture of fusion neutrons should therefore give a peak between these two, at 2224 keV.

Hoffman was under great pressure to work fast and took measurements over two full days. What made it worse was that he had a terrible cold at the time. His data showed clear peaks caused by the gammas from potassium and thallium (or so he thought, see Chapter 15)[12] and gave a hint of another peak between the potassium and thallium peaks. Interpolating between these two markers gave the energy of the peak as 2.5 MeV (2496 keV) which is approximately the energy that neutrons produced by *dd* fusion have.[13]

The range of peaks, as portrayed months later in *Nature*[14], is shown in figure 6. The sharp peaks are what Hoffman identified as the markers for 1461 and 2615 keV; the signal that interested Fleischmann and Pons consists of the dotted structure around 2500 keV. When they submitted their paper and Fleischmann gave his first public presentation on 28 March it was this peak alone that

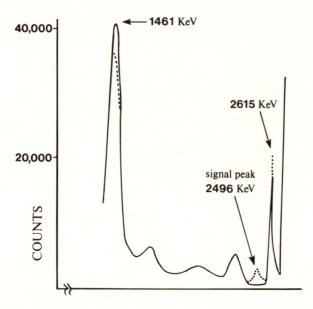

Figure 6a. Portion of gamma-ray spectrum as measured by Fleischmann *et al.*

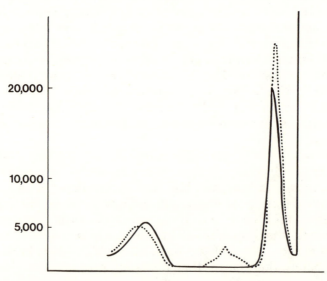

Figure 6b. The region near the signal peak in more detail. The solid lines are the background spectrum and the dots are the spectrum measured over the cell.

was exhibited; it is shown enlarged in figure 7 with the number of counts in the vertical axis and the energy scale along the bottom. Such was their purported evidence for a gamma ray from a neutron capture, an essential piece of the case for fusion taking place in their test-tube.

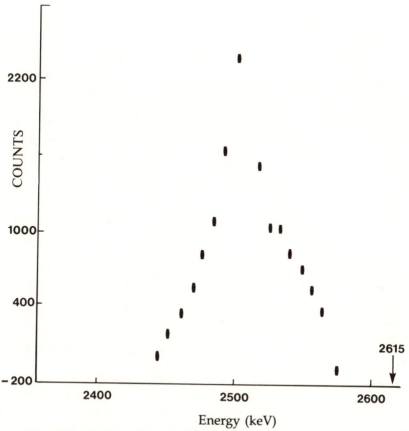

Figure 7. The signal peak as measured and originally presented by Fleischmann and Pons. The position of the nearby peak at 2615 keV has been added in this figure for reference.

Then fate played its hand with the entry of the second player in the drama of that week.

On Friday, 10 March, by which time Pons was becoming convinced that their hopes were at last being confirmed, Ron Fawcett, the US Editor of the *Journal of Electroanalytical Chemistry*, called him on a personal matter. The journal is published in Europe and its editor there is based at Southampton, Fleischmann's home institution. Pons mentioned the manuscript on test-tube fusion that he and Fleischmann were planning to write. Fawcett said that this sounded very important and that if Pons sent it to him he would rush it through. Pons sat down and quickly wrote a manuscript for a 'preliminary note' entitled 'Electrochemically Induced Nuclear Fusion of Deuterium'. This was written so quickly that several errors occurred, one being the omission of the name of the person who

97

had done much of the work since September: Marvin Hawkins, the graduate student.

The paper (published in volume 261, page 301 of the journal) contained a table of numbers for excess heat from the few cells that had measured 'positive' but gave no details of how many cells had balanced, or in their definition were deemed to be 'dead'. Careful reading showed that at best there was 10 per cent heat unaccounted for, but big numbers appeared because Fleischmann and Pons had 'extrapolated to rods 10 cm long' (their actual rod in this example had been only 1.25 cm) without saying what rule they had used to make the extrapolation from *measured* to *assumed*. And they had also made guesses as to what *might* be possible if the deuterium and oxygen gases that were bubbling off from the cell could be recombined and their latent energy recovered and used as heat. But very little of this was actually *measured*, much was theoretical and, it later transpired, much or all of the claimed evidence for radiation was incorrect.

This paper was one of the controversial centrepieces of the affair. Was it properly refereed or passed through at the urgings of the editor? Did it break the agreement of joint publication? Did its appearance signify a scientific publication and so excuse a *subsequent* press conference or was it, as many scientists claimed, a badly written uninformative document of limited scientific credibility that did little to support its authors' claims to have found fusion?

There was another feature of the accumulated data at this time that would cause problems later, namely their mistaken belief that the 2500 keV peak was evidence for neutrons; a true neutron capture peak should occur at 2224 keV. Had they checked with a knowledgeable nuclear physicist—and there were several in the nearby physics department who could have helped—or held a technical seminar before the press conference, they would have learned this before events overtook them. But they did not, and the 2500 keV peak went into their initial submissions to journals.

Optimistic that they were on the right track, but aware that much still had to be done, first thing on the morning of Monday, 13 March Fleischmann sent a fax to Harwell and then talked on the phone the next day remarking that their information was incomplete, that they had much more to do before they would be confident enough and expressing irritation that they were being 'rushed into premature publication'. Hoffman's measurement of the gamma spectrum was their third signature for neutrons: first there had been their own relatively crude neutron measurements with the dosimeter, then they had seen Jones' suggestive neutron spectrum and now this (erroneous as it turned out) gamma peak. These gave them some confidence that they were indeed seeing neutrons from fusion, but they were by no means definitive. A complete energy spectrum of all neutrons was needed, but Fleischmann's earlier hope that he could send cells to Harwell for them to make measurements on had been thwarted by customs and excise regulations. So on 13 and 14 March the best that could be done was for him to give Harwell full details and let them get on with it themselves (*see* Chapter 6).

Events were by now overtaking the scientists almost daily. Also on 13 March Fawcett received the 'preliminary note' at his editor's office and gave it top priority. Within a few days he had received some opinions on it which he transmitted to Pons immediately. Pons revised the paper over the weekend and rushed the new version back to Fawcett for publication. So by that weekend of 18–19 March Pons knew that not only was the DOE happy to fund the proposed research programme, but that the preliminary note had now received some measure of approval, which gave confidence in the fusion hypothesis. Moreover he knew that Harwell were looking for the crucial neutron spectrum which, surely, would clinch everything.

During the week leading up to 18–19 March the University of Utah authorities started increasing the pressures. Renewed accusations had been made against Jones on Friday 10 March by James Brophy, the Utah vice president for research who made allegations of piracy to Prof. John Lamb of Brigham Young University, and this was followed up the next week by Pons making general accusations against Jones to Dr Gajewski at the DOE. Pons was also concerned that Jones had enlisted the aid of Doug Bennison, chair of the Brigham Young University chemical engineering department and a leading electrochemist. Pons says this was as a result of seeing his and Fleischmann's work, whereas Jones insists that the DOE suggested it. Pons said 'I had no idea when Jones was going public.'

I asked Fleischmann why there was so much concern. I summarised the conflict as follows. 'The essence of the patent for you is surely that you believe the *heat* is there and if it can be scaled up, as you hope it can, *that* is something that could have manifest benefits to everybody. However, Jones claimed to see only neutrons, and a few of those only at levels that might be interesting to science and geophysicists but of no relevance to making commercial fusion reactors.'

Fleischmann replied that with hindsight that might be true but 'in the situation we were then in, we were obliged to tell the university of the work that we had done and they perceived that they were obliged to go for patent protection at that time. We could not tell whether Jones had heat data or was planning to look for this. How could one tell? He was certainly thinking about fusion as a source of heat in the Earth. If he was going to say *that* in the paper, which was surely his intention to do, it would almost certainly destroy any possibility of patent protection.'

So it was Jones' motivation—the possibility that fusion was at work within the Earth and possibly responsible for its *heat*, even though he saw none directly, that the University of Utah felt could derail priority for their own heat measurements. Moreover, it would be very easy for Jones to check out the *heat* by collaborating with an electrochemist.

Also one cannot overlook the fact that by this time (18 March) the chemists believed that they had evidence for neutrons *and* that those neutrons had the *energies expected for fusion products*. The number of neutrons was not enough to account for all the heat (a fact that they had been aware of for a long time) and this was a puzzle still to be solved, but at least there seemed to be some evidence that fusion was happening, proving their original hypothesis.

The University of Utah now decided to go ahead with a press conference. They were getting worried that news was leaking out. Brophy said that he was being questioned about leaks, and that rumours and false information were circulating. The University had filed for patents, believing that millions of dollars could accrue if test-tube fusion were commercially practical, and the patent lawyers were concerned about what would happen if other groups found out the results and duplicated them. The paper was written and accepted so why delay and risk losing everything?

Pam Fogle, the university news director, advised against a news conference until the paper was actually published in a journal, this being the usual course of events and a prior condition of several journals, in particular *Nature*. Brophy however insisted that if test-tube fusion holds up it would have such a major impact on society that 'Sometimes one has more responsibility to society than to the scientific community.'

Getting in first with the patent application was what the university authorities regarded as the top priority. Imagine yourself in the position of one of the university administrators. You hesitate as you are not convinced of the quality of the scientific work; test-tube fusion later turns out to be real and your university, your employers, lose billions of dollars in royalties. How would you view your career prospects? And it was easy to believe that the evidence for test-tube fusion was fairly solid, good enough that it would be only a matter of time before it was proven definitively, especially with the BYU people already claiming to have some independent evidence.

Thus it was that on Friday, 17 March the university press officer interviewed Fleischmann and Pons, who were even then in the process of revising their preliminary note, and the office had a press release ready by the Monday morning.

A lot took place the next day, Tuesday 21 March. The decision to go ahead with the press conference two days later was finalised; Pons edited a note alerting his staff to this effect, called Jones on the phone to reconfirm the joint submission for Friday the 24th but, according to Jones, said nothing about the press conference planned for the day before nor of the 11 March submission by Pons to the *Journal of Electroanalytical Chemistry*. And Fleischmann received a fax from Harwell and learned that they were seeing nothing at all.

Fleischmann certainly was concerned about the momentum of events; the news from Harwell magnified this and he expressed to some colleagues his nervousness and wish that the press conference could be stopped. But apparently it was all out of the scientists' control. He tried to contact Sir George Porter, the President of the prestigious Royal Society of London, in the hope that he would be able to use his influence in the USA and have the conference called off due to its potential security implications, but Fleischmann was unable to reach him.

At this point Fleischmann decided that as the whole thing was going ahead, then he should make the best of it and then get away for home in England immediately. He booked a flight for the afternoon of Friday, 24 March with Delta to San Francisco to connect with the British Airways flight for London. His hope was that he would be able to spend the Easter weekend in peaceful anonymity away from the media, not foreseeing that the press conference would merely be a beginning to a story that grew and grew.

Fleischmann's home base was in England and Harwell were already in action on his behalf. He thought that it would be nice if the BBC could be present at the press conference in their own right instead of getting a 'home grown' story at second hand, and so he contacted Richard Cookson, a close friend and retired Professor of Chemistry at Southampton University, whose son Clive was science correspondent with the BBC Radio.[15] However, Clive Cookson was no longer with the BBC but had instead become the science and technology correspondent of the *Financial Times*. He immediately recognised the immense potential for the story but was faced with a major problem: the press conference was arranged for 23 March and the next morning was Good Friday—his paper would not be published that day. Here he was with the journalistic scoop of his career and everyone else would be able to run it before him! So he asked if he could publish it on the morning of press day; appearing in one paper in London would not undermine the American publicity. Fleischmann consulted with Pons and they agreed; Cookson got permission to carry the story. However the two chemists did not discuss this with the Utah press office and that is how the Utah authorities 'confirmed' their paranoia that word was leaking out.

On 21 March 1989 Mike Salamon, professor of physics at the University of Utah, had been in Washington DC meeting with officials from the Department of Energy. Salamon studies cosmic rays and is an expert in recognising and measuring nuclear particles and radiations. Neither he nor his colleagues in the physics department had any inkling of the events taking place in the chemistry department, nor of the impending press conference, still less of the way their own lives would be transformed by them. Late that afternoon he was *en route* back to Salt Lake City by plane and among the passengers he noticed Chase Petersen, the president of the university.

Petersen knew Salamon only by sight and so there was no contact until, by chance during the flight, they were standing together at the galley getting drinks. They exchanged a few words, as one does in such circumstances, and then Petersen asked Salamon what his speciality was. When Salamon said 'physics' Petersen suddenly became interested and started asking him lots of questions about neutrons and tritium, radiation and muon catalysed fusion.

For Salamon there was no obvious reason why Petersen, university president and trained as a medical doctor, should be interested in these topics. Salamon recalled to me: 'I did not attach any particular significance to it at the time other than to come away really impressed that our president, a medic, seemed to be so conversant with the language of nuclear physics and such an esoteric topic as muon catalysed fusion!'

The Department of Energy received notice of the press conference on 22 March which stated that test-tube nuclear fusion producing net energy had been achieved at the University of Utah and that a reviewer of the proposal had confirmed the result! People at the DOE called Jones immediately. It was the first that he knew of it. Shocked and disappointed,[16] Jones and the Brigham Young University authorities communicated their feelings to Chase Petersen and James Brophy at Utah. In particular Prof. Grant Mason, dean of the College of Mathematical and Physical Sciences at Brigham Young University, spoke to Brophy and let him know that if the press conference went ahead the Brigham Young University people would interpret this as a violation of the agreements between the two universities.

Cookson's story in the *Financial Times* was indeed a scoop, appearing on the morning of the 23rd. Early morning in London is still before midnight the previous day in Utah, seven time zones around the globe. So even before the 23rd had dawned in Utah, the BBC and the international news services, such as Associated Press and Reuters, had already picked up the story. Many of the world's media had been alerted before the press conference began. Over 200 reporters, cameras and television crews crammed into the university; a large number for a press conference in New York or the White House, and unheard of for Utah.

There were still several shocks in store for the Brigham Young University team. It was intimated that a paper on Fleischmann and Pons' work had already been submitted, though they didn't say to which journal. In Jones' opinion this seemed to go against the agreement to submit jointly the next day, so immediately after the press conference Jones sent his group's paper to *Nature*. This was late afternoon on 23 March on a long day that had begun in time zones far to the east of Utah with Cookson's story in London. But now the time zones were against them as it was already 24 March in Europe and their paper, sent by fax to *Nature* in London was 'received 24 March'.[17] The University of Utah authorities subsequently criticised Jones for sending in his paper that day instead of waiting until the 24th. Indeed, one of them went to the agreed meeting place on the 24th expecting to send off the joint package; on learning that Jones had already submitted his paper, Fleischmann and Pons felt that *Jones* had reneged on *them*!

Whatever the reasons for this misunderstanding of the needs and hopes of both universities, it is hard to explain one action at the press conference by one of the most senior members of the University of Utah. A questioner asked: 'Is this going on any place else, or is this the kind of process that is currently being developed by anyone else?' James Brophy, vice-president of research at the University of Utah, who had been interacting with Jones and the Brigham Young University authorities for several weeks and who had spoken with them only the day before replied, 'We're not aware of any such experiments going on.'[18]

Haven Bergeson, a physicist at the University, told me that he had learned of Jones' work about two weeks earlier when BYU people had called him for some technical advice. That was the first that he knew about the work on test-tube fusion and, at that stage, he knew nothing about the work that Fleischmann and Hawkins were then doing in the chemistry department. 'I was not present at the press conference but I heard the account on TV. Brophy's denial I found most astonishing.'[19]

In February 1990 I asked Martin Fleischmann why the scientists had not clarified this error and he replied that the remark had not registered with him in the frenzied atmosphere at the time. Indeed, tapes of the conference show that Brophy's remark could easily have been missed, appearing starkly when presented in isolation of everything else that was going on. At that time Fleischmann's mind, at least, was focused on getting through the press conference and returning immediately to England.

# 7

## HARWELL

On Friday, 24 March the news hit the headlines. All around the world people tried the experiment for themselves wanting to be the first to prove or disprove the claims. The basic fusion kit seemed to simple that almost anyone could set it up; and almost everyone did. Partly as a result of this, during the following weeks there was much commotion and claims of dubious scientific value were seized upon by advocates of test-tube fusion to bolster their claims.

The quality scientific research proceeded much more carefully over several weeks and months. Universities and laboratories with first-rate equipment formed teams with wide ranges of expertise. A typical large scale effort would involve electrochemists measuring heat, nuclear scientists looking for signs of radiation, and metallurgists and materials scientists who studied the after-effects on the palladium metal that they had prepared. Whether test-tube fusion existed or not was a question whose answer cut across scientific boundaries. Groups consisting only of electrochemists (as were Fleischmann and Pons) or whose expertise was concentrated on nuclear physics (as at Brigham Young University) were less able to do a complete definitive study than the groups with multiple experience.

The story of the major laboratories' efforts naturally begins in Britain with Harwell, who knew of test-tube fusion due to Fleischmann's visit of 14 February. Harwell set up their own experiments soon after Fleischmann's visit, and by the date of the press conference were beginning to have the first doubts about the claimed phenomenon.

### Before the press conference

Martin Fleischmann's visit on Valentine's Day, Tuesday, 14 February, to David Williams, an electrochemist at Harwell, appeared to be nothing out of the ordinary until his unusual request to speak confidentially with Ron Bullough, the Chief Scientist of the UKAEA whose office is also at Harwell, on the grounds that he had something 'very important that might require a major decision'.[1]

Thus it was that Bullough learned of the fusion experiments at Utah and of the tantalising hints that something very exciting was taking place.

Fleischmann told him about the work in Utah and the ideas that he and Pons had about test-tube fusion, but he also expressed reservations because one of the referees of their grant application to the US Department of Energy had said that it was nonsense. Nonetheless he believed that they were seeing genuine effects that were inexplicable unless fusion were taking place.

Bullough too was sceptical, but he knew that Fleischmann was a respected scientist and that unexpected things happen in science (only a year previously there had been the startling discovery that certain compounds become superconducting at moderate temperatures and so one had to keep in mind the possibility that here again some weird phenomenon was happening). He also realised that if true this could have major ramifications, that Fleischmann was from Southampton and a consultant at the lab and that here was a chance for Britain to get in on the ground floor.

As Fleischmann was claiming fusion on the basis primarily of his heat measurements, the simplest explanation of the 'effect' was that they had made an error in these measurements. However, if fusion was really taking place then there should be radiation in the form of gamma rays and neutrons coming from the cell. To help detect these radiations Harwell took out a health physics radiation monitor from their production line for Fleischmann to purchase, telling him to call them if any further help was needed. Williams told me that neither he nor anyone else at that time made any attempt to try the experiment for themselves: 'My principle is that you don't pinch someone else's experiment.'

Bullough discussed the DOE proposal with members of the Harwell theoretical division to find out their opinion on the possibility that here was some unexpected phenomenon being discovered. Did the idea make sense, was the proposal rational, did it violate anything that was already known by the experts?

The view of the theory group mirrored that of the sceptical referee. The claim by Fleischmann and Pons that the experiment simulated enormous pressures, of the order of billions of billions of atmospheres, building up as the deuterium entered the palladium, were summarily dismissed. But the most significant comments addressed the question of radiation measurements; there was hardly a mention of this crucial test in the entire proposal. It announced that they were seeing '25 per cent heat excess' and that this is 'accompanied by an increase in the background radiation'. However, there was no evidence that this had anything to do with their experiment.

As chemists, Fleischmann and Pons were voting for fusion because they could not understand a mismatch of around a hundred in the energies required to produce the heat compared to what the chemical processes would have allowed. The nuclear physicists at Harwell could not understand the heat from fusion as there was a shortfall in the neutrons by over a billion. On the pure numbers game, a hundred was easier to live with than a billion and the prejudice was already leaning towards chemical effects and/or measurement errors.[2]

Meanwhile Fleischmann was back in Utah with Pons, putting the Harwell radiation monitor to use. On 1 March he phoned Bullough and told him that they were seeing ten thousand neutrons a second coming from their cell. Everyone realised that this was very important to verify.

Bullough offered him hospitality and support at Harwell. 'We have here a secure lab, he could get on with the work away from the eyes of the press. It would be his and Pons' discovery but we would be able to help confirm it and develop it much faster than otherwise.' Fleischmann said he was willing to send all of his equipment—a small cell, electrolyte fluid and palladium specimen—by air to Harwell to see if the lab could repeat the experiment and verify the reality of the effect. The new feature that Harwell could offer would be to determine the detailed energy spectrum of the neutrons (the advantage that Jones had displayed on 23 February) and thereby check what their source was—fusion of two deuterium nuclei, or a deuterium and tritium, or what.

Bullough contacted Archie Ferguson, the Head of Nuclear Physics Division at Harwell, who said he could locate the experiment in an area that was radiation shielded and provide them with the neutron detection equipment. David Williams' task, meanwhile, was to set up the electrolytic cells and then to lead the experimental teams. Bullough now informed Peter Iredale, the Director of Harwell, saying, 'I must say one can't help feeling sceptical about all this but with the reputation that Martin Fleischmann has, and the unequivocal statement that a large flux of neutrons is being emitted, it is clearly worthwhile simply to check whether the effect is a real one.' He reported too that the theorists had advised that any explanation must await verification and that this would be 'relatively simple for Harwell to do. I suggest that we go forward and repeat the experiment'. He added that if there was anything in the electrochemically driven fusion process then 'it's awfully important'. Iredale replied that it made little sense to him too, but that it certainly had to be tried.

The plan to send a cell to Harwell came to nothing; no cell came. There were problems getting paperwork allowing the transport of such 'radioactive' stuff that didn't fit in with the standard classification. As Bullough said to me 'Of course, as it eventually turned out it wasn't radioactive, but if Pons and Fleischmann had been correct, it should have been!'

Thus two weeks later, on 13 March, Fleischmann sent a fax to Williams in which he sounded rather desperate, noting that their information was still very incomplete, that they had much more work to do before they would be confident enough and expressing irritation that they were being 'rushed into premature publication'. He added that he would call by phone the next day at '8 a.m. Utah time, 3 p.m. your time' and 'maybe Ron Bullough should be there too'.

By now Fleischmann and Pons were convinced that they were in a race against Jones and must have feared that they were losing it; any electrochemist at Brigham Young University would be able to add the heat measurements to the neutron work of Jones and write a complete report, identifying the source and fine details of the fusion process, whereas Fleischmann and Pons were stuck with the heat and were hard pressed with the neutrons. Thus they were concerned when they learned that Jones had enlisted the aid of Doug Bennison, an electrochemist who is head of the chemical engineering department at Brigham Young University.[3]

Later Fleischmann asked if Williams could fly out with an 'energy dispersive neutron detector' (i.e. one that counted neutrons and also measured their energies thereby obtaining a spectrum). Williams laughed and said that it would require a Hercules transporter as the detector weighed several tonnes. So the idea was dropped and instead Fleischmann gave Harwell full details of the Utah experiments so that Harwell could build their own cells and replicate the phenomena for themselves with the very best neutron detection equipment.

Williams' team started building the cell and were soon ready to begin. The neutron counter sounded ideal to Williams: 'I spoke on the phone and was told "yes, there's a fantastic neutron counter here with an axial [cylindrical] hole six inches across running through the middle." This sounded great because we could place the cell in the hole and the counter would completely surround it.' The next morning he went over to Harwell's nuclear accelerator where the neutron detector was housed and saw it for the first time. It weighed several tonnes. 'There was much gnashing of teeth as we discovered that the cylindrical hole ran lengthwise horizontally through the detector, not vertically as we had imagined. To nuclear physicists it seems that *axial* means *longitudinal* whereas to chemists it is *vertical*. So we couldn't simply drop the cell into the hole, nor could we turn the counter on its side as it was so immobile. We had to redesign the cell.'

'By the Thursday, 16 March, everything was OK. The detector worked, we checked it by bringing some radioactive samples into the laboratory, and so we put the cell inside it and waited. By Friday evening there was no sign of anything, nor anything through the weekend and all through the Monday, still nothing.'

If the cell had been producing fusion at the rate Fleischmann and Pons believed then neutrons should have been easily visible for Harwell. By Tuesday Williams decided they had to talk with Fleischmann.

'I sent Martin a fax first thing that morning, while it was still night-time in Utah. He received it first thing that morning in his time zone and called me by phone at once to tell me what he and Stanley had discovered since we had last spoken.'

'Martin told me that they had an 8 mm diameter rod that was "dead" (producing no excess heat) and one of 4 mm that was a heat producer. He said that the gamma-ray counter registered counts when they pointed it at the 4 mm "live" rod but that when they pointed it at the "dead" 8 mm rod it registered nothing.'

The implication was that there were neutrons coming from the 4 mm rod and that they were producing gamma rays upon hitting the water bath that surrounded the cell (*see* Figure 5 on page 76).

'This sounded really significant because it implied that they could measure a "null"; in other words they could take a dead rod and tell that it produced no neutrons and then move the counter to the live heat-rod and detect neutrons coming off.' If that was indeed the case, Williams was right to conclude that in Utah they now had good evidence for fusion, and that some cells could be duds while others worked, for whatever reason. Fleischmann and he decided that Harwell must have been using a 'dud' cell, like the 8 mm Utah example, and that is why they saw no neutrons.

They discussed ideas as to why certain cells might work and others not, such as 'poisoning' the rods with sulphur to help the deuterium concentration grow at the cathode of the cell. Fleischmann said that 'he couldn't sit on this much longer'; this was Tuesday, 21 March and the organisation of the press conference was already underway.

'The fact that they could move the counter from one cell to another, detect a signal at the one and not at the other really impressed me. I thought, "do enough cells and we will find some that work". So we took the dead cell out and put three new ones in.'

That was on the Tuesday, less than 48 hours before Utah went public. 'Up to this point Harwell had kept everything deadly secret.' That day Ron Bullough, the UKAEA chief scientist, was driving back from London with Mick Lomer, Director of the Culham Laboratory fusion programme. Culham Laboratory is only five miles from Harwell and the two knew each other well. Bullough told Lomer of their efforts and the news from Utah, since if what Fleischmann reported was true then this would have major implications for fusion research worldwide, and Culham, together with the European JET (Joint European Torus) fusion research centre, housed adjacent to Lomer's laboratory, should be aware of it. It was fortunate that Lomer received this news as the next day he was called by Clive Cookson, the journalist with the scoop, and knew already what it was all about!

The sensational impact of the news, if it was all true, decided them to call an emergency meeting for the next day (Wednesday, 22 March). This took place at Culham, and was attended by senior scientists and theoreticians from JET and Culham and the Harwell people. As Williams remembers it: 'When the scientists first heard the claim they reacted by asking if this was some elaborate April Fool's hoax . . . I defended Martin because of his remark that they could detect neutrons coming from a live cell while showing that there were none coming from the dead cell. That claim was impressive.'

The next morning, Thursday, 23 March, the news was published by Cookson in the *Financial Times* and was picked up by radio and television during the day. (Because of the time difference between London and Utah, the press conference was still hours away.)

Williams again: 'That day all hell broke loose. The laboratory was due to shut early for the Good Friday and Easter holiday; we were trying to get hold of more heavy water to keep topping up our supplies as it gassed off from the cells and a reporter from one of the tabloid papers wanted me to tell him that everyone will soon have a fusion immersion heater in their living room. By midnight we had four cells running, each surrounded by a water-filled bath with a gamma-ray detector pointed at the bath and a neutron counter over the top. We purloined apparatus from wherever we could get it. I worked all through Good Friday and got to bed at 1.30 a.m. I was awoken early on the Saturday by the BBC wanting me to speak on their *Today* dawn programme. Then a photographer from the *Daily Telegraph* came; he seemed well briefed and took photos of our apparatus. Next there was a freelance photographer sent by another of the tabloids who didn't seem too well briefed, so he just asked what had the photographer from the *Telegraph* done and then he went and took the same photos himself.

'By Sunday we had eight cells running. I made a roster for keeping the liquid electrolyte topped up,' (so as not to expose the palladium to the air).[4] He showed me his diary: 'I top up at 5 p.m. and midnight, then someone else does it at 8 a.m. We topped up four times a day.'

This was now Easter weekend which is traditionally a welcome break, the first chance to have a few days' vacation after the bleakness of winter. The media were hot on what they already perceived to be the story of the century.

The Utah experiment sounded disarmingly easy. Roger Highfield in the *Daily Telegraph* noted that BDH Ltd, which supplies chemicals and laboratory equipment, quoted costs for the small amounts of platinum, palladium, lithium chloride (in the electrolyte) and heavy water as £28.10, £31.20, £6 and £17.70. Add a few pounds more for batteries and a test tube, and 'for around £90 someone could attempt their own fusion experiment and rival the conventional approach to fusion at the Joint European Torus in Oxfordshire which costs £75 million a year.' Members of the public were asking Harwell for advice on how to perform the simple experiment; a spokesman at Harwell expressed 'concerns for their safety'.

Word was getting out of the Utah press conference with its dramatic claims where Fleischmann said 'We have found conditions where fusion takes place and can be sustained indefinitely.' Peter Iredale, the Director of Harwell, became very concerned about safety as reports began to emerge that Fleischmann and Pons were warning 'we caution people doing this experiment that we have run some systems above the break-even point and one above ignition which melted. We were not in danger when this occurred as it was contained and there was nobody there. Since the accident we have scaled down our experiments. This is our word of warning: please, if anyone attempts this do not do it on a big scale.'

Williams immediately took precautions. 'We set up trip switches so that the current would switch off if anything happened and had television cameras monitoring the cells and the meters recording the signal coming from them. We put marks on the meters and told the people monitoring the screens in the control room 'these are the phone numbers you call if the needle goes past the mark'.

'Everything was an absolute frenzy. We would need some trifling piece of wire or something but the stores were closed for the holiday. The Director then called up the police who man the main gate to the lab and told them to open up the stores for us. So we went along to the stores with this policeman standing there and we would say ''right, let's have one of these and two of those'' and he recorded it all on the back of an envelope so that the accounting could be properly completed when stores reopened officially on the Tuesday.'

111

## Martin Fleischmann visits Harwell: 28 March 1989

Martin Fleischmann left Utah after the press conference to spend a quiet Easter weekend in Britain, and then visited Harwell on Tuesday 28 March as soon as the lab opened after the Easter holiday. To make a fast getaway he flew from Salt Lake City on Delta to San Francisco to connect with that evening's British Airways flight to London, but the Delta flight was delayed and he arrived in San Francisco to see the British Airways flight departing. He rebooked on a flight for the following day and thought 'no one knows that I am in San Francisco so I can be anonymous here as easily as in Britain.'[5] He checked in to a hotel and as he entered his room the phone was ringing. 'I wondered who on earth could know that I was there and when I picked up the phone, it was [a well-known scientist] who said, "Do not leave your room"!'

They discussed test-tube fusion but Fleischmann was not specific on how the scientist knew of his whereabouts. Fleischmann believed that the national security authorities had had concerns about the possible military applications of the Utah work and that for this reason he was being followed. He himself has always been worried about military applications of this work and had tried to have the press conference cancelled for that reason. This, superimposed on youthful experiences such as the Nazis moving into his Czech homeland, now amplified by the call to his San Francisco hotel room, may be why he now believes that much of the negative reaction to his work on test-tube fusion has ulterior motives which are not merely scientific.

Fleischmann was rapidly becoming convinced that it was important to be cautious in what he said about the work. His nervousness was not helped by what he considered to be a very hostile reception to his first scientific seminar on the project, which he gave at Harwell on 28 March.

Early on that Tuesday morning he was interviewed on the BBC Radio, in the course of which he announced that he would be visiting Harwell to give an official talk and to discuss the ongoing experiments there. David Williams told me of his impressions: 'To gain access to the site you have to go through several procedures normally. As a result of Martin's radio interview, lots of people wanted to come on site and listen. We said, "No", and arranged to meet him at the main gate so that photographers and media could interview there.'

Ron Bullough had organised the meeting and it was attended by scientists from Harwell, Culham, JET, SERC (Science and Engineering Research Council) and the CEGB (Central Electricity Generating Board). Fleischmann first described his evidence for heat and this was well received with one caveat; in Bullough's summary of the meeting he noted that it was 'surprising and unfortunate' that no experiments had been done with ordinary water in place of heavy water (as fusion should only occur in the heavy water, failure to see excess heat in an ordinary water experiment while producing heat with heavy water would have been rather impressive. The question of why they had not reported results with ordinary water was asked repeatedly in the following weeks).[6]

The claims to have evidence for nuclear radiation, however, were less well received. First there was the evidence that had impressed Williams when he was told over the phone that the Utah chemists had measured neutrons coming from a heat-producing cell while finding none coming from the dead cell. The original impression had been that this was a measurement that was being done continuously—put the monitor above one cell and it triggers; move to the adjacent cell and it stays quiet. However, it now became clear that it was not like that. The measurements took place over one cell for 50 hours, and registered a number of counts; then the measurements were taken for a similar extended period over a sink some distance away and fewer counts were recorded. So a period of several days was involved, during which time there can be significant fluctuations in the cosmic radiation from outer space which, on striking the atmosphere, can cause neutrons to hit your detector but which have nothing at all to do with the neutrons from the cell that you are really looking for. Discriminating between 'background' neutrons and genuine signal was one of the scourges of many of the experiments reported in the subsequent months. With care it is possible to do this, but the Utah experiment had not done so; the difference in numbers of neutrons 'measured' at the two cells was not particularly significant and could easily have been due to changes in the background radiation and nothing at all to do with the nature of the two cells.

Even worse, though, was the possibility that the gamma-ray measurements, which were being cited by the two chemists as evidence for neutrons, were in fact wrong. When Fleischmann showed his data, his evidence for gamma-rays (and by implication, fusion neutrons) was instantly objected to. This was the first time that the work had been presented to an audience with nuclear expertise and it immediately received the response: 'That is wrong. The peak has the wrong shape and it has the wrong energy.'

The report of the ensuing discussion[7] notes 'A peak at 2.5 MeV . . . was claimed. The peak was at the wrong energy for gamma-rays arising from neutrons captured by hydrogen. The calibration was therefore suspect and it was difficult to see how it could be so far wrong.' And the failure to show the range of peaks was also

113

commented upon. 'The raw data . . . were not provided, and without such detail it was difficult to assess the reliability of . . . the energy of gamma-rays observed.'

Where were the raw data, the list of numbers from which the entire spectrum of gamma-rays could be reconstructed? Fleischmann did not have them and said that he would ask Pons.

One can only imagine the consternation in Utah when Pons learned that a true gamma-ray from neutron capture should occur at around 2200 keV, whereas their data were exhibiting a peak at 2500 keV. Matters were not easy for him as the data had originated with Hoffman and Hawkins and urgent action was needed as the data had already been sent to the journals. However within a couple of days a peak, whose data points had appeared previously centred on 2500keV, was offered a replacement for the original but now centred at 2200 keV (*see* Figure 8 and compare with Figure 7) and it was this that appeared in their paper published in volume 261 of the *Journal of Electroanalytical Chemistry*.

The figure had been modified so that 2500 keV (2.5 MeV) became 2200 keV (2.2 MeV) but there remained clues in the text to the urgent changes—equation (vii) still referred to the wrong 2.5 MeV (2500 keV) energy:

'The spectrum of $\gamma$-rays emitted from the water bath due to the $(n,\gamma)$ reaction

$$^1H + n(2.45 \text{ MeV}) = {}^2D + \gamma(2.5 \text{ MeV}) \tag{vii}$$

was determined . . .'

In following the chronology of test-tube fusion, one should remember that almost all outside observers were unaware of this history and only had the evidence as presented in the published article; it would not be until late June that a full spectrum would be published, by which time the gamma-ray evidence had been called into question as a result of detective work by scientists at MIT (Chapters 9 and 15).

The general opinion at Harwell following Fleischmann's report was that the nuclear data—the claim to have seen neutrons as evidence for fusion—were questionable. A Harwell report on the meeting[8] said that fusion of deuterium would be 'grossly insufficient' to account for the observations of heat release and 'under questioning it was also admitted that the nuclear measurements were preliminary'.

The Harwell assessment said: 'The most dramatic (although qualitative) claim was of an experiment in which a one centimetre cube of palladium *had been observed to melt* [my italics] and in part

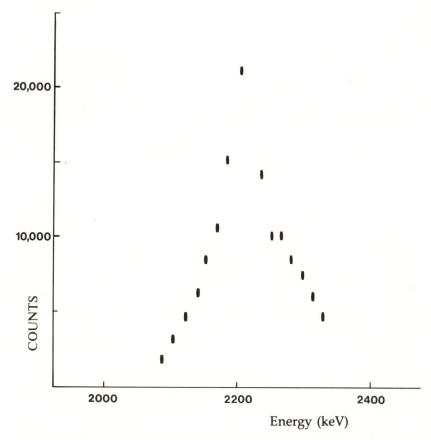

Figure 8. The signal peak data points in the form presented in the published paper.

vaporise. However, Professor Fleischmann indicated that not all cells showed an excess heating effect: this adds to the difficulty of either substantiating or refuting his claims.'

I emphasised the 'had been observed to melt' because this reinforces what had impressed Fleischmann months earlier, when Pons had called him by phone and told him of this event. For many people the vaporising of a lump of palladium was indisputable evidence that something beyond mere chemistry was at work. The unasked question though was 'Did anyone actually *observe* it melt?' or was it something that happened when no one was around? In fact the latter was the case and the lump could have exploded from the pressure building up within it from the deuterium, or because the level of electrolyte had fallen below the top of the palladium with the result that the gases started to escape and then recombined explosively due to a small electrical spark.

115

The opinion at Harwell at the start of April, as no neutrons were being seen, was that they should now put effort into checking the heat measurements as Fleischmann's talk made these appear to be very clear. In addition they decided to continue using the neutron detector as news had by now come through of Jones' experiments at Brigham Young University, where he claimed to see neutrons (though no dramatic heat phenomena such as melting) at a level slightly above that of the background radiation from cosmic rays and other materials in his lab. Jones' result, if correct, would not rewrite the world' energy budgets but could, nonetheless, be interesting for science, not least in their possible implication for geophysics and the source of the Earth's heat.

So began attempts to test Jones' work by duplicating his experiments with beakers filled with a 'witch's brew' of salts and which bore only a superficial similarity to those of Pons and Fleischmann.

The Harwell team designed a shuttle arrangement whereby two cells, one 'real' and one a 'blank', were supported on a rod. The idea behind the design was that a motor would drive the 'real cell' inside the neutron detector which would record if any neutrons were being emitted. Then the cell would be withdrawn and the 'blank' cell inserted in its place. Any neutrons recorded while the 'blank' cell was in place must be coming from the background. As the apparatus, background and experimenters were identical in both cases, any difference in neutron rate would be evidence that the 'real' cell was producing neutrons. With the apparatus motor driven, they hoped to be able to monitor both cells almost continuously, pulling in and out every few minutes.

Preparing the apparatus, cleaning it so as to remove any possible sources of contamination, was a careful and painstaking job which took several days. While this was going on Harwell received another of the many rumours that were flying at the speed of light around the electronic mail networks. This one was that Birmingham University had confirmed the Jones effect. Harwell knew that with their detector they would be able to check his results within a few days of experimenting and redoubled their efforts to get started in light of the Birmingham rumour. David Williams recalled that 'There were guys out there in the snow until 10 p.m. cleaning the apparatus to get it ready. The next day dawned beautifully clear. We spent six hours on the Friday and again on the Saturday pulling and pushing the cells in and out manually because we couldn't wait for the motorised version to be completed.'

Harwell had had an inside track. They learned about test-tube fusion before the rest of the world and did what may well be the most complete and definitive set of experiments. As in so much of the

116

## SUMMARY OF EVENTS; MARCH 1989
(Horizontal lines are between Sunday and Monday)
F: Fleischmann; P: Pons; FP: Fleischmann and Pons

1. F informs Harwell of neutrons. Arranges to send them cell (unsuccessful)
2. US DOE approve FP funding
3. Heads of UU and BYU talk
4.
5.

---

6. Major meeting at BYU
7–9. Gamma ray spectra
10. Editor of J. Elec. Chem. talks to P
11. Paper written and submitted by FP
12.

---

13. Harwell receive details from F
14. Harwell begin experiments
15.
16.
17. UU Press officer interviews FP
18. Revision to paper
19.

---

20. Press release ready
21. F learns that Harwell see nothing. He suggests to them that rods can be 'dead'
22. DOE receive press notice – Jones alerted by DOE
23. PRESS CONFERENCE AT UU. Jones faxes paper to *Nature*
24. FP send off paper to *Nature*. F departs Utah
25.
26.

---

27.
28. F talks at Harwell. Problem with gamma peak. F notifies P
29. University physicists meet with P. They tell university officials and P that gamma peak is erroneous
30. Fax copies of papers sent out. Gamma peak at 'right place'
31. F talks at CERN. P talks at UU

---

test-tube fusion story, rumours abounded, in this case to the effect that Harwell were confirming some aspect or other of the Fleischmann and Pons phenomenon. But these rumours were groundless; Harwell indeed fulfilled their promise of being a 'secure lab' as no detail of their work became known until May when they announced that none of their experiments had yielded 'statistically

117

significant evidence of fusion reactions taking place under electrochemical conditions'.

Meanwhile, ignorant of the Harwell work, on 24 March the rest of the world learned of the chemists' claims to have achieved cold nuclear fusion and lived by rumour for weeks thereafter. Whereas in private most scientists expressed scepticism, there was a reluctance to go public too soon and say that the claims were wrong. The earliest and sharpest refutation that I found was in the *Observer* of 2 April, where geologist Beverley Halstead noted that many science correspondents seemed already to have 'booked tickets to Stockholm to witness the Nobel ceremony . . . but it really does not make sense . . . Until tangible evidence is to hand and details of the exact techniques used, we should not have our time wasted by such tall tales.'

# II

# DEUTERONOMY

World reverberations after 23 March 1989: attempts to generate fusion
by deuterons

# 8

# THE FIRST REACTIONS

Everyone who is over 35 years old is supposed to recall where they were when they first learned of the assassination of President J. F. Kennedy. For many scientists and others the same will be true for cold fusion. I was in the arrival lounge at, ironically, J. F. Kennedy airport in New York having just flown from Britain. It was around 2 p.m., Eastern Standard Time, on Thursday, 23 March and it was not until some days later that I realised how bizarre this was; not only was cold fusion incomprehensible with respect to the laws of fusion, but it also seemed to have violated the theory of relativity which asserts that information cannot be transmitted at faster than the speed of light, backwards in time.

The Utah press conference, which first released the news, had not yet taken place and the news would not hit the American papers until that evening. However, it turned out that the story had been headlined already on the British evening TV newscasts, which had taken place while we were still in the air, and they had obtained the information from that morning's *Financial Times* where Clive Cookson, the science and technology correspondent, believed he had won the scoop of his career. Allowing for the time difference, the news was in the morning paper in Britain a full fifteen hours before the press conference took place. The British Airways plane had copies of the paper on board and others on the flight (but not I!) had read it and were talking about it.

In the furore during the months following, the University of Utah defended their calling of the press conference by citing this leak as proof that journalists had spies in the place and that the news was leaking out with consequent threat to possible patents for what, they believed, could be the greatest financial prize of the century. However, in reality there never was a spy leaking information to unnamed journalists to undermine the fusion project. If there was a mole it was Fleischmann who, with Pons' agreement, had passed the story on to Cookson.

## PARTIAL SUMMARY OF APRIL to MAY 1989
(Horizontal lines are between Sunday and Monday)
F: Fleischmann; P: Pons; FP: Fleischmann and Pons

1.
2.

3. P: Indiana
4.
5. P learns of Texas A&M
6.
7.
8.
9. Period of confusion over fusion in light water

10. Texas A&M announce heat. Georgia Tech announce neutrons
11.
12. P: Dallas; F: Erice
13. Helium experiments in process in P's lab.
14. Univ. of Washington announce tritium
    European Fusion Community meeting
    G. Seaborg advises President Bush
15.
16.

17. P claims helium found
18. Claims that heat output:input 8:1 achieved. ENEA Italy report neutron burst
19. US Dept of Energy meeting
20. Utah Governor signs cold fusion bill
21. Rep. Roe calls Federal hearings for 26 April. Caltech internal seminar
22. FP withdraw *Nature* paper: 'too busy'
23. Private discussions with nuclear physicists: Indiana, Oak Ridge

24. US Secretary of Energy instructs US DOE laboratories to study cold fusion
25.
26. Congressional hearings
27.
28.
29.
30.

1 and 2 May   APS meeting: Baltimore


Cookson has a high reputation in British science journalism. He has been science correspondent for the BBC as well as the editor for science and technology at the *Financial Times*. People instantly took note of what he had written. Roger Highfield of the *Daily Telegraph* was once an engineer, knew Fleischmann to be a reputable scientist and also respected Clive Cookson. He told me: 'I saw Clive's story and knew that it would be reliable. I couldn't easily dismiss it. Easter weekend it may be but I had to follow this up.'

That weekend many journalists in Britain and elsewhere were working late following up on what already was being perceived to be 'the story of the century'. I, meanwhile, had heard only verbal details in the Kennedy airport baggage hall and dismissed it as nonsense. Releasing millions of volts of energy from a nucleus by means of a few volts from a car-battery made no sense at all and I, like some of the Harwell scientists, thought that it was an April Fool's Day story run a week early. I dismissed it from my mind and spent the weekend on the Long Island Coast with no radio, television or newspapers. That Easter holiday I was working on a very different story unaware of the circumstances that would intervene, or that the garbled nonsensical news of the chemists would deluge the world of science.

Columbus landed in America and it was a year before Europeans knew; the Eagle landed on the Moon and the whole world watched it a mere heartbeat after it happened. A news-story in a British paper one morning is common property as the Sun rises around the globe: *Financial Times* in the morning, British radio at high noon, and Associated Press wire it to news services round the world within the hour. Some 6000 miles to the west it is dawn in California and a Japanese language radio station seems to have been the first off the mark and broadcast the news while English-speaking Americans were still unaware. Ryoichi Seki was listening in to his native language and the news interested him; Seki is a physicist and was visiting the California Institute of Technology—Caltech—in Pasadena, a Los Angeles suburb. He went in to work and spread the word around the scientists at Caltech's Kellog Lab—a nuclear physics laboratory named after the cereal magnate who was its benefactor (and who thought that the giant X-ray tubes were going to cure cancer).

Charlie Barnes, a professor there, had spent many years investigating fusion, the reaction that powers the stars. He called Bob Finn in the press office who almost immediately obtained a copy of the AP press release based on the *Financial Times* article. So before the Utah press conference had taken place, the news was already in the hands of nuclear professionals. The trouble was that there were not enough details to make use of.

They spent Thursday listening to every newscast trying to decipher what Fleischmann and Pons had done. They knew that if deuterium fusion really was happening then there must be neutrons produced and that they were expert at detecting them. So began a collaboration that would eventually provide headlines of its own, on the front of the *New York Times*, six weeks later (Chapter 11).

Also in California, but on vacation, was Van Eden, a physics student from the University of Washington at Seattle. He was totally cut off from the news and the first that he knew of cold fusion was when he read the *New York Times* on the Tuesday after Easter. A student leading an ordinary life that weekend, he had no hint of the turmoil that would soon surround him and that within a few weeks he would be patenting a scientific discovery and appearing as a lead item on the NBC Nightly News (*see* page 133).

## Oak Ridge, Tennessee

On Easter Sunday, 26 March, I returned from New York to Oak Ridge National Laboratory in Tennessee where I was working. Having learned of test-tube fusion and dismissed it in J. F. Kennedy airport on the Thursday, it was coincidental that I should be in New York's domestic La Guardia airport and pick up a discarded *Wall Street Journal*. It was two days old—Friday, 24 March—but the headline caught my immediate attention. 'Taming H-bombs? Utah Scientists Claim Breakthrough in Quest for Fusion Energy' and the confirmation of what I had previously dismissed as mistaken rumour: 'Batteries and Palladium Wire'.

Certainly it seemed to be legitimate. The chemists claimed to have triggered a fusion reaction in a test tube that continued for a hundred hours. This eliminated my first suspicion, namely that some dubious glitch of a counter dial was being heralded as significant. 'Four watts produced for every one watt put in'—that certainly impressed, and it was that statement, coming from James Brophy, vice president for research at the University, that was the flag for the experiment in the following weeks but whose credibility was later to be questioned. The name Pons meant nothing to me, but Fleischmann I recalled having been made a Fellow of The Royal Society only a couple of years before. This suggested that perhaps he was unlikely to be fooled easily and was the reason why now I began to take note of what was going on.

People make mistakes in all activities, and science is no exception. The following weeks would see examples enough of that as attempts to be the first to prove or disprove the astonishing claim got underway. The experiment turned out to be disarmingly easy to set up and equally easy to misinterpret. Just as the experienced adult is less likely

124

to get burned than the naive child, so will a senior researcher with a proven track record recognise the many pitfalls that the unwary can fall into. 'An established scientist is one who has already made all the mistakes.' Martin Fleischmann had nearly 40 years of research experience behind him.

There were two other remarks in the first *Wall Street Journal* article, one so obvious that it did not unduly strike me at the time. The paper noted that in science an experiment's results weren't accepted until duplicated in a second laboratory. University of Utah officials emphasised that the claims had to be subjected to 'the world court of scientific opinion' before being believed.

The second piece of information which impressed me and, I learned, many of my nuclear physics colleagues too, was the news that at Brigham Young University a team led by Steven Jones had been carrying out experiments on cold hydrogen fusion 'similar to those in Salt Lake City'. I well knew that Jones had been studying fusion induced by muons but this bore no obvious relation to what the chemists had been doing. What was afoot? 'We cannot verify their results', Jones was quoted as saying. What did that mean—that they disagreed, that they weren't in a position to say one way or another, or that he was saying nothing until his results were published in the scientific press in the time honoured manner? The paragraph concluded with the cryptic remark that his team's work shouldn't be interpreted as casting doubt on the Salt Lake City claims.

If the intention had been to arouse excitement and anticipation, these titillations could not have been bettered. The magic word was *fusion*; the dream of the ages was apparently being fulfilled. It was that word that was the light around which the media moths were fluttering; if instead the chemists had claimed to have discovered a new form of storage battery, no one outside a limited circle would have paid any attention.

Why were Pons and Fleischmann claiming *fusion*?

Their cell was quite small; small enough to hold in the hand. We know how much a single palladium atom weighs and so we know roughly how many such atoms there are in their apparatus. The overall heat output claimed by them implied that each individual atom was putting out at least one hundred times as much energy as would be possible if chemical processes alone were at work.

Chemistry involves the redistribution of electrons in the periphery of atoms—sharing among atoms to build up molecules, jumping from one quantum state to another, accepting or releasing energy in the process. The warmth of your hand is due to the heat released by chemical processes taking place in your body. The energy release is faster in some processes than in others; strike a match and it bursts into flame. In extremes it can be explosive as in TNT or terrorist devices. Even in the most extreme of these each atom is releasing less than 10 eV (electron volts) of energy, yet the net heat output in the Fleischmann and Pons experiment seemed to imply that each atom was releasing many hundreds of eV—the scale of the output appeared utterly to be beyond that of chemistry. However, such energies could easily be tapped within the central atomic nucleus. Either Fleischmann and Pons had miscalculated or they had a nuclear effect; and the only candidate for this was fusion of the deuterium nuclei from the heavy water as it entered the palladium.

Palladium is not particularly radioactive; it is not like radium for example whose spontaneous fissioning causes it to glow in the dark. Palladium has been used for years by jewellers, dentists and hydrogen transporters, none of whom go around wearing radiation monitors. The nuclear energy, if that is what it was, had to be due to the deuterium packing into the palladium. Somehow the palladium induced them to fuse without the thermal encouragement of a million degrees input, as advocated by the hot fusion researchers. If so, we physicists were going to look rather foolish.

But fusion of deuterium nuclei is well understood. There may well be a mystery as to how the palladium encourages them to fuse, but the aftermath of fusion is standard physics. Roughly half of the fusions between the two deuterium nuclei produce neutrons immediately; essentially the rest produce neutrons eventually. The direct production of helium occurs less than one part in a thousand. Translate apparent claims for power production of the '4 watts' into a rate for producing fusion neutrons and you find that there should be a thousand billion neutrons pouring out of the apparatus each second!

So Fleischmann and Pons appeared to have made a table-top reactor and instead of surrounding it with a protective shield of concrete blocks and lead against what would be a lethal dose many times over, there they are holding a cell in their hands, in jackets and ties, proudly in front of what may be the most concentrated nuclear source west of Chernobyl. And they are manifestly alive.

If a picture is worth a thousand words, then it was that photo that made physicists, at least, somewhat puzzled. There was more going on than met the eye. The need for real information, a detailed scientific paper, became paramount.

At Oak Ridge Jones' paper arrived by fax, and showed that he only measured neutrons and that the evidence for fusion was a small shoulder in the energy spectrum. Local experts immediately queried whether he could be sure that these were not simply neutrons that cosmic rays had knocked from materials in the environment. A fax then arrived of Fleischmann and Pons paper, but this added little more than frustration. The tables in it contained no information on errors, numbers were quoted to several significant figures without details of where they had come from. The only place that support for the sensational claims of '4 watts out for 1 watt in' could be found was in 'column C' of one table which did not deal with measurements but with some theoretical estimates of what *might* be achieved in some idealised circumstances (this point was made clear only when Fleischmann remarked in his testimony to Congress that these numbers were 'projected as is commonly done in fusion research', (Chapter 11)). One scientist even sent around a plea on the computer network for guidance: 'Is this real? Is it a joke?' A graph purporting to show evidence for tritium detection had part of its labels missing, and if what was shown was correct, then the missing label suggested a negative number for a quantity that should be positive. The text had several errors and contradictions all of which suggested to experienced scientists that the paper had been prepared in a rush.

In the office just beneath mine at Oak Ridge was Don Hutchinson, a nuclear engineer and leader of a group developing advanced diagnostic devices for fusion. He had first heard the news on the Saturday morning when a colleague had called him, and as a result he had watched the newscasts all weekend, becoming more and more excited and amazed. On the Monday morning at the lab he talked about it with his group which included Roger Richards, also a nuclear engineer, and Chuck Bennett, a physicist. Bennett suggested that they should call around to see if the lab was going to make any response and start to work on it, so that afternoon Hutchinson called Mike Saltmarsh, head of the Oak Ridge National Laboratory fusion programme, asking what the laboratory planned to do and expressing a desire to be involved. Saltmarsh said that he did not yet know but promised to call back, and did so at 8.30 a.m. on the Tuesday morning, telling Hutchinson that he would be in the primary effort and to proceed. It was during that Tuesday lunchbreak while I was sitting outside the building in the sunshine discussing the news with the other nuclear theorists that Hutchinson came over and gave us the news that he had been told to start experiments.

The story was for me suddenly becoming alive and real.

I then had to return to Britain for a month and did not interact with Hutchinson until early in May. J. D. (David) Jackson, who 35 years earlier had made the first detailed calculations in the west on muon catalysed fusion (*see* Chapter 4), was visiting Oxford University. I showed him the papers and within hours he had calculated the possibilities, evaluating what would be needed to generate fusion at the rate Jones' experiment would require. These showed that the deuterium nuclei would have to encroach at least a factor ten nearer to one another than experience with gases in palladium suggested occurred. Already the claims seemed rather improbable and the heat claims from the chemists were even more bizarre. Jackson wrote up his notes and sent them around the local laboratories, Rutherford Lab, Harwell and others, as an open document.

News then broke that Fleischmann had spoken at Harwell on 28 March. David Williams gave a report about this at a meeting of the Institute of Physics one week later, and it was from this that people at large learned about the concerns expressed, namely the impossibility of the gamma-ray peak cited as evidence for fusion and the lack of usual control experiments. Sir Denys Wilkinson, a leading nuclear physicist, had come to talk with me about some other aspects of nuclear physics but, as was the norm in those early weeks, the conversation soon turned to test-tube fusion. I told him about the heat and the failure to find neutrons at the required level. He immediately wondered if the production of tritium and proton, which is normally produced at the same rate as the neutron channel in hot fusion, might be responsible and relatively enhanced as a result of electrical disruption. However, this seemed unlikely to help by a factor of billions, even allowing for the fact that the rate of fusions required to make watts of heat was way beyond comprehension. He was surprised to hear the reports that there had been no control experiments with ordinary water and promised to ask for clarification of this at a special meeting that had just been arranged for 12 April in Erice, Sicily.

Rumours were now beginning that cold fusion was being seen elsewhere, but there was little clear information. We waited for the first definitive announcement in the race to be the first to prove or disprove the phenomenon.


## 10 April: Texas heat and Georgia neutrons

Fleischmann and Pons' claims seemed to show the promise of a bright new horizon. If true, then this could turn out to be one of the most seminal discoveries since that of atomic structure. In the best traditions of scientific research, all around the world groups set up experiments to check it for themselves.

The first stage in this research process would be to replicate the phenomenon at different places using a range of apparatus and thereby verify that it is a true natural phenomenon and not an artefact.

Once several groups have independently confirmed a phenomenon's reality under subtly different conditions where some experiments prove easier than others, one forms a range of experience from which the *essential* secret gradually becomes known. No one wants simply to reproduce what Fleischmann and Pons or Jones had done—there is little or no scientific kudos for coming in second and merely replicating what has gone before. The aim that drives experimental science is to understand the origins of the phenomenon and from this base to *improve* on the techniques and advance the state of the art.

Personal ambition and natural curiosity each played its part. If Fleischmann and Pons' claim stood up to these tests then here was probably the beginnings of a new industrial revolution and those who learned the techniques the fastest would be leaders in the new field. But natural curiosity was the prime mover: could it *really* be that the way to harnessing fusion was via not an intricate technological maze but a direct route once the door was found—and that that door was now open?

Like children enthralled by the magician's sleight of hand, we all wanted to see this trick of nature and then perfect it for ourselves.

In those first few days it was hard to be sure quite what the evidence was and whether or not Jones' work related to Fleischmann and Pons'. In the USA people were at this stage unaware of the Harwell work and no news of their results became known until May. The news reports suggested that the experiment was easy and thousands set it up. Indirect clues to the amount of activity showed up in strange ways. There was so much sudden demand for palladium that the price soared and Ontario Hydro, major suppliers of heavy water who usually deal with orders of a thousand tonnes for use in power stations, were inundated with requests from all over America for lots of a few kilogrammes.

## Texas A & M

Charles Martin had known Stan Pons professionally for some time and, being an electrochemist, he set out with colleagues Kenneth Marsh and Bruce Gammon at Texas A & M University to see if they could produce the phenomenon. Martin was one of a few electrochemists who had talked in some detail with Pons in the first hectic days after the 23 March press conference, and they designed the cells with every hope of success. Early in the first week of April they began to get positive results, seeing large amounts of excess heat being produced in their heavy water cells with palladium electrodes.

The excess was up to 90 per cent over and above the amount of power that was entering the cell from the electricity supply. This was in line with the magnitudes that the Utah press conference had led them to expect: if Utah after years of experience were claiming 4 watts out to 1 watt in, then a near doubling in the first two weeks of trying after taking advice from Pons has the feel of being 'on the right track', and a clear replication of the phenomenon. Martin called up Pons who was delighted to hear the news, telling Martin how great it was to learn that he had replicated the phenomenon.

By this stage the trio felt that they were succeeding, but before writing a paper there were important checks to be made. If the heat was indeed due to the fusion of deuterium nuclei—dueterons—in the heavy water ($D_2O$), then there should be no such effect if they ran the experiment with ordinary water, $H_2O$, in place of the $D_2O$. And so they set up this 'control' experiment, but found that heat was apparently being generated in this experiment as well. They double checked and there seemed to be no doubt: excess heat with *both* the $D_2O$ and the $H_2O$. Confused, Martin called Pons at the end of the week of 8 April to tell him this and ask for Pons' opinion.

Pons gave an astonishing reply, saying that 'yes'—he too had seen heat with ordinary water, that now he understood the reason why and that this was the most exciting part of the whole story. This was the 8 or 9 April; on the 7 April Pons had received on the electronic mail a theory 'paper' by Steven Koonin and Michael Nauenberg which pointed out that at low temperatures fusion between protons and deuterons (*pd* fusion) occurred faster than that between pairs of deuterons (*dd* fusion) (though still far too slowly to generate significant heat). What Pons had decided was that small amounts of deuterium in the ordinary water (present at about 1 part in 6000) were fusing with the copious protons and, because the *pd* fusion is favoured, astonishingly the net effect was that there is measurably fusion heat even with ordinary water!

No wonder Pons was excited. In March he had believed that fusion can take place in a beaker with heavy water—itself an amazing discovery—and with the most 'exotic' commodity being the $D_2O$. And now, by the weekend of 8 April, he believes that you do not even need the heavy water: ordinary water is enough; literally fill the beaker at the kitchen sink. Pons believes this because heat had been seen in his lab in experiments using light water,[1] Koonin and Nauenberg give him what he believes is a possible reason for it and now Martin is confirming it.

Then Pons told Martin a strange thing. He said that he (Pons) was not allowed to make this news public but he encouraged Martin to do so, and that it would be Martin's discovery.

For Martin, naturally, this was exciting news. Pons was confirming that the results from Texas were indeed as they should be since light water also could give out heat. Earlier the BYU and the Fleischmann–Pons teams had drawn mutual encouragement from each other's results while being rushed into the helter-skelter race to publish; now Texas A & M and Pons were doing likewise. The decision by the Texas team to go public was fuelled by Pons' assurances that he too saw heat with light water. Pons had also told Martin that he knew of dozens of experiments confirming the Fleischmann–Pons effect that had not yet gone public. The sense of competition, the desire to get an important result such that they had been working 24 hours each day on the experiment and the spur that dozens of others were in the wings apparently in agreement, plus Pons' confirmation and encouragement, all combined in deciding them to go ahead. They submitted a paper outlining their results so far and the University put out a press release. The *Wall Street Journal* of Monday 10 April announced 'Cold Fusion Experiment is Reportedly Duplicated' and that 'School officials said they would hold a news conference today to announce that their researchers had achieved the same kind of cold fusion'. A Texas A & M spokesman said that they were announcing it as soon as possible 'as a gesture to the researchers in Utah'.

## Georgia Tech

If the palladium and heavy water were indeed producing fusion as Fleischmann and Pons were claiming, then there should be radiation and tell-tale products—tritium, helium, protons and neutrons. Jones also had reported seeing neutrons and so several groups set out to look for these as their initial strategy.

At Georgia Tech in Atlanta a team of physicists and metallurgists was also in the race and came in almost at a dead heat with Texas A & M. James Mahaffey directed a team including Bill Livesey, a physicist and metallurgist, and Rick Steenblick who set up a cell and looked for neutrons by using the traditional device—a $BF_3$ counter. The $BF_3$ counter works because there is a large affinity between neutrons and boron (the $B$ in $BF_3$). The fast neutron hits the massive boron nucleus, which ejects a fast alpha-particle ($^4He$), the basic reaction being written

$$n + {}^{10}B = {}^4He + {}^7Li$$

In effect the neutral neutron has been converted into an *electrically charged* alpha particle which ionises the surroundings, ejecting electrons which pass down wires to a counter where the amount of charge arriving is in proportion to the number of initiating neutrons.

Rick Steenblick was in the lab on the night shift and noticed that, if the detector was to be believed, lots of neutrons were pouring out from the cell. James Mahaffey had been at home in bed, but thinking about the experiment. As was the case for most people in the test-tube fusion race, there was the feeling that they were involved with what, in generations yet unborn, would be seen as a seminal moment in human culture: the discovery of the key that solved the energy problems of the species and saved the world from pollution death. When he arrived at the lab next morning he found Rick Steenblick in a state of high excitement. Neutrons were coming from the cell; the detector was recording thousands of them.

As a check, Mahaffey decided that they should move the neutron detector to behind some shielding bricks that surrounded the cell. The bricks would intercept the neutrons coming from the cell and so none should be seen outside the protective fence. And this is indeed how it appeared to be: outside the bricks the detector registered almost nothing whereas in front of them, where the cell was in full view, the detector registered neutrons. They repeated this over and over and found that there were apparently ten times as many neutrons recorded adjacent to the cell as when the detector was taken all the way out behind the protective bricks. They tried every way they could think of to check the detector and there seemed to be no doubt that neutrons were coming from the cell: a signature of fusion is neutrons.

Now, Fleischmann and Pons had reported both heat and radiation products as evidence. The Georgia team measured the temperature of the cell and found that the hotter the cell was, the more neutrons the detector recorded. This seemed to imply that whatever was causing the neutrons to come out was also causing the temperature rise. If so, this really clinched the proof: fusion producing heat and neutrons in concert.

The Georgia group called up the local press and the media poured in for the press conference. Fleischmann and Pons had been verified; test-tube fusion was real. Georgia Tech was suddenly on the map. The press conference was at 4 p.m. on Monday, 10 April.

The power of the media in the test-tube fusion story was already apparent when Livesey remarked in a phone interview with the *Boston Globe* that he had not yet seen a scientific paper describing the Fleischmann and Pons experiment but had 'obtained all of the information he needed from newspaper accounts'. The scientists stressed that the results were preliminary and needed checking but this subtlety didn't come across much to the public who saw this as confirmation. Texas A & M confirmed the heat and Georgia Tech saw neutrons: the dream was coming true.

During the next four days the media heralded more reports of cold fusion being replicated. There were the first reports from India that a group at the Indira Gandhi Atomic Research Centre saw a sudden rise in temperature in a cell, though the more cautious scientists pointed out that it was not clear whether this proved excess heat. The next day, 12 April, Tass reported that Dr Runar Kuzmin, a physicist at Moscow University, had replicated the Fleischmann and Pons phenomenon. However, there were no details of what he had done or seen, or of how he had measured it. Indeed, there was no proof that anything had been done at all, but this nonetheless added to the conviction of test-tube fusion supporters that there was something real.

On the 14th in the USA came the third major announcement of the week. This seemed to complete the cold fusion trinity of heat, neutrons and tritium: Texas A & M University reported that they replicated the heat, Georgia Tech had reported neutrons and now Van Eden and Wei Liu, two bright graduate students at the University of Washington in Seattle, announced that they saw tritium.

## 14 April: the Seattle tritium

For the two Seattle students the adventure had begun on 28 March. Van Eden had been on vacation in California during the Easter weekend and out of contact with the news, but in the *New York Times* on the following Tuesday he had learned about the chemists' claims. Eden and Liu had been using ultra-high vacuum equipment in the course of their thesis work, which has no obvious connection with electrochemistry, palladium or fusion. As was the case at MIT and Caltech, it was the enthusiasm of youth that drove the initial effort.

While cleaning out the laboratory they found some bottles of heavy water and then had the fortune to learn that the departmental glassblower, Bob Morley, had supplies of hollow rods of palladium which he used as a hydrogen filter.[2] He let them have a 5 mm diameter broken piece. A successful strategy in science, as in life, is to play to your strengths, use the techniques with which you are most familiar and apply them in novel circumstances. That is what distinguishes Eden and Liu's attack on cold fusion. Their expertise was with ultra-high vacua, and fortune had presented them with a *hollow* palladium rod. So they set up their electrolysis cell with the heavy water surrounding the rod, but with a vacuum pump connected

133

to the rod's hollow centre. Their novel idea was to force the deuterium into the rod and if fusion took place in the rod, any gaseous products could be extracted from the hollow centre and analysed in a mass spectrometer. One possible product of cold fusion was tritium which would likely desorb from the inner surface of the palladium rod as a molecule with deuterium, since deuterium would be in great abundance. The mass of this compound, $DT$, is five (on the atomic scale where hydrogen is one). Other fusion products, such as helium-3 and 4 have the same mass as $HD$ and $D_2$ molecules respectively and so could not be identified unambiguously within the resolution of their mass spectrometer. As they were measuring atomic masses and, from that, wanted a clear signal for a fusion product as distinct from materials such as hydrogen and deuterium which were there to begin with, they decided to concentrate on mass 5. There is no stable atomic nucleus with this mass and the only way to make it in molecular form from a pair of nuclei would be from $D$ and $T$, mass 2 and 3 respectively, which involves tritium—a fusion product.

After several hours of electrolysis, mass 5 and 6 signals began to appear and after about ten hours reached as much as a factor of hundred to a thousand times bigger than the background amounts, and only 100 times smaller than the $D_2$ signal. The deuterium being forced in by electrolysis gave a huge signal and their results made it appear that a significant percentage of it was fusing and producing tritium, since they recorded signals for $DT$ at mass 5, and $T_2$ at mass 6. To check if this was indeed the case they replaced the heavy water with light water and found that the mass 5 and 6 signals dropped down into the background levels after about a day. This showed that the products were only present when deuterium, from the heavy water, was used. They returned to heavy water and the mass 5 and 6 signals reappeared.

By now they were sure that they were onto something, but there was still the possibility that the signals were some artefact of their mass spectrometer and not caused by the action of the palladium. To test this, they closed off the palladium from the system and introduced deuterium gas directly into the spectrometer. There was some signal for mass 5 and 6 but less than one per cent of that measured with the full electrolysis working. This was certainly suggestive that the palladium-deuterium system was important in producing the large signals that they were observing.

This must have been a wonderful moment for them. Every student who is serious about a life in scientific research dreams that some day they will have the joy of being 'the first to know', or to be in at the crest of the wave. Here were Eden and Liu who had done a clever thing—introducing a new technique (as far as I know no one else has attempted to extract products from a hollow rod) so that they are making a genuine contribution, more than merely a replication; they have done control experiments with light water and checked out their appartus in other ways. For them the wave was a tsunami.

When Sam Fain, their research supervisor, arrived in the physics department on Wednesday, 5 April he found a message from the students awaiting him: 'Please come and see us as soon as you get in.' They talked about the experiment, to see if they had done all possible checks to see if the apparatus could be deceptive. They discussed whether there was a question of a patent for the idea of the hollow rods and the vacuum pump for tritium extraction; and then went to the university technology transfer office, obtained patent forms and spent the day completing them after informing the head of the physics department and the Dean of the Faculty of Science, each of whom had to approve the application. Each agreed to keep quiet. On Friday the 7th Eden and Liu worked with an attorney on the patent. The lawyer had had experience in patenting biochemical developments where he had learned that one tries to patent everything imaginable, covering any and all possibilities.

The attorney also confirmed their suspicions that if tritium really was being produced then there was a chance that the whole venture would be shut down: if this was a process for producing and extracing tritium, they might not be able to tell anything. (Ironically a faculty member at the university, John Cramer,[3] had earlier written a novel about two graduate students who make a discovery which the military then try to shut down; as soon as they realise that they might be shut down they publish by electronic mail (*Bitnet*), to tell everything they know.)

On Saturday, 8 April Fain left Seattle to attend the American Chemical Society meeting in Dallas. Eden and Liu called him on Saturday evening and again on Sunday, and Fain urged that they write up a report. Fain talked to David Lazarus, his former adviser, who told him that the rules of the American Physical Society are that once you have submitted a paper you can talk about it. Fain did not want the students talking about it until it was written up. However, on Tuesday, 11 April the chair of the physics department received a call from the head of the OTT (Office of Technology Transfer), who had been out of town during the previous Thursday and Friday, and also from the University Information Services. The physics chair remarked to Eden and Lei that 'the cat is out of the bag'.

There was much pressure on the University Information Services to make a big splash at this juncture. On Monday, 10 April Texas A & M and Georgia Tech had announced that they had confirmed cold fusion; reports were beginning to come in from abroad; Pons was about to receive adulation at Dallas where, according to some media reports, he had to change hotels and register under an assumed name in order to avoid the reporters, and the media understandably were giving all of these events headline prominence. Test-tube fusion was the hot news, and on top of all this the state legislature was meeting to debate the university budget. In such circumstances any university administrators worthy of their salaries would recognise the importance of going public lest they be beaten to it by someone else. Who knew which institutions would be the next to make the headlines, and if one of those had detected tritium using a hollow palladium rod then all would be lost.

Now one had conflict between opposing pressures. The press office wanted to have a press conference; people in the physics department were wary that there was some overlooked problem as they believed that tritium could not be produced in such amounts by these room temperature processes. The physics chair wanted Eden to give a talk in the department about their work so that other experts could evaluate it. It turned out that the first opportunity for a presentation was at one of the regular informal meetings held between some of the physicists and chemists who specialise in surface science. The scheduled speaker at the meeting of Friday, 14 April allowed them to speak instead at very short notice.

The University of Washington has a very strong nuclear physics group who knew that *dd* fusion should produce copious neutrons, not merely tritium. If tritium was emerging in the amounts that Eden and Liu believed then there must also be lots of neutrons being radiated by the cell. A graduate student in nuclear physics had helped them by bringing neutron detection equipment into the laboratory to watch the apparatus on the first full day of work (29 March), and the health science department measured the tritium in their $D_2O$ and provided them with health monitor radiation badges: 'if this is working you could be dead already'. So here we see the co-operative nature of open science at work: more expertise is brought to bear, advice, support—'keep at it', additional apparatus and expertise provided, and suggestions made on what other tests one might do.

Eden and Liu had completed their paper on Tuesday the 11th—the same day that the OTT broke the news within the university—and faxed a copy to Fain at Dallas, also sending it to *Physical Review Letters* for publication. There was pressure being put on them to give a press conference; the public relations people stressed that if there was no press conference the students' phone would be ringing all the time, half of the departmental phones would too, no one would get any work done, press would be waiting on their doorsteps and

136

they would be hounded until something was said. The talk in the chemistry department was scheduled for Friday and the public relations people said that the press would all be there if there had been no release before then. They held off as long as possible, but the press conference was arranged by the university for Thursday and took place in the Administration Building. As English is Lei's second language, the students decided that Eden should do most of the talking as it was important that they not be misquoted.

A report sent out on *Bitnet* by an attendee at the press conference described how there were four television stations represented plus reporters from ten other agencies. 'Eden handled the leading questions very well and was quite pointed about the press making a premature sensation about the work. He repeatedly gave a cogent description of how science works.' Eden told me his own impressions. 'The press are very good at presenting what *they* want, but I felt it was difficult to get our message across to them. You can fall into the trap of using quotable quotes, which is what they report. But they also report the meaningless jokes, the things that you include to keep the atmosphere light.' He felt that their hopes to avoid being misquoted came to nothing.

In the USA the NBC News announced the claims of Georgia, Texas A & M and of the Seattle students over pictures of tubes with dramatic bubbling liquid and flashing lights, giving an aura of 'energy' throughout. The headlines blazoned that 'In Seattle the results mean that three American labs have duplicated part of the experiment . . . An increasing number of scientists now agree that this experiment does produce at least small amounts of nuclear fusion, the reaction that powers the Sun.'

This weekend of 15 April was arguably the high point of test-tube fusion.

Van Eden showed me his notebook in which he had pasted a cartoon taken from the 1989 Gary Larsen calendar for 13 April, the day of the press conference: it showed a devil at home waving goodbye to his wife as he sets off to work saying: 'Well, down we go.' On 14 April, the day after the conference, the Larsen cartoon showed a small boy standing outside a tank of sharks which peer down at him hungrily. The sharks are saying: 'So near and yet so far.' Eden told me 'That is exactly how we felt on those two days.' He added, 'People kept coming into our lab, people I had never seen before. They would come in, look around and leave without saying a word. It was incredibly rude.'

A rumour that spread on the 14th was that a test-tube fusion experiment at the Lawrence Livermore Lab in California had 'blown up'. This again sounded significant and test-tube fusion supporters

137

took it on board. First, this was a National Lab with professional researchers, first rate equipment and lots of know-how so it couldn't easily be dismissed as naive error—'blowing up' is not easy to misinterpret! Second, it had the hallmark of Fleischmann and Pons' original 'discovery'—their vaporised block of palladium which gave rise to Pons' warnings that fusion was a dangerous business. No one had actually seen Fleischmann and Pons' block vaporise—only the after effects were there as evidence. Had a Livermore scientist replicated this, the first link in the Fleischmann and Pons phenomenon?

It is quite possible that they did, but in doing so showed that it had nothing to do with fusion. It seems that a palladium rod had been loaded with hydrogen and then left on some 'handy wipes'. The hydrogen gas started leaking out into the air and, releasing this strain, the palladium began to get hot; very hot. This is a phenomenon known since early in the nineteenth century, a chemical effect not fusion, that has been used in the 'Dobereiner Cigarette Lighter'. The temperature rises rapidly and is sufficient to light a cigarette.

At Livermore the 'handy wipe' was like the cigarette, and caught fire. But the heavily loaded palladium kept on releasing hydrogen and a miniature *Hindenberg* disaster took place as the flame ignited the hydrogen which combined explosively with the oxygen in the air. Dramatic to be sure, but it was chemistry, not nuclear fusion, and possibly a verification that Fleischmann and Pons' vaporised block was equally irrelevant for fusion. It is interesting to compare this 'mundane' explanation, that I received from one Livermore scientist, with Fleischmann's belief that someone had fired a laser beam at a loaded block of palladium. Rumour and anecdote were as rife as fact.

### 14 April: at the White House

On 13 April, the day after Pons' address in Dallas, officials at the Department of Energy summoned the physicist and Nobel Laureate Glenn Seaborg[4] from the University of California at Berkeley to a briefing in the White House. Seaborg flew immediately across the country and by the following morning was in a DOE car in Washington DC, accompanied by Admiral Watkins the DOE Secretary of State, en route for an 11.15 a.m. appointment with John Sununu, White House Chief of Staff, and a number of his aides.

The meeting in Sununu's office lasted about twenty minutes, during which Seaborg briefed him and his aides on the test-tube fusion announced by the University of Utah chemists and also mentioned that there had been the observation of low levels of neutrons during the electrolysis of heavy water, in particular by Jones at Brigham Young University. He told them of the tremendous reception that the work had received two days earlier at the annual meeting of the American Chemical Society in Dallas, but added that a number of nuclear physicists took a 'very dim view' of the development and tended not to believe in it.

Most of the conversation was between Seaborg and Sununu. Governor Sununu has an engineering background and this helped him to understand the issues: Seaborg's notes recall that Sununu raised 'a number of penetrating questions'.

Expressing his own opinion Seaborg indicated that he was sceptical, but that he believed that the phenomenon had to be investigated and that he was recommending that a special panel be created to look into it. He had discussed this with the DOE Secretary, Admiral Watkins, earlier and Watkins indicated to Sununu that the appointment of such a panel was underway.[5]

In the more general conversation after the main business was concluded Seaborg mentioned to Sununu that he had been one of the discoverers of the radioactive iodine that had been used earlier that week in treatment of the President's wife, Barbara Bush. This interested Sununu very much and he left the room briefly to pass the information on to the President, returning after a few minutes with the news that the President would like to meet him. They walked down the hall to the Oval Office where Bush greeted them and pictures were taken.

Seaborg told the President about the radioactive iodine and how it had worked a miraculous treatment on his own mother years before. He touched on the problems of pre-college education in science and mathematics and then described briefly the test-tube fusion story. Seaborg indicated that it was something to be viewed dispassionately, that there were some signs that it may not be nuclear fusion but that, nonetheless, it must be looked into. He also told Bush that the DOE was about to create a panel to do that.

The meeting in the Oval Office lasted about ten minutes and Seaborg's notes concluded that 'The President seemed very interested and encouraged us very much to go ahead with an investigation of this intriguing situation.'

By noon the business had been completed. That afternoon the DOE began to plan out the strategy for the test-tube fusion investigations, the first actions of which would be felt early the following week. Meanwhile Seaborg contacted Pons by phone in order to clarify some rumours that he had heard at the DOE to the effect that Pons was seeing heat not just with heavy water but also with light water.

Seaborg asked Pons if this was true, and 'Pons ummed and ah-ed and then said "Yes, that is true." '[6] Pons then explained that he believed that this was due to fusion between protons in $H_2O$ and the deuterium which is present in ordinary water in small traces. This was essentially the same as he had told Chuck Martin of Texas A & M the previous weekend, and which had encouraged Martin to go public, in turn raising the excitement at Dallas.

Pons also told Seaborg that he believed that the absence of copious neutrons in their experiments with $D_2O$ was because the $dd$ fusion produced helium, and to support this he referred Seaborg to Cheves Walling, professor of chemistry in Pons' department. Seaborg then called Walling who confirmed what Pons had said.

## 17 April: the Utah helium

During that Thursday and Friday experiments were in progress at the Utah chemistry department which led to another news conference—on Monday, 17 April—proclaiming evidence for test-tube fusion. Two of Pons' colleagues, Cheves Walling and Jack Simons, had suggested that the reason for neutrons and tritium being found in too small amounts to explain the heat was because the dominant $dd$ fusion produced helium, $^4He$ (this route is normally ten million times suppressed relative to the neutron and tritium production channels). As apparent proof of this theory, Pons and Hawkins informed Walling and Simons that they had used a mass spectrometer to analyse the gases coming from a cell that was delivering 0.5 watts of excess power.

They said that they had found evidence for helium-4 being produced at a rate of about $10^{12}$ atoms per $cm^3$ per second. The *Salt Lake Tribune* reported Simons as saying 'The helium results are important [because] you can't get helium without a nuclear reaction. Chemistry doesn't make helium.' Furthermore, this rate of helium production was commensurate with the amount of excess heat coming from the cell. This appeared at last to be the smoking gun.

Pons confirmed the claims at a press conference on the Monday and propagated them nationally on the PBS programme *Science Journal*.

A complete range of evidence now seemed at hand: heat, neutrons, tritium, gamma rays and helium. But one by one the pieces of nuclear evidence were undermined.

The news of the helium was being cited in *Fusion Facts* later in the year as evidence favouring the Fleischmann–Pons phenomenon, yet the experiment was not formally published. With helium being detected in amounts that would explain the heat, this would seem to be the definitive proof. What had happened?

There was a series of flaws in the claims that were unearthed primarily by a team at Caltech (whose major contribution to the test-tube fusion story is documented in Chapter 11).

When Fleischmann and Pons began their work electrolysing water in the presence of palladium they were unaware of the historical work from the 1930s. There had been detailed study of the diffusion of hydrogen in palladium for many years and a vast amount was known to materials scientists. There had also been much study of the behaviour of helium with palladium. Its behaviour is rather different from that of hydrogen: whereas hydrogen freely diffuses through palladium, helium stays put—helium generated within metals has been observed to stay there for over ten years. Thus if helium had indeed been formed by fusion within Pons' palladium it should have stayed within the metal and not emerged in the gases. To find helium would require analysis of the palladium electrodes themselves, not of the gaseous emissions.[7]

So where had the helium come from? Almost certainly from the air, the same source as bedevilled Paneth and Peters, as noted by Chadwick, Ellis and Rutherford 60 years ago. In the present case the cause was identified by a group from Caltech led by Nathan Lewis, professor of chemistry.

Lewis had been in contact with Walling about the helium results. The first question he needed to ask was whether checks had been made to ensure that the helium was not from air contaminating the apparatus. Air contains about 5 ppm (parts per million) of helium and other inert gases in trace amounts along with the dominant nitrogen and oxygen. Lewis asked if they had double-checked for the absence of air by looking at higher mass elements to measure the relative amounts of helium, argon and nitrogen, and was told that this had not been done.

If this was indeed true, the potentially crucial piece of evidence for fusion was being presented without all necessary controls having been made. The proof that the helium was indeed from the air came when a Caltech team replicated the measurements using the same model of spectrometer as had been used in Utah.

They confirmed that the device could register helium in the air and distinguish helium from $D_2$—which is also mass 4—so long as helium was present above 1 ppm relative to the deuterium. However, the amount of heat claimed by Utah, even if it were all associated with helium, would involve the production of much less than 1 ppm relative to $D_2$ and so be unmeasurable. Lewis summarised: 'They could not have detected the correct amount of helium with this

141

instrument; if they found some, it obviously was too much, and probably came from an air leak. We knew something wasn't right.'

Here was the first clue, though known in detail only to a few people at the time, that claims to have evidence for cold fusion in the University of Utah chemistry department were based more on enthusiasm than on well-controlled science. As the months passed and more details emerged of what had been going on in the chemists' experiments so did concern grow about the quality of some of the 'evidence', and even whether it had been obtained within the usually accepted procedures of science. However, this was still in the future; by the start of the week Monday, 17 April, cold fusion was looking assured to outsiders.

### 19 April: at the US Department of Energy

Peter Bond is chairman of the Physics Department at the Brookhaven National Lab in New York. On Tuesday, 18 April he was attending a meeting at the lab when he received an urgent message: he was to go to Washington DC early the next morning, plane reservations had already been made and the Director of the Laboratory would brief him on what it was all about.

A similar scene was being played at other Government laboratories across the country. Senior scientists on the west coast found themselves booked on overnight flights to the headquarters of the Department of Energy (DOE) in Washington, where they arrived next morning red-eyed, such was the haste with which the meeting had been put together.

On Friday the 14th Glenn Seaborg and Admiral Watkins, the Secretary of State at the DOE, had met with Governor Sununu and President Bush in the White House and the recommendation had been to form a special panel to investigate test-tube fusion. Seaborg had been chair of the US Atomic Energy Commisson, the forerunner to the present DOE, and Watkins as the present Secretary had already begun to take action on this. That was the Friday afternoon immediately prior to the weekend. By Monday morning events were moving fast forward with the latest announcements emanating from Pons' laboratory that helium had been found and in amounts that appeared to explain the amounts of heat being claimed.

The DOE had been presented with a problem. Before forming an ad hoc panel it was important to be sure how its own laboratories were responding to the events which seemed to be making fresh headlines each day.

Bond, a nuclear physicist, found that his fellows, all summoned with similar dramatic urgency as he had been, were the leaders of materials science or fusion programmes in the other major national laboratories with broad based energy research programmes. They had come from Argonne Illinois, Oak Ridge Tennessee, Berkeley and Livermore in California, Los Alamos, Pacific North West and the Ames Lab in Iowa plus the Sandia Lab, a non-profit research and development organisation in New Mexico which works closely with the DOE. These were the collective Goliaths whom David, in the disguise of the chemists, seemed to have felled. A fine mess indeed.

The DOE felt the pressure welling up from the public and from congress because it was pouring billions of dollars of taxpayers' money into the fusion business, into big scientific labs the length and breadth of the continent, and now two chemists were claiming to be able to do it with a car battery and stuff from a dime store. There was even speculation around that the major laboratories were stonewalling, reluctant to admit that cold fusion was true, scared that their major funding would be cut off. The group was confronted directly: 'Why are you doing nothing?'

Bond knew that his lab was already looking into the chemists' claims and had set up experiments. It rapidly transpired that the story from the other labs was similar; roughly five teams per laboratory had been working day and night with state of the art equipment. It was too soon for them to go on the record and write scientifically reputable papers, but the trend was already clear—there were no signs of cold fusion. This didn't mean that they saw *nothing*, in fact they *were seeing things*, the same sort of things that others had seen and had called press conferences about. The difference was that the major laboratories contained a wide range of expertise, knew many of the pitfalls that the unwary could fall into and had already pointed out to some groups, such as Georgia Tech, where they had gone wrong.

One of the representatives at the DOE meeting pointed out that his group too had seen what at first they thought to be neutrons, just as the Georgia people had done. But his group ran some checks and discovered that the signal was spurious. They found that when they moved the detector away from the 'fusion' cell, the apparent neutron signal stayed as strong as ever. If it was coming from the cell then it should have faded away as they moved further from it. They eventually discovered what was going wrong, realising that the detector was sensitive to changes in temperature and that it was acting more as a thermometer than as a neutron detector. Usually this subtle effect is swamped by the signal coming from genuine neutrons but in the case under study the neutron signal was too feeble to pick up; the apparent 'signal' had nothing to do with neutrons. So when they heard of the Georgia Tech news conference they called up the University and asked if they had checked to see if the signal died away with increasing distance from the source. It turned out that Georgia had not done this, and were then informed about the subtle temperature sensitivity of the device that they were using. Georgia then ran a second series of tests, using the new wisdom gained from the DOE laboratory, which resulted in them calling a second press conference and withdrawing their earlier claim (page 152).

The retraction by Georgia Tech was greeted with amusement in some quarters, but in reality was scientific credibility of the highest order and short of having a correct positive result was the best thing anyone could do. Having found their error they immediately announced it so that others did not repeat this mistake, and thus they helped to make a positive contribution to the scientific record. Within the DOE it also carried an important message: the professional laboratories were indeed active and trying to limit the damage from the developing bush fire as the media reported vague and incorrect claims about cold fusion.

The damage limitation was a serious concern. As Bond commented to me: 'For the media and the public this was the US dream; the little guy beats the big battalions. Heat with no radiation. You want it to be true. It would be wonderful. No wonder that everyone is head over heels. But wanting it and it being true can be two very different things.' It was a dream for those who had nothing to risk, but for the major laboratories it could turn out to be a major embarrassment. For congress and the funding agencies it was a major problem because the budgets for the large scale hot fusion were in the process of being renegotiated; no one wanted to sign a cheque for a few hundred million this month if next month they would be emblazoned across the front page headlines as profligate bureaucrats because it could all be done with a flask of heavy water.

'Where do we go from here?' was the question from the DOE at the end of briefing. One suggestion was that a major lab be given access to a 'working cell', one that was purportedly producing heat. Either the cell be brought to a lab or the lab be allowed to take its own detection equipment to the site and train it on the cell.

Los Alamos reported that it had been negotiating with the University of Utah chemists to get a cell, but that discussions had come to nothing. (The news of this leaked around the scientific community via the electronic mail and the apparent reluctance of the University of Utah to collaborate did not help the credibility of the chemists.)

The main result of the meeting came a few days later with a letter on 23 April from Admiral Watkins, the Secretary for Energy. The

144

instruction was that the labs were to redouble their efforts, report to the DOE weekly on progress and to get an answer urgently.

This was the first acknowledgement of the large scale efforts under way in the major laboratories; the public was essentially unaware of it and was still subjected to a diet of sensational claims by press conference and anecdotal remarks, often emanating from Utah, to the effect that several groups around the world were seeing fusion phenomena and would soon be confirming the chemists' claims. The desire for it all to be true was so great that suspicion of the silent labs was already beginning in some quarters. When Peter Bond arrived back at Brookhaven he was called by a news reporter seeking comment on a claim that 'The big labs *are* seeing fusion but are keeping it secret because the big oil companies have bought them off'!

# 9

# THE PARTING OF THE WATERS

In October 1988 the chairman of the University of Indiana chemistry department had sent an invitation to Pons to give a talk on electrochemistry, dateline 4 April. Pons had agreed, but in the meantime had been so busy as a result of the race with Jones that he had not been in touch with them. So on Monday, 20 March Dennis Peters, Distinguished Professor of Chemistry at Indiana, called up Pons to finalise plans for his visit and to ask him for a title so as to announce the talk. Pons replied that he 'couldn't tell (Peters) the title until after a press conference on Thursday, 23 March'. Peters was amused and later recalled that 'I have never had to call a press conference to decide what the title of one of my talks was going to be'! With the sudden eruption of interest in cold fusion Peters decided to make admission to Pons' talk ticket only.

This was a wise move as the celebrity status of test-tube fusion was rapidly rising. At Pons' only other talk on the science, specially arranged at the University of Utah Chemistry Building on Friday, 31 March, the line of would-be attendees went around the courtyard as about a thousand people turned up. Originally this talk had been open to the public but the authorities then restricted it to faculty and College of Science graduate students with any extra seats being 'first come, first served'. In the event they attempted to restrict it to faculty and provided an overflow room with audio visual displays. Campus security was tight and a couple of TV crews were removed from the halls. The decision to hold the talk in a relatively small room rather than a large venue, such as the basketball arena, was an attempt to preserve some scientific decorum and not have 'a circus'.

No recording was permitted, but four days later Pons' talk in Indiana was video-taped. The tape shows that although he was competent and confident with the chemistry, he was very unsure with the physics and wrong on some basic points. It was the performance of an expert chemist who has not had time to brief himself on the physics, reinforcing the impression that while five years may have been spent

146

on the investigation, the physics had been 'discovered' only at the last moment and had not yet been assimilated.

A nuclear physics colleague of mine, John Cameron, told me about reactions to the talk. 'We were impressed with the style: it was low key, honest. He didn't seem to be trying to sell anything and admitted that there were lots of open questions.' However, the doubts began during the question session following the talk. There were lots of questions that he couldn't answer to the audience's satisfaction: 'this generated much scepticism'. 'Chemists were unhappy with his heat measurements, the physicists were not convinced by his reports of his neutron and tritium detection.'

These concerns emerged more sharply at a get-together at Indiana University of physicists and chemists on the next Friday afternoon, 7 April. 'The chemists had decided that the effect must be physics; the physicists had decided that it must all be in the chemistry. When they got together it became clear that no one was prepared to take up the defence. Someone said: "If it's not physics and it's not chemistry then it's not happening!" '

If their heat was really due to fusion of deuterium nuclei–deuterons—then there should be no such heat if the heavy water, $D_2O$, is replaced by plain water, $H_2O$. Thus a convincing proof would be if they had made similar experiments with $H_2O$ and balanced the books, found no excess heat. Pons was asked about this and replied, 'We have not made controlled additions of water. I would imagine that there would be some killing of the process'—a response that is in line with Fleischmann's given at Harwell on 28 March.

Yet later that week Pons received news from Texas that Charles Martin and his team were seeing heat with both heavy and light water and he assured Martin, in no uncertain terms[1], that he too saw excess heat with light water, encouraging Martin to go public. Texas and Georgia Tech both made claims of replication of aspects of the cold fusion on Monday, 10 April and so when Pons checked in to his hotel in Dallas, where he was due to speak at the American Chemical Society meeting on the 12th, he was already being mobbed like a rock star.

Pons was on friendly ground as chemists and media formed the bulk of the 700 audience in the basketball arena. The general flavour was that chemists were beating physicists at their own game. The ACS president Clayton Callis opened the special session on cold fusion and remarked that the physicists' efforts at hot fusion using tokamaks and lasers were 'apparently too expensive and too ambitious to lead to practical power', and added, 'Now it appears that chemists have come to the rescue.'[2]

At this the crowd burst into laughter and applause. Pons summed up the different approaches of chemists and physicists with a single slide that parodied one of the Princeton tokamak that had been shown in an earlier talk by Harold Furth. Furth had shown a picture of Princeton's machine covered with pipes and wires that is still some years away from a breakeven fusion reaction. Pons' picture was of his beaker of water in a plastic washtub which, he was claiming, produced breakeven fusion energy. 'This is the U-1 Utah tokamak' he said, and the chemists were ecstatic.

Before his talk Pons had given the (by now compulsory) press conference in which he appears to confirm his remarks to Martin by stating that he had seen heat in an experiment with plain water. Most of the questioning in the hall after his talk was friendly but he was asked the crucial question about the light water and replied, 'A baseline reaction run with water is not necessarily a good baseline . . . We do not get the total blank experiment we expected'.[3] In addition to the news from Texas he had also learned that proton–deuteron fusion (*pd* fusion) could exceed *dd* fusion and so plain water, which contains traces of $D_2O$, could give measurable fusion, and hence was 'not a good baseline'.

Meanwhile Martin Fleischmann had been making presentations before more critical audiences who were highlighting problems with the chemists' claims, in particular at Harwell on 28 March when he was told that 2.5 MeV is the wrong place for a $\gamma$-ray ('gamma-ray') coming from neutron capture.

On Friday, 31 March Fleischmann was at CERN—the European Particle Physics Centre in Geneva. Fifteen years previously I had seen Samuel Ting speak to a packed auditorium on his discovery of a new subatomic particle for which he won the Nobel Prize; Fleischmann's audience exceeded even that, with an overflow audience watching on closed-circuit TV in the CERN council chamber. One of the audience in the main hall was Douglas Morrison, an experienced experimental high energy physicist, who sent out on electronic mail an extensive newsletter about the occasion. This spread around the world in days and over the next months Morrison became a conduit for much of the news, providing regular bulletins and contributing his own commentary, notably comparing aspects of the developing saga with 'Pathological Science'—where belief in a claimed phenomenon causes one to see in marginal data evidence where there is none, to explain away contrary evidence and, above all, to believe in effects which are not reproducible and are seen, if at all, at the limits of experimental detectability. Morrison's report of the meeting included the news 'There were some questions . . . of null tests. Carlo [Rubbia] immediately asked if [Fleischmann] had repeated the experiment

148

with $H_2O$ instead of $D_2O$—they had not.' Fleischmann said that $H_2O$ experiments were 'in progress'.

The $\gamma$-ray peak, shown at Harwell centred on 2.5 MeV, had been replaced by a peak at the right place, 2.2 MeV, with 20,000 events recorded at the top of the peak (Figure 8, page 115). This was in the right place but it looked too sharp to be a true $\gamma$-peak from neutron capture; the second worry expressed at Harwell was still present.

The International Centre for Scientific Culture in Erice, Sicily, hosted the first real investigative workshop into the cold fusion episode on 12 April. Physicists, chemists, materials scientists and the media arrived in the cobbled streets of this Sicilian hilltop retreat to debate the new phenomenon. Apart from Jones and Fleischmann, there were several present who had been attempting to replicate the phenomenon, without success, and who were keen to learn what they 'were doing wrong'.

As the months went by the answer became clear—'probably nothing'—but on 12 April the cold fusion phenomenon was near its apogee. That day, as Pons was being lauded by the chemists in Dallas, Fleischmann was raising the first doubts in the minds of many of the more critical audience at Erice.

His description of the electrochemistry was articulate and confident, a clear expert utterly on top of his subject. There was not a single hesitation, no 'umm'-ing or 'err'-ing; a transcription of the presentation would find itself with all the punctuation in place and without any need for polishing. The evidence for nuclear effects, the neutrons, was briefly skimmed over but, as he pointed out, he was not a nuclear physicist.

As a chemist it was clear that he believed in a nuclear source because the heat was '100 times that from any conceivable chemical reaction', alluding to a production of 5 MJ/cm$^3$ in 100 hours. He added that 'we are convinced that if we run for 1000 hours we would get 50 MJ but we have not had time to do such experiments.' Several people were surprised to hear this as the University of Utah publicity kept making a point about the five and a half years' duration of the experiments, yet they had 'no time' to run one cell for two months.

In the long question and discussion session afterwards he received a detailed cross-examination, and it was here that concerns began to be aired. He was asked a long series of questions about the electrode preparation, electrolyte and currents, and answered each precisely. In answer to a question from Matthijs Broer, of AT&T Bell Labs, who wanted to know if there were any details being held back, he confirmed that 'I have left nothing out.' He seemed confident and in control. But there were two questions where the certainty

149

evaporated; these dealt with the γ-ray peak and whether they had made control experiments with light water.

The γ-ray evidence was effectively a *third* peak: after the 2.5 MeV at Harwell and the 20,000 event peak at 2.2 MeV shown at CERN (and portrayed in the first version of their paper), at Erice he showed the peak, still at 2.2 MeV, but announced the scale showing the numbers of events to be 'totally wrong. It should be around 1000 (not 20,000 events at the peak)'. This peak was a smooth curve passing

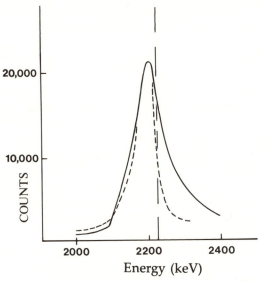

Figure 9. The solid line illustrated the purported (n, γ) peak in Fleischmann and Pons' original paper (compare with Figure 8). The dotted curve outlines the peak in their erratum (the vertical scale has been magnified by a factor of 23 for this latter case, for which the true value is about 900 only). The vertical dashed line denotes 2224 keV where a true (n, γ) signal should peak (compare with the MIT curve in Figure 11). It was the abnormal thinness of the dotted peak that, in part, alerted Petrasso's group at MIT.

through some small marks, such as are used to direct a curve and rather unlike real data points with vertical errors which vary from place to place.[4] The following exchange took place between Fleischmann and Richard Garwin of IBM (who wrote a resumé of the meeting for *Nature* and had been one of the organisers of the workshop).

Garwin: 'Your γ-ray peak seemed to have no experimental points. They were just plotting ticks.'

Fleischmann: 'They were all about one standard deviation.'

Garwin: 'No, they all have the same size.'

Fleischmann: (hesitating) 'There is a background [pause] I'm not able to answer without access to the data. The background count is about 400 and we get [in our peak] about two and a half times the background.'

But then an astonishing thing happened. Fleischmann stopped and said 'Er, there is a problem there. There is a problem.' He paused and then repeated 'There is a problem. I think you may be looking at an incorrect diagram.'

There was some nervous laughter at this, but Fleischmann continued with 'Is this the one that shows the peak at 2.5 [sic]?', to which Garwin said, 'No, it is at 2.2.'

Fleischmann then expanded by saying that they had measured the γ-rays up to about 3 MeV, and saw the expected signals from background radiation (such as a peak at 2.6 MeV coming from decays of naturally occurring thallium), and that, 'The count rate is identical everywhere except in the region of 2.225 MeV.'

Garwin then alluded to the 'wrong shape' and commented that it would be nice to see the whole spectrum. (In the USA a team from MIT was already chasing hard on this and were able to discredit the data, as we shall see in Chapter 15.)

Following this, Denys Wilkinson asked some questions, which Fleischmann debated, and then Wilkinson asked the central question, 'What happens when you use ordinary water?' Fleischmann responded tartly, 'I'm not prepared to answer that question.'

There were some incredulous gasps, a hubbub of comment around the hall and a clear sound of someone saying 'Oh no, no' on the tape of the meeting. Fleischmann could not ignore this reaction and repeated, 'I do not wish to go into experiments with ordinary water.'

Steven Koonin, the nuclear theorist from Caltech, then followed this up, saying. 'This question is the same, but let me ask it again to make sure that I heard the answer correctly. I would have thought that standard scientific procedure would ask you to run a control experiment with light water to prove that deuterium was really important. How can you make such a claim without making such a control?'

Fleischmann was by now sounding irritated. 'I've said quite specifically that I'm not prepared to answer that question at this moment', he said, and as more murmurs of dissatisfaction came from parts of the hall he added: 'You can read into that whatever you like.'

Richard Garwin, who also talked with Fleischmann extensively over dinner, read into it that the light water control had not been done. Whether this is the only interpretation that one 'can read into that' will be evaluated in Chapter 16.

## Retractions

The most significant though less prominently reported news story of the weekend of 15 April concerned Georgia Tech. On the 10th they had started the sudden chain of 'confirmations' of fusion, but in the subsequent days had learned much more about the subtleties of their experiment. They called in the media a second time, but this time it was to announce a significant retreat.

The original news announcement had made them famous. The director of the team, James Mahaffey, learned that his next door neighbour had been in London and was going round telling everyone that he was *his* next door neighbour: such was fame. In the US President Bush had received a personal briefing from Glenn Seaborg, Nobel Laureate in chemistry and one time chair of the Atomic Energy Commission, who said he was very sceptical. Even so, the authorities, nervous at the possibility that test-tube fusion had something to it and that it was seemingly being confirmed at various universities, decided that the Department of Energy (DOE) should make a committed effort to get to the bottom of it all. The news also was pored over by many others in the US and around the world who were themselves already attempting to replicate test-tube fusion. These included universities, national labs and Caltech (Chapter 11), some of whom had been using the same type of neutron detection kit—the $BF_3$ counter—as had Georgia Tech, and had discovered a flaw with it.

Usually people using $BF_3$ counters are counting large fluxes of neutrons, for example from nuclear fission where many thousands of neutrons pour out each second. If the Fleischmann and Pons effect had really been fusion then there should have been large numbers of neutrons here too; the trouble was that there weren't. If there were any neutrons at all they were similar in number to those in the environment. As one's deficient eyesight is exposed in dim light, so were deficiencies in $BF_3$ counters shown up when people tried to use them to see dim fluxes of neutrons. The $BF_3$ counter was designed for measuring large bright neutron fluxes; in dim fluxes, as for many other detectors, it can mislead.

When your eye sees something, a lot of complicated and indirect electrical signals have flowed to the brain which decodes them. This is second nature to us and we tend not to think about the logistics and philosophy of what is involved in such a 'simple' thing as seeing. Similarly, when a neutron counter counts neutrons it does so in an indirect way. Neutrons induce electrical signals to be passed along some contacts; the more electrical signal so the more neutrons there must have been.

But other things, such as faulty electronics, dropping the apparatus on the floor or even vibration, can produce electrical impulses in the detector and be misinterpreted as neutrons. If there are thousands of genuine neutrons hitting each second, then these other spurious sources do not affect you—ten or twenty miscounts in a thousand are neither here nor there. However, when you are looking at a dim source where there are only one or two neutrons insted of a thousand, then the ten or twenty is suddenly the dominant effect. You think that you are counting neutrons when in fact it is nothing of the sort.

What Georgia Tech had thought so significant at the time was that the hotter the cell was, so the more neutrons the detector recorded. This suggested that whatever gave the neutrons also gave the heat and as such correlated two manifestations of test-tube fusion—namely radiation and heat. They had monitored the experiment for several hours, taking data at hourly intervals, and then let the experiment run down, watching the temperature fall and the neutron rate drop with it. But after the press announcement a major national lab and also Caltech contacted them and said that they too had done exactly the same experiment, using very similar instrumentation, and they too were getting data that seemed to track with temperature. Then one of them had happened to notice that when he took one of the detectors out of the experiment and held it in his hand, the neutron count rate suddenly doubled. When he let go, the neutron rate fell. The heat of his hand caused the detector to react as if it was measuring neutrons; it was sensitive to temperature, acting as a thermometer. You are measuring temperature but think that you're measuring neutrons!

So the Georgia team tried a test to see whether it was neutrons or temperature that they were measuring. They emptied the water from the experiment, which was now cool, and took a reading. Next they filled the cell with hot water and count rate jumped up as if neutrons were coming from the experiment. The horrible suspicion began to form that they had blundered.

They ran more tests and found that the supposed neutron detector was so sensitive that they could literally use it as a thermometer. They calibrated it to see how many 'neutrons' were counted at any chosen temperature and then removed this number from number measured when the cell was running. To their dismay there was little or nothing left. They had not been counting neutrons at all.

What many people outside science probably didn't realise is that it is quite usual in the course of an experiment to discover that your instruments do not behave quite as expected. You run the experiment for a long time, become aware of the idiosyncracies of the apparatus and get to know it as one might come to be familiar with the quirks of your car. And it is normal to feel frustration that you've lost a week's work because of a problem that was only discovered later. The test-tube fusion was such a media event and the pressure to be first with the glory so great, that the moment a pointer on a dial moved, a press conference would be called and only later would the subtleties be discovered.

$BF_3$ detectors are quite adequate for monitoring large fluxes of neutrons such as come from fission, when you're counting many thousands a second. As that was the sort of flux that people had been led to believe would be coming from the test-tube fusion cells, then it made sense to run a modest experiment in a small department without spending many millions of dollars to run a more sophisticated experiment. Georgia's intent, as was the case in most initial set-ups, was to see if the phenomenon was real; only if it was would a more ambitious longer term and detailed study be merited.

Where they were caught short was that there turned out to be only few neutrons at best. The $BF_3$ counter was quite inadequate to measure in such a regime. Georgia told the media that they were now planning to use better neutron detectors. Not just Georgia Tech but a group in Calcutta, India, and others too were caught out this way. Georgia had the scientific credibility to announce their error, and this retraction was actually scientific honesty at its best. Had more groups given equal prominence to the negatives as to the positives the true picture might have been determined more rapidly. If truth is to be determined, both prosecution and defence must be heard with equal voice.

At their second press conference, where the claims were withdrawn, a reporter sagely asked a crucial question: 'You discussed the $BF_3$ detector. Have you been able to learn if Dr Pons and the Utah people were using the same kind of counter, the same model as you?'

It turned out that of the many calls that Georgia had received, one had been from Pons. He had just been at the Chemistry Society Meeting in Dallas where 7000 had hung on his every word. He called Georgia on the Wednesday evening, 13 April, by which time they already knew of the blunder but had not yet announced it. Pons called up believing that Georgia had a positive result, confirming part of the Fleischmann and Pons effect, indeed seeing neutrons which was the part that Fleischmann and Pons were the least confident about. Pons had been excited by the Georgia claims and asked details of the experiment, the type of palladium, its shape and size, how much electrical currents they had used and so on.

154

Then Steenblick told him that they were having difficulties with the $BF_3$ counter and asked what counter Pons had used on which to base his data.

Pons replied that it was a $BF_3$ counter but that he was sure that he had the temperature effect under control.

Steenblick recounted this conversation at the press conference. A reporter then asked 'And *what* were Pons' temperature controls?'

Steenblick replied: 'He didn't describe them.'

On that same day in Seattle Van Eden and Wu Lei were giving a technical presentation of their work and the first hints of a possible problem were beginning to emerge as a question was raised about triatomic molecules. We are used to hydrogen molecules containing *two* atoms, $H_2$, and similarly $D_2$, and hence $DT$ and $TT$ with mass 5 and 6. However, it turns out that *tri*atomic molecules and ions, such as $H_3$, can form in a variety of situations such as gas discharges (an example is the formation of ozone, $O_3$, during lightning in thunderstorms). Neutral triatomic hydrogen molecules are unstable, but the positive ion—where a single electron is missing—is stable. It turns out that in the set-up that Eden and Liu were using, $HD_2^+$ and $D_3^+$ ions, with mass 5 and 6 respectively, could form.[5] Hence these masses, detected in their spectrometer, did not necessarily imply anything about the presence of tritium.

The spectrometer that they had been using up to that time could not discriminate between $DT$ and $DDH$, nor between $T_2$ and $D_3$. About a week after the press conference they took a palladium tube to the medical chemistry group who had a higher resolution device. Half a day's work showed that there was no big $DT$ signal. Then they returned to their original equipment to make various tests which took a further six weeks of careful measurement to complete. These confirmed that triatomic hydrogen molecules were being formed and that the earlier results had been due to this, not tritium. On 25 May 1989 they sent a note to the Information Services at the university which reported: 'In particular the signal seen at mass 5 is likely due to a triatomic molecule composed of one hydrogen and two deuterium atoms ($DDH$) and not due to molecules of deuterium–tritium ($DT$). These latest results therefore, do not provide support for the work on cold fusion by Fleischmann, Pons and Hawkins.'

It was bad luck for Eden and Liu; their idea and experiment had been clever. There is a parallel to the work of the Utah chemists in that things went wrong when the research went beyond the local expertise. However at Seattle, the moment others were able to dissect the work, the errors were detected, careful refined experiments were done and the erroneous claims were clearly and explicitly withdrawn.

That should be the end of the story, although some still cite Eden and Liu as having provided support for cold fusion. Their withdrawal was explicit—the university press office sent copies of the statement to everyone who had previously been mailed the original announcement—but you cannot force the media to issue it: there were no headlines to compare with the original announcement. '*Fusion Facts*', a publication from the University of Utah Research Park which attempts to excite corporate interests into the possibilities of cold fusion, cited Eden and Liu as confirming test-tube fusion, in *Fusion Facts* 'Scorecard on the F–P effect replication' (selling for $35 in September 1989 for a single issue, '36 issues $900'). However, it did not mention that Eden and Liu had withdrawn this.

Meanwhile at Texas A & M Charles Martin and his colleagues continued developing the experiment during the two weeks after the press conference using platinum and carbon electrodes in order to check: even if $H_2O$ fused in the presence of palladium and gave heat, there should be no heat produced with other electrodes. (Fleischmann always insisted that the best control experiments were with platinum electrodes.) They tried an experiment using $H_2O$ and platinum which should have been blank, but they found that this gave excess heat. Then they tried with carbon electrodes, and found heat here too!

At this point they became seriously worried. Then by chance they stumbled on what was going wrong.

To ensure a uniform temperature throughout the cell they were stirring the liquid. Electrolysis causes lots of bubbles in the liquid and so stirring caused the bubbles to disperse all over the volume. One day, for no particular reason, the stirrer was temporarily turned off and they noticed that the bubbles were not coming solely from the electrodes, where electrical current flowed, but also were forming at the metal casing of a thermometer in the cell. This showed that the thermometer had not been properly earthed—electrical current was flowing into the cell via the thermometer, adding power and hence heat that had not been accounted for and consequently was causing the temperature to rise. The 'excess heat' was entering from the outside, not being generated from within.

About two weeks after the initial press conference they duly announced that the experimental claims were flawed and withdrew the paper that they had written earlier that month.

156

They properly earthed the thermometer and immediately most of the excess heat disappeared. There was still a 10 to 20% excess—about half a watt unaccounted for—and for a while there was some confusion as to whether or not Texas were claiming excess heat. However, this was never reproduced. Martin told me that 'the experiment had been compromised, the heat was never reproduced in any of our subsequent attempts, and the summary is that we have never conclusively demonstrated excess heat.'


## Physics doubts

Following the meeting at the DOE Bond had become aware of the amount of effort that was going on, not just in the government labs but in many universities too. In April, when individual groups had begun their exploratory experiments but had no real knowledge of what was going on elsewhere, Bond remarked to me: 'I have talked with nuclear physicists whom I have known for years and whose work and integrity I respect and who are either directly involved in or are near to people who are into this business. No one sees anything. Now, your first reaction when this happens is that you must be wrong. Consequently everyone is double checking, triple checking to see what, if anything, they can have overlooked.'

Across the country individual groups were at work and only now, towards the end of April, were they confident enough to start talking to one another about their work and comparing notes. One of the first large scale gatherings of nuclear physicists took place on Sunday and Monday, 23–24 April. It was there that for the first time I began to piece together the way things were developing. It became clear that while there was much trumpeting of positive claims (which were then withdrawn as errors appeared), there were also many careful experiments going on with nuclear physicists seeking fusion products, such as neutrons, and seeing nothing.

The meetings had nothing to do with cold fusion and had been arranged weeks earlier to discuss the planning of the next five years of research in various aspects of nuclear and high energy physics. Specialists in different areas were meeting at different venues that weekend. I was due to speak at a meeting in Bloomington Indiana on the Sunday, and then went to Oak Ridge Tennessee to hear the concluding summaries of a different meeting on the Monday.

The conversations during coffee breaks, dinner and late into the night had little to do with the meeting's official agenda but instead dealt with cold fusion and provided the first picture of work across the continent. What emerged was the impression that someone who was an expert in detecting neutrons would describe how it is much more difficult to use the sophisticated instruments than at first one would imagine; and that he disbelieved the results for the following reasons. Then would follow a list of technical details which made one realise that you could no more go out and buy a neutron detector off the shelf, switch it on and claim to be detecting neutrons than a tone deaf person could go buy a violin and start playing a concerto. Neutrons, $\gamma$-rays, helium—detecting these is as much an art as it is science. To say that the chemists didn't understand what they were doing was not intended as an insult to chemists—the physicists for their part would not have expected to be taken seriously if they had suddenly claimed to be able to measure subtle effects in chemistry.

Also I heard criticisms of the chemists' claims which became voiced more and more widely in the following weeks. For the first time I learned of a group that was doing experiments and had interacted with Pons through a talk that had, by chance, been prearranged at their University. This was the University of Indiana and my first contact was Charles Goodman, a distinguished experimental nuclear physicist.

Goodman has some things in common with Martin Fleischmann. He is a similar age, enjoys skiing, and having had the pleasure of skiing with him from his cabin above Telluride, Colorado, I can testify that he is a good cook too. Fleischmann's reputation among chemists parallels Goodman's in the field of nuclear physics. Goodman has discovered subtle effects in nuclear physics for which he won the prestigious Bonner Prize of the American Physical Society in 1984.

During his career he has spent many hours detecting neutrons from all sorts of sources. When the news broke on 23 March it was natural that his colleagues turned to him. 'If the chemists had produced four watts of power from a fusion reaction then they would have been bathed in so many neutrons that they shouldn't be here to tell the tale. As they seem to have taken no precautions against radiation they are lucky that whatever it is that gives the heat is NOT fusion, at least not fusion as you and I understand it. But whatever else is going on, if there is fusion at all there must be some neutrons with energies of around a couple of MeV. I have never measured neutrons with energies in this energy region, but as I have spent a lifetime in nuclear experiments, someone suggested "You have all this equipment set up and know all about neutrons, why don't you give it a try".'

Goodman's experience already forewarned him about the difficulties and pitfalls. He told me, 'If you see what appear to be a few low energy neutrons, it isn't always easy to be sure where they are coming from. You have to be careful. For example, at Berkeley someone set up a palladium cell, turned on the electricity and they saw some neutrons. But at Berkeley the labs have been used for nuclear experiments a great deal and there has been, and still is, a lot of radioactive material about. It's no hazard to you or me but it makes the environment dirty in the sense that there are a few neutrons around which can trigger your detectors and have nothing at all to do with the effect that you are trying to look for. It turned out that someone had brought some curium (a radioactive element) into the lab, and it was irradiating the detector with neutrons.'

'Now Berkeley are wise to this sort of thing and soon tracked down the source. But you can imagine how someone less experienced could have been led astray and called a press conference. You have to be careful. It is easy to get spurious signals and I predict that lots of people will do so. People who ''see'' something, erroneously, will announce it; no one is going to be telling the media ''hot news; we don't see anything''. The public is going to get a very unbalanced impression and will believe it is all true even while we are still trying to get at the bottom of it.'

'Either way I am not optimistic about the outcome. There will be egg on scientists' faces if it turns out all to be wrong because you can be sure that all over Japan factories are tooling up to see what they can develop, all based on hype. The danger if it's all true is the tritium production. Fusion will produce tritium. It is the difficulty of doing this that has limited the spread of thermonuclear weapons; finding out even basic information about tritium is often not easy because it is classified it's so sensitive. But if cold fusion is real, then anyone can do it. Any tinpot dictatorship with some water will be able to make the stuff.'

Goodman's neutron detector was ideal for identifying neutrons whose energies were large, about 100 MeV, but not good for low energy ones, at about 2 MeV that would be coming from fusion. He had a system suitable for catching speed merchants whereas the search here was for pedestrians. Consequently he did not personally do any test-tube fusion studies but his colleagues at the Indiana Cyclotron at Bloomington, Indiana, set to work immediately after Easter, within days of the news breaking. Two nuclear physicists, Les Bland and John Cameron, joined forces with Tim Ellison, an engineer, and Bill Lozowski who built targets for irradiation. Their nuclear tradition, as Goodman's, said 'Neutrons, that is easy for us', and then later they discovered how hard it really was.

When they spoke to me in April they had been working for four weeks 'on and off'. The 'off' is because the Indiana Cyclotron Facility was already committed to other tasks including preparing radioactive samples for medical use, and there was a lot going on that required

159

their expertise. Test-tube fusion had to work its way into a busy schedule.

They tried several ways of forcing the deuterium into the palladium. They used electrolysis like the chemists, and also tried heating and cooling the palladium in the presence of the deuterium gas. This is a much faster way of getting the deuterium in; the chemists took months with electrolysis, whereas the 'gettering' technique, as it is known, can do the equivalent in a few hours.

At this stage people had been misled by the large effects that the chemists had claimed and were supposing that the fusion would be easy to reproduce; it seemed therefore to be simply a matter of forcing the deuterium in as fast as possible. Later people realised that if there was any effect at all it seemed to be highly sensitive to conditions and supporters of the chemists could make excuses that 'by trying to do the experiment different to the way that the chemists did it you changed the magic ingredient and so the fact you see nothing is because you did it incorrectly'.

Indiana had been lucky in that Pons gave a talk there on 4 April, and although it was only a few days after the press conference, the Indiana physics group had hoped already to be able to have enough data to be able to say something at his talk, maybe to confirm or refute the claim.

But then the difficulties began to show up. It is hard to detect a low flux of 2 to 3 MeV neutrons, and you also need to look for them on top of whatever else is about. Their neutron counter was a liquid scintillator: essentially any neutrons coming into it scatter from protons in the liquid and emit a flash of light of characteristic energy. This would be the 2.2 MeV $\gamma$-ray that the chemists had also claimed in their paper as evidence for fusion. The Indiana team were swamped by signal. Natural radioactivity in the environment, in particular the concrete in their laboratory, was producing $\gamma$-rays tantalisingly like those they were looking for. Once you use a lab it gets contaminated. And there are cosmic rays hitting us all the time. Our eyes don't see them so we think everything is clean and simple. But if our eyes saw $\gamma$-rays we would be living in a perpetual light.

They realised that for the task in hand their neutron detector was too crude. With this discovery they decided that Jones' group at Brigham Young University, with a more sophisticated detector than theirs, could have detected neutrons as Jones claimed, but the chemists' use of a radiation monitor struck them as useless for distinguishing signal from the noise. 'It was as if Fleischmann and Pons were trying to pick up a transmission from a spy in the field while using a crystal set and his frequency is the same as the local pop music station. Even with a good radio we can't pick it up. Jones

160

with his hi-fi may, but even that seems to us to be on the borderline.'

That was the situation at Indiana when Pons arrived on 4 April.

My notes of the conversation on 23 April read: 'Two weeks have passed and they haven't seen any neutrons yet.'

'So what can one conclude at this time? They feel in their hearts that if there were a real phenomenon taking place they would have already seen it, given their years of experience with these sorts of measurements. "Time will tell," John Cameron told me. "We want to be sure that we are doing everything right. Measuring zero is hard; how near to zero can you get?" They are putting all their time into this now as the American Physical Society meeting is next Monday, 1 May, and they would like to be able to have some results. "Maybe by then" were his parting words.'

The APS meeting turned out to be the first public airing of widespread criticism about the claims for heat and copious radiation products. Two central players were teams from MIT and Caltech whose investigations went directly to the heart of Fleischmann and Pons' claims and exposed serious concerns about their quality.

My impression at the end of April, four weeks after the saga began, was that here were teams of people who had a lot of experience in making technically sophisticated measurements and who were uniformly failing to see anything. Even so, they were trying every possible way of finding what they could possibly be doing wrong. They had not gone into this with a preconceived notion that it was crazy and that they would destroy it. They still felt that it could be real. There was no sense of a witch hunt—at least, not yet.

## The MIT story

At MIT in Boston Massachussetts there is a large centre studying fusion with its own experimental programme and with its own dedicated set of municipal power generators and its own cable connected to the Cambridge Electric Power Company. The experiments draw so much power that when there are disturbances in the local power supplies people often blame MIT. This is high quality big science which attracts some of the best undergraduates to its campus.

MIT's interest in cold fusion followed hard on the original announcement on 23 March and, as at Caltech, it was the students who reacted first to the news. They were being trained for a scientific career and their chance of making a mark in an increasingly competitive subject was still years away, so the apparent simplicity

161

of the experiment seduced them into believing in the possibility of being able to do definitive work, of 'being present at the start of a new technology, of being able to verify the claims of another scientist' as one undergraduate described it later. At 11.30 on the night that the news broke, a group of students went scavenging for apparatus. 'It was a very simple experiment that a chemistry student might do in their first year. All we needed was a jar, a battery and some palladium. We didn't really believe that it was going to work but it seemed worth a try.'

They found nothing but their excitement did lead to a more professional research effort in the Plasma Fusion Center. The next day was Good Friday and that afternoon people were chatting in the hallway preparing to go off for the Easter Weekend. Marcel Gaudreau, an engineer on the MIT faculty, was shown a page of notes from one of the undergraduates, indicating how on the day of the news conference they had set up a cell and measured the temperature with all the correct ingredients. It was all very quick but it impressed the faculty—the enthusiasm struck home. Gaudreau recalls: 'This got me excited. I thought ''these guys are really rolling and we ought to try and figure this thing out''.'

'We had spent years trying to understand controlled fusion and now the chemists believe that they can do it in a little jar, so I thought we ought to try and figure out how this works.'

A group of scientists led by Stan Lockhardt at the MIT Plasma Fusion Center began the real experiments that weekend. As at Harwell and Caltech, theirs was a multidisciplinary team containing physicists, engineers and chemists. There was no published paper to go on and they had no success reaching Pons and Fleischmann by phone—that day was absolute bedlam in Utah with *Fortune500* companies, news reporters and cranks all calling one after the other and tying up the phone lines. So the MIT team ended up getting their information from TV videos of the Utah laboratory that had accompanied the newscasts. They could see how many cells there were and how they were wired up. They spent many hours trying to understand the wiring of the batteries as they would need to know how much total energy had been put in to the system and compare it with what was being measured as output. In fact this was wasted effort because the batteries were only there to power the cells and their details were irrelevant, though MIT did not know that at the time.

The next day the father of Martin Schloh, a chemist in the collaboration, arrived from London together with a copy of the *Financial Times* containing the story of the cold fusion announcement. This article was special in that it had a diagram of the cell and with this new help the MIT team thought that it now might be possible to reproduce the experiment. They worked round the clock over the Easter weekend with materials that they had available. On the Monday they called around trying to get hold of larger palladium rods as one of the news reports had given the impression that this was important. This turned out to be harder than they had anticipated; as at Caltech, and all over North America, palladium was rapidly becoming hot property. Schloh recalled that they eventually got what he believes was the last available palladium rod in New England and they had to drive up and get it themselves. That evening they were so excited that they stayed up all night with the experiment, half expecting to see the apparatus start to glow.

The team contained fusion experts who were all too aware of the potentially lethal amounts of radiation that would be released if, as they expected, the experiment worked. Yet at the same time they were aware that as Fleischmann and Pons were still alive there was an upper limit to the amounts of radiation to expect, and this did not seem to fit with the amounts of heat being reported—if the heat was due to fusion that is. As one scientist wryly commented: 'One of the interesting things about the CNN (Cable News Network television channel) report of the conference was that the man explaining the experiment to the reporters was apparently touching the glass bulb containing the active elements and yet none of his bodily parts fell off.'

Even so the MIT team took precautions and set up the experiment in their own (hot) fusion lab where six-foot thick walls of boron-impregnated concrete can shield the surroundings from blasts of a million billion neutrons per second.

After a week there had been nothing dramatic; no explosion, no flashing lights, no glowing rods. They decided that although they knew what the apparatus in Utah looked like, and had designed their own as a copy, there were still too many unknowns. They had no idea what power was needed, what voltage or currents were used, or what chemicals were in the solution.

Eventually more details became known to them and, as at Caltech, they produced a report that they had seen neither heat nor fusion. In this they were 'merely' another well known institution contributing to the swell of 'No'. However, MIT made a unique contribution to destroying Fleischmann and Pons' claims to have seen the fusion neutrons.

Among the evidence that Fleischmann and Pons claimed for fusion was a spectrum of $\gamma$-rays purportedly generated when the fusion products—neutrons—are absorbed in the surrounding water. Their 'fusion cell' was surrounded by water which slows down neutrons and captures them converting some of the hydrogen in the $H_2O$ into deuterium. As the mass of a deuterium nucleus is less than the

163

combined mass of a free neutron and proton, the capture liberates energy—a $\gamma$-ray of 2.224 MeV (mega (million) electron volts) is emitted giving a characteristic autograph for the presence of neutrons.

The more fusions that have taken place in the cell, so the more neutrons emerge to be captured in the water and the more 2.224 MeV $\gamma$-rays will be seen. You count the number of $\gamma$-rays coming at you, measure each of their energies and plot the number for each energy on a graph. You will end up with a figure that looks like a range of mountains, a set of peaks on top of a high landscape. A peak at 2.224 MeV energy will be the signal that neutrons have been captured, and if fusion has occurred, producing neutrons, then such a peak must be present—we shall call this 'Mount Fusion'. There are other peaks which are due to $\gamma$-rays coming directly from other naturally occurring radioactive decays of elements in the environment. They are there because everything is radioactive, if only mildly so, and many of these radioactive decays emit $\gamma$-rays. Not just uranium but many mundane elements are active, including potassium which is common all around us and even within us. The presence of the human in the laboratory adds to the $\gamma$-rays already flooding into the detector; these $\gamma$-rays from decays of potassium occur at 1.46 MeV, easily distinguishable from the 2.224 MeV fusion peak.[7] There are so many of these 1.46 MeV $\gamma$-rays that they form a peak that is like the Mt Everest of the range in the average environment (it is peak number 3 in figure 10). The peak is so easy to see that you can use sight of it to verify that the scale on your detector showing the energy of the $\gamma$-rays is calibrated correctly, analogous to the way that you can find your bearings in a range of real mountains by orienting on a prominent one. Radioactive decays of naturally occurring elements also produce thallium, and this in turn gives a $\gamma$-ray at 2.6 MeV which is easy to see too. So the fusion peak at 2.224 MeV should be partnered by this pair; 1.46 MeV potassium and 2.6 MeV thallium.

The analogy of being among the mountains has an important message in the present story. In the mountains suppose that you are looking for Mt Fusion. You might sight it by relying on your compass alone, if you were sure that your compass was working properly and that you understood how to use it. But in a strange mountain range you would be unlikely to rely solely on the compass. A good guide will convince you of the sighting of Mt Fusion by pointing out other features in the landscape, noting the huge peak of Mt Potassium on the left and the small Mt Thallium to the right. That is what nuclear physicists do in real experiments. They show the whole range of peaks and from their relative heights can determine the relative abundances of the various sources.

If cold fusion had occurred in such a way as to be commercially

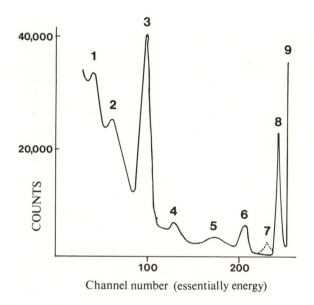

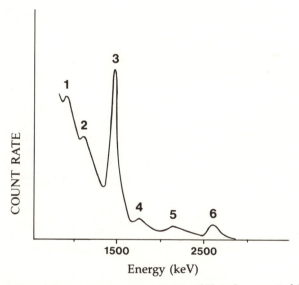

Figure 10. Comparison between gamma-ray spectra of Fleischmann *et al* (upper) and the MIT group (lower). Note the similarities 1 to 6 which are all peaks familiar to nuclear physicists, and the extra features 7, 8 and 9 in the upper spectrum. Reproduced courtesy of *Nature*.

viable then Mt Fusion would have dominated all the other peaks and even made Everest seem puny. One would have had no need of other topography to be sure of what one was seeing. This was the impression that many people got when they first saw the figure because in Fleischmann and Pons's paper, faxed around the world following the infamous press conference, there was only one peak and they asserted this to be Mt Fusion. There was no way to be sure other than to take their word for it. However, as soon as people who knew this range of peaks saw the picture of the isolated one they said 'That is wrong', as at Harwell, for example, where this was pointed out to Fleischmann at his talk on 28 March, and he transmitted this information to Pons.

Rich Petrasso, experienced with $\gamma$-ray spectra at the MIT plasma fusion laboratory, noticed that the data looked wrong, in particular that the peak was too narrow and did not have a characteristic shoulder called a 'Compton edge';[8] as Petrasso stressed 'that edge has to be there as surely as if I let go a ball it must drop.'

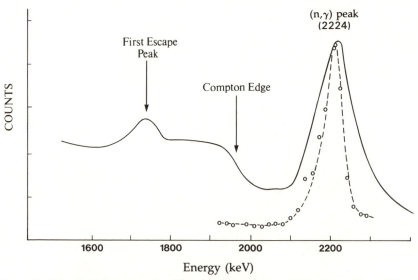

Figure 11. The dotted peak is after Fleischmann, Pons and Hawkins. The solid curve is as measured with a genuine neutron source at MIT. Note the absence of the Compton edge, the narrow width and the slightly displaced maximum in the FPH data.

In *Nature* on 18 May Rich Petrasso and five colleagues published their seminal paper entitled, with scientific understatement: 'Problems with the $\gamma$-ray spectrum in the Fleischmann *et al.* experiments.'

166

Though published in *Nature* on 18 May a preliminary version of the report had been circulating at the APS meeting in Baltimore on 2 May. The tension behind the scenes were apparent as one of the authors, Ronald Parker, denied a report in the *Boston Globe* that he had called the chemists 'fraudsters and shlocks'.

The manner of the MIT exposé was itself extraordinary in the experience of scientists in that whereas the chemists had publicly displayed only a single peak, the MIT team had managed to obtain a full mountain range from the local Utah television station.

Fleischmann and Pons had been media stars throughout April, culminating with international exposure in *Time* and *Newsweek* on 8 May. With megastars in their town the KSL-TV station in Salt Lake gave them a lot of attention in newsreel of their laboratory. This was the first sight of the place that most scientists had seen. The spy in the lab showed a VDU with a range of peaks instantly recognised by nuclear physicists as the spectrum of γ-rays that are constantly being given off in nuclear processes all around us. The MIT team obtained a copy of the picture from CNN and confirmed with Marvin Hawkins—the previously forgotten co-author—that it was indeed taken in the Pons laboratory. Then later in April they obtained every news-clip from Utah News Inc. and scanned them to confirm that they only showed one γ-ray spectrum from Pons' lab amid views of the lab, further confirming the source. (These are shown as photos A, B and C below.) Petrasso's group then made a spectrum in their own lab and compared the two. They were almost identical but for slight differences—minor to the casual observer but devastating to the experts who knew these peaks so well (*see* Figure 12).

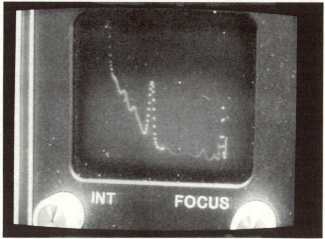

A. Gamma-ray spectrum on VDU, shown on television during item on test-tube fusion and Pons' laboratory. Reproduced courtesy of J. Dismore and R. Petrasso.

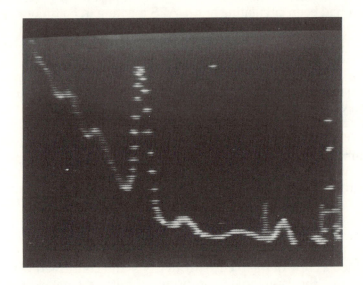

B. A reproduction of the spectrum as used by an MIT group in *Nature* (*see* text). Reproduced courtesy of R. Petrasso.

C. Television spectrum as in A, but highlighting region to the right of the finger (in shadow).

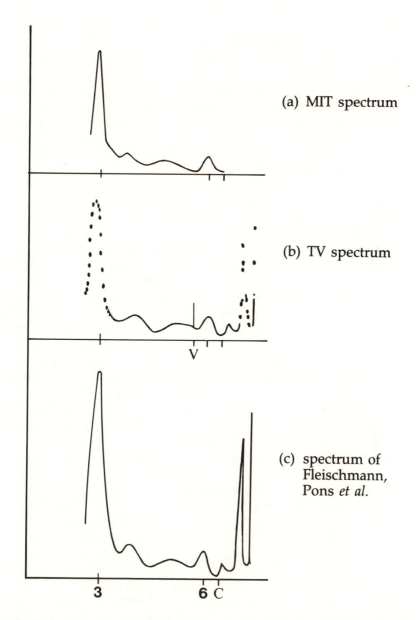

Figure 12. Comparison of a (standard) gamma-ray spectrum from MIT, the spectrum shown on television emanating from Utah and that from Pons' lab. 3 and 6 identify the peaks shown in Figure 10, C denotes the endpoint of the MIT spectrum. Compare the structures to the right of C in (c) with those in the television spectrum (b). V is a cursor. *See also* note 2 in Chapter 18.

Although this picture captured the attention, the technical destruction of the chemists' claims did not depend on it. The MIT group showed that it was impossible to see such a slim peak as the chemists were claiming with the apparatus that they reported they had used, and the absence of the Compton edge eliminated it as a viable candidate.

Petrasso had extensive phone conversations in late April with both Hawkins and Hoffman who had taken γ-ray spectra in March, trying to understand what radioactive backgrounds might have contaminated the signal, how the measurements were done and what the actual readings were. From this it became clear that there was no 2.2 MeV peak and that the peak that Fleischmann and Pons were advertising had been moved there from elsewhere. However the MIT team felt unable to say this explicitly as they had not officially seen the data, so in their paper they merely concluded by reminding everyone how important it is to identify the energy of the feature claimed by Fleischmann and Pons to be the fusion γ-ray; 'this can hardly be overemphasised'. And in understatement, 'It is extremely unfortunate' that the chemists chose only to display a single peak with no supporting evidence of the γ-rays from potassium (at 1.46 MeV) and thallium (at 2.6 MeV) 'which must be present in their spectrum in order for their identification to be correct'. They concluded that whatever the chemists were displaying, it had nothing to do with the 2.224 MeV γ-ray from fusion and that it is 'probably an instrumental artefact with no relation to γ-rays' (i.e. nothing to do with nuclear processes whether fusion or not).

The paper of 18 May showed that the peak was false; that was all that it did. It did not imply that there was no fusion taking place, merely that the displayed peak was not evidence.

What MIT did not say, although they strongly suspected it (and, as we shall see, these suspicions were correct) was that the lone peak that Fleischmann and Pons were advertising at 2.2 MeV had really been measured at 2.5 MeV and moved without explanation (compare Figures 7 and 8 on pages 97 and 115). Meanwhile the rest of the world was presented with the 2.2 MeV peak as if it proved fusion and as if there were no doubts about its energy and status.

The chemists had to make some sort of public response to MIT's paper if they were to retain any remnants of credibility. They wrote a letter to *Nature*, co-signed by Hawkins the student and by Hoffman, the radiologist who had taken the measurements that had appeared under Fleischmann and Pons' names alone. Far from answering the critique, it presented the whole range of peaks, confirming the so far unstated suspicions, and enabling MIT to make a further paper showing what the chemists' had done. Moreover, it destroyed any

170

remaining credibility for the neutron-capture data and showed that there are no neutrons. This story is in Chapter 15.


MIT were beginning to uncover these details only towards the end of April and they were not yet clearly known, even to them, during the weekend of 23–25 April when I was gathering together the first news.

In Utah, April was the time for exploiting the cold fusion news to the full. A major push began to get money into the state. While the public, fed by positive stories and statements emanating from Utah, believed cold fusion to be almost confirmed, the scientific community was growing sceptical as informal news, such as that I gleaned at the meetings on 23–25 April, began to spread.

During that weekend I also talked to Doug Beck, a nuclear physicist from Caltech. He had flown in directly from a seminar at Caltech, Los Angeles, which had been attended by 1000 people who had heard the first reports of a multidisciplinary effort. The team had a wide range of expertise. Nathan Lewis, an electrochemist, knew Pons; Charlie Barnes is a nuclear astrophysicist who is expert at studying the sort of processes that ought to be present if cold fusion was real, and Steve Koonin is a theoretical nuclear physicist who had been re-evaluating the theories, looking for hidden loopholes.

What none of us foresaw were the events that would occur during the next ten days. The University of Utah authorities, with the chemists, went to Congress seeking millions of dollars of support for cold fusion. The first public statements came from Ron Ballinger of MIT who announced that groups at MIT were seeing no signal, and he also hinted at the more serious worries that Petrasso and others were beginning to uncover. Indeed the claims that some of the scientists and administrators associated with the University of Utah were making in Congress astonished members of the Caltech team, who also were beginning to understand what Fleischmann and Pons had done as distinct from what the media reports were suggesting. These events so disturbed the Caltech team that they decided to make a public statement at the American Physical Society meeting in Baltimore on 1 May in which they denounced the 'incompetence and possible delusion' of Fleischmann and Pons. In order to understand this chronology better I shall delay their story until Chapter 11 and go first to the Congressional hearings.

171

# 10
## MONEY

On 7 April the University of Utah had told the State Legislature in Salt Lake City that if they didn't put five million dollars in to support test-tube fusion 'the discovery of the century will be developed by Mitsubishi'. The State approved the money by a vote of 96–3 and on 20 April Governor Bangerter signed the bill. A letter the next day in the local *Deseret News* asked, 'If the governor thinks fusion research is so wonderful, why doesn't he put his own Bangerter family fortune into it as a good example to us all', to which Robert Deltman of Salt Lake City replied, 'Why indeed. Because it's so much easier and convenient to spend the public's money.'

With this cash in place the University's ambitions moved upward. The following day, 21 April, Federal funds were the new target as Robert Roe, the Chair of the US House of Representatives Committee on Science, Space and Technology was impressed enough to call a special hearing to evaluate cold fusion. Twenty five million dollars was the sum being asked for, with 125 millions being a longer term goal.

The hearings were set for 26 April and the media circus came to Washington: Fleischmann, Pons, Jones and an entourage of supporters and detractors were in town. C-SPAN and CNN television channels and the other media carried the proceedings live. Representatives from major laboratories—Oak Ridge, MIT and Princeton—gave their first public statements, the first clues for many of what the eventual outcome would be. The testimony of the supporters of test-tube fusion was given to a full house; some of the negative contributions received less immediate attention but did make more impression in the serious papers.

The testimony was politics at its most transparent: it was all about money. In one corner were the chemists and the University of Utah making a direct pitch for Federal aid for cold fusion, supported by Ira Magaziner, a business strategy consultant, who appealed 'on behalf of [his] children and all of America's next generation that we are not the first of our nation's ten generations to leave its children a country less prosperous than the one it inherited' and warning everyone that if the US didn't develop cold fusion, even though not yet proven, Japan would beat them once again. Some measure of scientific support came from Prof. Robert Huggins, a materials scientist from Stanford University, whose final plea was to remind the committee that the Federal budget for research into metals and energy had been reduced and how small science (which includes his field of research) had made more contributions to the US economy than had large scale science. There were representatives of some areas of 'large-scale science' in the form of Mike Saltmarsh of Oak Ridge National Lab and Harold Furth of the Princeton Plasma Fusion Research Program. They urged caution and advertised scepticism, and were later accused by some supporters of test-tube fusion of wishing to preserve the status quo, and of protecting the funding for their own big research progammes. There were contributions from the Brigham Young team who stated that they were seeing a few neutrons, but that although the magnitude of the effect was interesting scientifically it would never make any contribution to solving the world's energy problems; they were later castigated by some in Utah as being unsupportive of the State. And finally Ronald Ballinger from MIT went so far as to express 'distrust of the reported results'; this was the first clue that a group from MIT was in the process of exposing flaws and even fabrication in the Fleischmann and Pons paper.

The hearings were set up in a climate that perceived Fleischmann and Pons to have saved the world from impending environmental disaster. The supporters appealed to patriotism and geopolitics more than to science. To the media this was Fleischmann and Pons's show; by the time the 'negative' testimony of the scientists appeared, many of the attendees had left.

Although it was still very early for a clear scientific opinion to be formed, it was interesting that some of the comments from the US laboratories as to the likely errors and lack of regular scientific methods, and the general scepticism about the work, turned out to be almost identical to opinions being expressed independently within European labs such as Harwell. No one knew that at the time, and although the emotional appeals may have swayed the uninformed towards believing that energy from test-tube fusion was simply a matter of time it was ultimately a credit to the US process and to cool heads that the lack of substance in the evidence managed to come through; that facts rather than gloss won the day. To many watching at the time it was by no means clear that this would be the case. Indeed, in California Nathan Lewis (see Chapter 11) was extremely concerned; his own data showed that some of the claims of Fleischmann and Pons were most suspect and even wrong, and it was the nature of these congressional hearings, that Lewis watched on television, that decided him to go public with his own results and even to visit Washington to help put some of the record straight.

The detailed testimony shows a cross-section of perceptions prevalent at the time. Fleischmann and Pons were widely believed to have found heat and also fusion products—the refutation of some of these as other scientists discredited them was still in the future. The Brigham Young team clarified the relation of their work to that of Fleischmann and Pons; up to that time the media had portrayed them as confirming the two chemists' claims but here for the first time was a clear statement that 'the only similarity between [the Fleischmann and Pons] experiment and the work at BYU is the use of an electrolytic cell. The electrodes, the electrolyte *and the results* are completely different.' Support for the effect was mainly emotional: 'We want this to work. Energy is the lifeblood of a nation and fusion energy would be an enormous step towards energy independence.' And for the first time there were official statements from the big laboratories who had so far only made presentations in private.

The seed for the hearings was sown on 21 April by Democrat Robert Roe, the chairman who called the meeting 'to review the dramatic developments that have occured with Dr Pons' and Dr Fleischmann's discovery of cold fusion.' (Note here that no mention of Jones was made; the official perception as in Europe was that the question at issue was Fleischmann and Pons's claims to have harnessed fusion energy of world significance.) Roe outlined the reasons for the committee hearings as: 'We want to understand these startling developments and gain some insight into the potential implications. We feel that we may be standing at the door of a new regime of scientific knowledge.'

The ranking Republican member of the committee, Robert S. Walker of Pennsylvania, a strong supporter of fusion energy, was very enthused by the hearings: 'The committee is extremely excited by the opportunity to examine the promise of this new potential source of energy. Cold fusion may eventually turn out to be one of the most significant sources of energy for the world in the 1990s and beyond, and I look forward to hearing from its discoverers and other experts on the implications of this intriguing breakthrough.'

That was all on 21 April. Within five days the protagonists had rearranged their schedules and come to the Capitol in Washington DC.

Roe and the Hon. Marilyn Lloyd opened the proceedings. Roe set the temperature of the times: 'In recent weeks an atmosphere of high excitement and anticipation has permeated the scientific community

174

as startling possibilities for sustained nuclear fusion reactions at room temperture have emerged. The potential implications of a scientific breakthrough that can produce cold fusion are, at least, spectacular.' Marilyn Lloyd picked up the theme and echoed much of the public's hopes for test-tube fusion as a saviour when she noted that her subcommittee on Energy Research and Development dealt with matters like preventing disasters such as Chernobyl and Three Mile Island, cleaning up nuclear waste and dealing with acid rain and the greenhouse effect. Test-tube fusion was good news in such a world: 'Your presence here shows that good news can make headlines', and, reminding everyone of the good news, she told of the 'hope of developing the power of the Sun, a power that runs on limitless sea water and leaves no dangerous radioactive waste. This has put research projects all over the country on hold.'

If there was any doubt that it was Fleischmann and Pons rather than Jones who were the focus of attention, this was dispelled by both Lloyd and Roe. Lloyd said, 'Gentlemen, the world awaits the crucial details of your amazing claim. The amount of energy that your experiment produced is larger than any chemical reaction we know can justify and the production of *so much heat energy without a corresponding number of neutrons* is novel to say the least. Roe referred only to Fleischmann and Pons by name and said: 'We are extremely pleased to have assembled here *the two professors who may have discovered cold fusion . . . additionally we have with us several recognised experts in the fields of fusion energy research and materials research from across the country.'*

And the summary of all the excitement 'We all want this to work'. All wanted it to work for the good of mankind; earlier fears that we would die from nuclear destruction were being taken over by a new growing awareness that we were creating an ecological desert on the planet. The perception that the environment was important was no longer the concern only of fringe ecologists and 'greens'. *Time* magazine had chosen no one as its 'Man of the Year' but instead had awarded 'Planet Earth' this honour as it heralded the impending disaster. The environment was becoming a cause that politicans wished to be identified with. In Britain Mrs Thatcher announced 'I am a Green' and the US Presidential Election had touched upon the issue, but Governments were finding that it was one thing to recognise the problem but quite another to act on it with so many vested interests in industry to be considered. So the claims of pollution-free test-tube fusion offered them an exciting way out.

The wish to be identified with test-tube fusion took on many forms. For those at the centre of the new technology there was the hope of great riches. The University of Utah authorities were naturally jealous

175

to protect the fledgling research that was not yet patented, not yet developed, and not yet confirmed.

Chase Petersen, the president of the University, took the microphone. He introduced the listeners to fission, with its intense long-lived radioactive waste products, and to fusion, practical use of which had defeated investigators for 40 years. He compared the huge cost of these endeavours with the simplicity of the Fleischmann and Pons experiment, with the now familiar rhetoric that nuclear waste is no longer a problem, the cost of heavy water fuel is moderate and since there is one molecule of heavy water in every 6000 molecules of sea water, the supply is inexhaustible. Goodbye carbon dioxide pollution, goodbye greenhouse effect, goodbye acid rain.

Then he addressed the question of what had led to the Utah experiments, and noted the humour and bite of the scientific controversy. 'A capacity to see an old problem from a new perspective was required: chemists looked at a problem traditionally reserved to physicists.' He then proposed that it might not be mere chance that it had happened at Utah, 'a University which has encouraged unorthodox thinking . . . The USA has prospered and innovated at the frontier and the University of Utah is still a frontier that attracts faculty who value highly their intellectual freedom. As these studies are confirmed they will need to be moved rapidly to developmental and commercial phases or we will lose their harvest. In a real sense we are obliged to build the first floor of engineering and the second floor of commercialisation in this edifice at the same time we build the foundation of scientific understanding.'

Then Peterson revealed his agenda: 'The state (of Utah) has appropriated five million dollars to assist [this project]. 1.1 million has already been raised privately with the promise of much more. We are prepared to build a novel consortium of Federal, corporate, state and University resources if you choose to join us. Without Federal participation the race for competitive leadership will be handicapped.' Roe had asked Fleischmann earlier what the government could do to speed up progress in test-tube fusion. Fleischmann had answered that it should support technological development but that it would be expensive. Now it was Peterson's turn. Asked how much money was required he replied, 'The figure that comes to mind is 25 million.'

In fact officials at the University of Utah had met privately with some of the committee members the previous day to suggest that Federal seed money was needed: 25 to 40 million dollars in the coming year, 1990, to help create a 100 million dollar cold-fusion institute in Utah to speed commercialisation of the work.

To keep the appeal for cash moving along, the University then introduced Ira Magaziner, the business strategy consultant, whose principle was to follow the Pennsylvania lottery principle, namely that if the pot is big enough why worry about the odds.

Magaziner announced that he was not from Utah, nor would he know palladium if he saw some; he was there because 'I am concerned about my three children and the future prosperity of their generation in America.' The crucial ingredient in achieving success here lay in 'America's ability to lead the world in pioneering the technologies and products of the future.' Then followed a moving inspirational plea for the US to save itself from an economic disaster, in which he used facts that were less than totally established, and he gave little consideration during most of the speech as to whether or not there was any scientific basis for his claims.

'From its early days this nation has prospered in great measure because we have led the world in taking the scientific knowledge of the day and bringing forth commercial products which we made more efficiently and in greater abundance than anyone else. This ability made us the most prosperous nation in the history of this planet. Over the past fifteen years, however, we have been losing this ability. We have lost the knack of converting scientific inventions into products to create jobs for our people.'

He then introduced the Japanese threat. He cited a range of products and discoveries that had been initiated in the US and were now 80, 90, even 99 per cent Japanese or East Asian dominated. These included the inventions of microwave ovens at US Raytheon, RCA's colour television, Ampex and the VCR, Texas Instruments' memory chips, Control Data and Cray's Supercomputer, and AT&T Bell Lab's solar cell. The result of all this was 'a 135 billion dollar deficit despite a 48 per cent drop in the value of the US dollar over four years forcing the US to borrow from its competitors, selling them our land, buildings even our productive companies to finance our current living standards.'

A crucial cause of this, he argued, was that foreign companies converted the science into commercial products better than did the US. They do so, he asserted, because they were prepared to develop marketable products 'even before the science is proven.'

This led him to suggest that the US should change its business strategy and follow the Japanese and European approaches where the Government worked closely with the Universities and industry in the early 'pre-development' stage. With examples such as the European Airbus—where Government backing had brought it from nothing to 25 per cent of the world's commercial jet market, surpassing Lockheed and MacDonnel Douglas—and Japanese new electronics devices for the 1990s that were being developed with billions of Governmental resources. Clearly the US should be supporting research and development with public money to match the Japanese and European effort.

Now he made his pitch for Utah and brought in the perennial demon, the clichéd imagery that permeated much of the test-tube fusion propaganda literature. 'As I speak to you now, it is almost midnight in Japan. At this very moment there are large teams of Japanese scientists in University laboratories trying to verify this new fusion science. Even more significantly, dozens of company engineering laboratories are now working on commercialising it, thinking of products which can be created if the science works. The Ministry of International Trade and Industry is already in the process of forming a committee to implement its plans for a coordinated push on this new industry. "Fusion Fever" has already gripped Japan's scientific *and* commercial communities.'

'Similarly a project team is being formed at the Eureka programme in Europe. A number of European Universities and companies are already at work to develop a European 'Cold Fusion' capability.'

While these claims may have been literally true the images were far from the reality: the official reaction in Europe and the resulting efforts to replicate test-tube fusion there were not essentially different from those already underway in the USA sponsored by the Department of Energy. And, we now know, their conclusion that test-tube fusion had little if any basis in fact and certainly was not going to solve the world's energy problems was essentially the same as that arrived at by the DOE laboratories. What's more, by 26 April, when Magaziner was painting his picture before Congress, the Harwell work in Europe and the Caltech work in the USA were already showing that test-tube fusion was unlikely to be the bonanza that others were claiming. However, nothing of this was yet public. Magaziner then offered his remedy; America should 'fight to win this time' and the way to do this was to form a 'research institute around this new science . . . which can be funded with money from the University, state . . and Federal Government. The state of Utah has already committed 5 million dollars . . but to match the competition in Europe and Japan there must be Federal support.'

He continued, 'But wait a minute, you say. This science isn't even proven . . . Wouldn't it be prudent to wait until we see whether there is really something of value here? We could all wind up with an extra large egg on our faces and waste the public's money in the process . . . I have no idea whether this is the most important discovery of the century or whether it is nothing. I do know that some very serious and accomplished people think it is real and I do know that if it is the implications are dramatic for the world and in particular for the nation that pioneers products based on it.'

'Suppose that a week or a month from now scientists find that there really isn't much to test-tube fusion. If we move aggressively ahead and invest as I suggest we will lose a few thousand dollars if it is discredited next week, a few hundred thousand if it is discredited next month and a few million dollars if it is discredited next year. But suppose that this science is real and it does open up a new energy source in the next decade and becomes a multi-billion dollar or even hundred billion dollar industry in the next few decades. If we dawdle and wait until the science is proven . . . we will be much slower off the blocks than our Japanese and European competitors because they won't run the race that way. If we fall too far behind at the beginning we may never catch up. The downside of wasting a few thousand or even a few million dollars is far less risky than the downside of losing this possible future industry to foreign competitors. The right decision is pretty clear.'

'I have come here today to ask you to lead so that we will not be the first of our nation's ten generations to leave its children a country less prosperous than the one it inherited. I have come here to ask you, for the sake of my children and all of America's next generation, to have America do it right this time.'

Fleischmann and Pons' testimonies were equally positive. Pons said 'we are as sure as can be' that the experiment showed nuclear fusion rather than a chemical reaction.' Fleischmann too said that other scientists' statements that the results were chemical, not nuclear, 'just don't stand up'. They announced that they had cells producing even more heat than previously, one of which was boiling. He said that advancing the work to design a high-pressure steam reactor would cost between one and ten million dollars. 'A vast amount of research is needed,' he said, adding that money was critical to commercialising the technology. 'Any scale-up is limited by cash flow. You cannot do engineering on the cheap.'

Fleischmann was asked if he was aware of any attempts by Japanese or European scientists to develop viable test-tube fusion. He replied, 'This is being researched around the world and I have confirmation of our results from . . . (he paused, grinned and raised an eyebrow) . . . far afield. *Very* far afield. You can guess where that might be.'

Magaziner's argument that you can't wait until the science is proven may have some merit in 'usual' circumstances, but these could hardly be claimed to be such. Discovery and 'proof' tend to be gradual cumulative processes rather than sudden breakthroughs, and so one can have a fair idea of how things are likely to work out some time before you are satisfied that they are proved, assuming that the science has been done to the usually accepted standards. But in the present case this seems not to have been so. The whole claims were based on supposed evidence for fusion, namely heat, neutrons, gamma-rays and tritium production. However, the last three pieces of 'evidence' were already being questioned by knowledgeable scientists, some data would soon be withdrawn by Fleischmann and Pons (but only after others had queried them) and the claim for heat was based on experiments for which incomplete controls had been made. 'Science not yet proven' may make sense *if* the science has at least been done but that cannot really be claimed to be the case here for there was already reason to suspect that the evidence was flawed (as became clearer in the second part of the proceedings). When Fleischmann and Pons made their announcement on 23 March it was the news on the 24th that Jones also had been doing similar research that had formed the first favourable impressions in people's minds. The perception had grown in the media that Jones confirmed Fleishmann and Pons' claims.

But now it was the turn of the Brigham Young University representatives to state in the full glare of the cameras precisely what they had, and what they *had not*, seen. It rapidly became clear that the Brigham Young University experiments were rather different to those of Fleischmann and Pons in detail and certainly very different in conclusions.

Daniel Decker, chair of the physics department, reviewed the history of the Brigham Young University research, from muon caatalysed fusion to solid state fusion, for which Jones had coined the name 'piezonuclear fusion'. He recapped how there was a year of encouraging but inconclusive results that showed that 'if any fusion was taking place *it was at a very slow rate* and its detection would require a more elegant detection system'. By late 1988 an improved detector was completed and preliminary studies were made on palladium, *tantalum, nickel, aluminium, iron and lanthanum* electrodes loaded with deuterium by electrolysis *and by gas pressure*. (The italics are mind to highlight the extensive differences between Jones and Fleischmann and Pons who did not do these). 'The results were tantalisingly positive and those on electrochemically loaded *titanium* were considered in February to be publishable.'

And what Jones had measured was also very different to that by Fleischmann and Pons. As Decker pointed out, the best proof of *dd* (deuterium) fusion would be to detect the 2.45 MeV neutrons that are emitted. The way that Fleischmann and Pons had tried to detect them was by 'moderating' them, slowing them down in a water bath until their energy was typically the same as that in the warmth of the room—hence called 'thermal neutrons'. They then detected them with a conventional thermal neutron detector, but the problem is that the low rate of neutrons was in competition with background thermal neutrons coming from cosmic rays.

The Earth's surface receives a constant rain of neutrons. Cosmic rays from outer space hit the upper atmosphere, shattering its atoms and ejecting the constituents downwards. Those of us who live at low elevations are protected from these particles by the intervening atmosphere which slows them down and captures them; mountain dwellers experience more radiation. At low altitudes the neutron flux is around one to ten neutrons per square centimetre per minute. In addition there are a few neutrons around from the spontaneous fission of natural uranium and the break up of thorium and its decay products. The quantity of neutrons you receive from these processes depend very much on the minerals in your vicinity, or whether someone has left a radioactive source in your office.

The neutrons ejected by cosmic rays vary in intensity with barometric pressure too. This is because the higher the pressure the more air there is for the cosmic rays to bump into—producing more neutrons—but the fraction of those neutrons that are captured by the air before reaching ground will also be affected by this. This changes the 'background' neutron rate slightly, but if you intend to look for a few neutrons coming from some feeble radioactive source, it is important first to be very sure just how big the cosmic ray background is.

Not only do cosmic rays and radioactive materials affect the number of neutrons, but people can too! We contain a lot of water, and water is adept at 'soaking up' neutrons. The hydrogen nuclei (protons) in the water are the lightest nuclei of all and when neutrons hit them, the neutron transfers its energy to the hydrogen much as a billiard ball gives up energy when it strikes another. The neutrons are slowed and captured in the water, a property that is used in reactors, for example (and was also exploited as a way of detecting neutrons by the 2.2 MeV gamma ray emitted in such capture). Water is an effective neutron shield, and if you lean over a neutron detector you will prevent some of the neutrons reaching it.

Cosmic rays may have penetrated all the way to ground without ejecting neutrons and then at the last moment hit the outer casing of your neutron detector, ejecting neutrons from it which are then instantly recorded as 'incoming' neutrons. Lead is particularly adept at producing neutrons this way. Indeed, scientists often use lead in demonstrations for the public. If your local college physics department wants to display a neutron detector at work it needs a ready supply of neutrons for it to detect. The best source is a natural neutron emitter such as the nuclear isotope Californium-252. The problem is that this can be difficult to get hold of, so the trick is to cheat by hiding a lead brick inside the detector. Cosmic rays hit the lead and produce a shower of neutrons. Moreover, lead is almost transparent to neutrons, so it lets through any that were passing anyway and adds to them from cosmic ray collisions! For demonstration purposes this is ideal, but if you are seriously wanting to count 'genuine' neutrons from some source, keep lead out of the way or you will get more than you bargained for.

If nuclear fusion was occurring in the Pons–Fleischmann cell, then it should have been producing hordes of neutrons whose number vastly exceeds the natural background. If there had indeed been 4 watts of nuclear heat coming from the cell then there should have been over a billion neutrons pouring out each second. There would have been no need to worry about how many were coming from background sources as the signal would have been so strong. However, what happened was that the number of neutrons that were being detected was so small that it left the question of whether or not any were from the cell, or whether all of them were coming from the various background sources, such as those just mentioned. It is because the measured rate was so near to that of the background that many people questioned the reliability of the measurements, and in particular the way in which the chemists measured the strength of the background, for it was necessary to know that first if one was going to claim something 'genuine' on top.

A crucial ingredient of the Brigham Young University experiment was that they developed a detector that measured both the number *and the energy of* the neutrons; this was the missing link for Fleischmann and Pons that they had hoped to get help with from Harwell. It was the observation of excess neutrons, albeit a small number, at the crucial energy of 2.45 MeV that had been the convincing feature for the Jones team, though even here there was the question of how much contamination was coming from cosmic rays.

A feature in common to both Brigham Young University's and to Fleischmann and Pons's presentations was that they claimed that other groups had evidence supporting their separate results; in both cases agreement was presented as supportive and disagreement argued away. Decker said, 'Several other labs have undertaken to check on these results. Some have good facilities and have reported detection of neutrons at levels similar to ours; some have not good enough counting systems and they see no neutrons above background.' However, there were also labs with *good* neutron detection facilities that saw *no* neutrons above background, such as Harwell, Caltech and Yale, but he did not know of them: announcements of the negative results from careful painstaking experiments were still weeks away. But his main point was a statement about the scientific process that was also an implicit criticism of the

Fleischmann and Pons work: 'None of these reported results constitute verification for true verification requires publication of a refereed paper showing method, technique, results, and analysis. If those laboratories have verified our results, then such papers will be forthcoming. It is too early to tell.'

Then he made an explicit comparison with the Fleischmann and Pons work that spelled out that it had little or no relation to that of Jones' team, and that Jones' work could in no way be used as support.

'The world is also very curious about another experiment [in which] Drs Fleischmann and Pons . . . argue that the amount of heat is greater than one could explain by a chemical reaction. There is no present evidence that this heat is related to fusion; even by their estimates of fusion rates the number of neutrons or of tritiums is deficient by many orders of magnitude. Rather than consider the possibility of some heretofore unknown chemical reaction taking place, they prefer to suggest that a violation of various laws of physics is necessary to explain this Fleischmann and Pons effect.' The crucial comparison then followed:

*'The only similarity between this experiment and the one at Brigham Young University is the use of an electrolytic cell. The electrodes, the electrolytes, and the results are completely different.* In the Brigham Young University experiment the emission of neutrons began just a few minutes after the application of current rather than after many days. In the University of Utah experiment the objective and detection was calorimetry. Other labs have tried to reproduce the Fleischmann and Pons effect . . . it is very difficult to sort truth from rumour by what one observes in the media so, in the final analysis, I must conclude that these results have also not yet been duplicated.'

He appended scientific comments on the current status by Kent Harrison, a physics professor at Brigham Young University. Harrison had written these on 18 April and they are interesting in that they could equally well be tabled today as the state of play. His essential points were: 'A nuclear fusion source for the energy production at the level claimed has not been demonstrated and indeed is unlikely. Nuclear fusion at much lower levels has been demonstrated, but cannot account for the energy. A clear demonstration of high level fusion in the present experiments would require a positive identification of the actual reaction(s) taking place.'

'If chemical reactions are the source of the energy, would this constitute a large new energy source? No. The virtue of nuclear reactions is that a sizeable fraction of the rest mass of the nuclear particles is converted to useful energy (by Einstein's famous equation $E = mc^2$). However, in chemical reactions this effect is negligible; such reactions are similar to ordinary combustion. It is possible that in the

183

current experiment that the charging of the cell for many hours—which the researchers said was necessary—simply stored energy which was later released; in effect then the cell would simply be *an ordinary storage battery, not a source of energy.*'

The first public discussion of careful experiments then followed, and slowly the chasm between Fleischmann and Pons (heat) and Brigham Young University (neutrons from fusion) began to open.

The only support for Fleischmann and Pons came from Robert Huggins, who had been Director of Stanford University's Department of Materials Science and Engineering for seventeen years. He had been working in 'solid state ionics' in which his group used electrochemical techniques to study solids which show unusual properties due to the rapid motion of atoms and ions within them. He had particularly been interested in metals which can rapidly incorporate hydrogen and 'thus also deuterium' into their crystal lattices, and had been studying the idea of solids as storage materials for hydrogen. He was already thinking ahead, realising that there is a large number of metals that have properties similar to hydrogen-doped palladium, and so he conjectured that if cold fusion in solids was a reality, there might be other materials less expensive than palladium which could be used. This was an important consideration, and was also noticed in official comments in Britain (*see* Chapter 13).

Huggins had taken on this project following the long session at the American Chemical Society meeting (12 April) where he realised that the validity of the claims would be more credible if hydrogen and deuterium with palladium showed different behaviours. He stressed that 'such experiments would subtract out any contributions from spurious effects because they would be present in both'.

He claimed that his results supported Fleischmann and Pons but did not give details pending a formal presentation of the data that night at a meeting in San Diego. However he did report that he saw different thermal effects with deuterium from those with hydrogen, but this did not of itself really mean much as deuterium weighs more than hydrogen and so will in general behave differently.

He also said that he had not made any radiation measurements. Though few realised it at the time, this remark was the beginning of a systematic trend. No single experiment saw *both* heat and neutrons or gamma-rays from fusion: Fleischmann and Pons claimed that they did, but everything other than the heat was soon discredited and withdrawn. There were experiments, such as those by Jones, that saw neutrons at a low level on the borderline of discrimination from background, or in random low level bursts, but no heat measurements were made. Other experiments looked for both and saw neither (e.g. Harwell), or at best saw heat unaccounted for at the ten to twenty per cent level and then not in all cells; and if they looked for neutrons at all, they saw none.

It was possible that there were two distinct phenomena: ten per cent heat imbalance (chemical or miscalculation but certainly not fusion) and low level neutrons (interesting for physics but irrelevant for practical energy production).

In this regard Huggins stated that in one of his experiments the internally generated power was 1.14 times that supplied, a number typical of that later measured at Oak Ridge and consistent within errors of *one* of Harwell's cells. However, there were many many cells that showed no effect at all, which tended to be ignored by cold fusion aficionados but which should really be included in the census when attempting to gauge the overall significance of the 'effect'. As one scientist said, 'It's like tossing a coin that you predict will always land head up; you then ignore the occasions when it comes down tails and claim that your prediction works.'

Huggins noted that if the reason why there were no neutrons was because the fusion gave helium-4 and heat (though no helium-4 production was ever seen), then it would be very attractive because it implied useful heat without radiation. With this upbeat swing, based on hope rather than evidence, he followed a similar theme to Ira Magaziner, though whereas Magaziner's pitch had been for a Cold Fusion Institute in Utah, Huggins' was more wide ranging, and was geared towards his own mainstream interests as he reminded the committee how severely the Federal budget had been reduced for energy research and that 'maybe this is a good time to give this new consideration'. Then he continued more controversially: 'I am sure that you will also give some attention to . . . the distribution of effort and funding between the few large and very expensive efforts and the possibility of many somewhat smaller, yet perhaps more innovative, efforts. I need not also point out that essentially all of the major advances in the types of science that may have some relevance to our national technological welfare have been in what is sometimes called "small science" rather than in "big science".'

His examples of innovations from 'small science' were warm superconductors and the tunnelling electron microscope which, while legitimate as examples, had little to do with the issue at hand which was energy from fusion. 'Small science' here meant Fleischmann and Pons whom Huggins' data seemed to support: he dismissed experiments that disagreed by asserting that they were using 'wrong materials'—though precisely what were 'right materials' was not made clear. 'Large science' was the large scale hot fusion research. Fleischmann had been asked about this after a seminar at CERN in Geneva on 31 March, and to his credit replied, 'It would be a total

185

disaster to cut back on other fusion research. Ours is small scale, theirs is large scale generation of electricity.'

Huggins' work was reported on later in the day of the hearings, at a meeting of the Materials Scientists in San Diego, California. This was the nearest that it had yet been to peer review and here again one saw some of the dangers in presenting to Congress results that the scientific community had not yet evaluated. There were many questions that night whose answers revealed that Huggins had not performed standard calorimetry, in the sense that he didn't have an absolute measure of the heat in any one cell but only differences, so, for example, one cell could have been losing heat more than the other rather than excess heat being generated. Another point that came out was that both cells were charged to the same voltage but as one contained ordinary water and the other had heavy water they had different electrical resistances, and hence different currents flowed in each; but different currents with the same voltage means a different power balance, so this also would naturally give different heat characteristics. In addition it was important to know, for example, how much power was being removed by gas evolution; Huggins was asked about this and replied that he didn't know. Many other questions were asked about the different boiling points of heavy and light water and about the diffusion of deuterium and hydrogen into the palladium being at different rates, all of which would cause differences in the heat behaviour and had to be accounted for. His 'confirmation' of Fleischmann and Pons that night seemed rather less certain than in the less scientifically critical Congress earlier that day.

Huggins' testimony advertised the wider aspects of the hearings; the University of Utah people wanted money for a Cold Fusion Institute in Utah, but the Administration had a more general concern—what was the future for *hot* fusion?

The effect on the hot fusion programme also concerned George Miley, Professor of Engineering at the Univerity of Illinois. He listed his credentials as Director of the Fusion Studies at the University, editor of the journal *Fusion Technology* and he claimed 'I read the technology section of the *Wall St. Journal* every morning to get the latest news.' He realised that there would be an impact whether test-tube fusion was confirmed or not. If the results were positive, then present hot fusion programmes would face competition for what was already a very tight budget; if negative he feared that the whole fusion community would be accused of false optimism.

The three remaining testaments were from Princeton, MIT and Oak Ridge, big labs with an interest in fusion and in preserving the large budgets. The politics was in the background, but in the scientific arena

186

there was no doubt: they had sophisticated equipment and first rate expertise—and saw nothing.

The first representative of the big laboratories was Harold Furth, Director of the Princeton Plasma Physics Lab, who repeated the difference between the Fleischmann and Pons and the Brigham Young University experiments and developed it further. First, since the neutron radiation or tritium concentrations are typically one billion times smaller than required to explain the heat, they clearly could not be used as experimental proof of the test-tube fusion interpretation of heat. However, there was a catch: the possibility of *dd* fusion without neutron emission, though hitherto unknown, was 'still under debate', so the absence of neutrons didn't necessarily disprove test-tube fusion either.

But whatever was going on, there had to be *some* end product that would give the essential clue. If it was a *chemical* process, then the various elements remain the same but join together in new ways making *new compounds*. If it is a *nuclear* process then the neutrons and protons rearrange, seeding *new elements* from the deuterium. Deuterium has the simplest nucleus after hydrogen and there is only helium or tritium that can result from *dd* fusion. So if *dd* fusion were taking place, its products had to be there somewhere; $^3He$ and $^4He$ had to be looked for. If it was fusion that was liberating the heat, then there is a precise relation between the amounts of heat and the concentrations of the fusion products. *The proof of test-tube fusion would be when its products turned up in amounts corresponding to the excess heat.*

He advertised the helium measurement as being the most effective discriminator against natural background since when cold fusion produced helium, the amount of helium would accumulate in proportion to the reaction time and to the level of the 'excess heating power'. 'Since there is little hope for a theoretical model that accounts for deuterium-fuelled fusion power without any production of helium (or tritium), failure to observe the appropriate accumulation rates would constitute a clear-cut disproof of the test-tube fusion interpretion of "excess heat".'

Furth then offered some cogent comments on the role of ordinary water versus heavy water experiments as controls. If excess heat still occurs it could be due to hydrogen fusing with deuterium, as there are small concentrations of the latter in water (about 1 part in 6000). In such a case the decisive control would be to mix heavy water into the light water and see if the excess heat rose as the fraction of heavy water was increased. And if it really was due to fusion, then neutrons would be produced but with energy of 5.5 MeV each in the *HD* fusion, instead of the 2.5 MeV characteristic of *dd* fusion. This was what

Princeton had started to do—trying to reproduce both the heat measurements and the nuclear *dd* experiments and initiating a search for *HD* fusion too, but it was too soon to tell the outcome.

Michael Saltmarsh, Associate Director of Fusion Energy at Oak Ridge National Lab, also emphasised that the Brigham Young University results had nothing to do with the claims made by Fleischmann and Pons. 'Despite the apparent similarities between the two experiments, it should be noted that there are substantial differences between the two experimental set-ups, between the two sets of reported results and between the interpretations of the data.'

The possible importance of the results for world energy 'triggered an immediate and concentrated effort to duplicate the reported effects at . . . ORNL and all of the DOE's other major national laboratories . . . Thus the normal scientific process of duplication and experimentation aimed at understanding the new results has begun. Despite initial high hopes and the apparent simplicity of the experiments, it has generally proved very difficult to reproduce the reported results. It is clear that experimental details are important, but because the mechanisms which produce the reported results are not known, we are not sure which aspects of the procedures are crucial. Thus the task of duplication has been hampered by lack of detailed written technical information on the precise details of the original experiments.'

Efforts to reproduce the results had started at ORNL immediately following the initial press release. Saltmarsh reviewed this and made the telling comment that 'most of the other institutions with whom we are in contact have a similar status to report. At a meeting last Wednesday with representatives from all of the DOE's major national laboratories, they all reported similar efforts and similar results' and, what's more, at least for the neutron searches, with 'detectors with overall sensitivities 3 to 5 times higher [than used even by Jones].'

But the most damning testimony, the full background to which only emerged later (Chapters 9 and 15), came from Ronald Ballinger of MIT.

Ballinger specialised in the environmental effects on materials with applications to nuclear safety. His field covered corrosion in light water reactors, the effects of radiation on chemistry and stress corrosion cracking. He was an expert on hydrogen, metals, electrochemistry and nuclear phenomena—the full range in question. And he was only one in a multidisciplinary team involving electrochemists, metallurgists, fusion and radiation physicists based on MIT's Plasma Fusion Center, Nuclear Engineering and Materials Science and Chemistry departments.

Initially his statement mirrored that of Saltmarsh: MIT had started work soon after the announcement and, as had others, immediately realised that the Brigham Young University results had very little to do with the claims of Fleischmann and Pons. They were unable to verify any of the results; 'This is in spite of the fact that we are employing calorimetry and radiation detection methods of even greter sophistication and sensitivity than those at the University of Utah.'

But now his testimony went further and for the first time was voiced what were the growing frustrations of the scientific community.

The failure wasn't due to lack of effort, far from it; since the Easter weekend MIT too had been labouring around the clock, but they 'and the other teams have been handicapped by lack of enough scientific detail to guarantee that we are actually duplicating these experiments.'

'In the scientific community the soundness of experimental or theoretical research results is evaluated through peer review and duplication. For results such as those reported, whose potential impact on the scientific community and the world are so great, this review process is absolutely essential. Unfortunately, for reasons that are not clear to me, this has not happened in this case, at least so far. The level of detail concerning the experimental procedures, conditions and results necessary for verification of the Fleischmann and Pons results have not been forthcoming. At the same time, almost daily articles in the press, often in conflict with the facts, have raised the public expectations, possibly for naught, that our energy problem has been "solved". We have heard the phrase "too cheap to meter" applied to other forms of electric energy production before. And so the scientific community has been left to attempt to reproduce and verify a potentially major scientific breakthrough while getting its experimental details from *The Wall Street Journal* and other news publications.'

'Experiments conducted in haste and based on insufficient detail coupled with premature release of results have often resulted in retractions and embarrassment on the part of the scientific community caught in the heat of the moment. I guess we are all human.'

'The result of this unsatisfactory situation has been that a healthy scepticism and, in some cases, distrust of the reported results has developed. We at MIT share this scepticism.'

And his scepticism was well founded, because not only had MIT failed to verify Fleischmann and Pons's claims, but they had even found results that questioned the reliability, even the credibility of some of those claims. He stated: 'We have found that the results reported in the few available published documents from the University of Utah are inconclusive or unclear. For example with respect to the detection of neutrons, critical products of the fusion reaction, the reported results either do not agree with or are not presented completely enough to show that they are consistent with what one

189

would expect from the emission of neutrons in a deuterium fusion reaction. Specifically *the reported gamma ray spectrum produced by neutron emission does not exhibit a shape and intensity that demonstrates an increase in the number of detected neutrons above normal background.* The tritium levels are also consistent with natural background. Yet the results have nevertheless been reported as ''significant''. Until such time as (a full disclosure of the experimental details) occurs we feel that the data are insufficient to demonstrate the presence of neutrons.'

Even these remarks were muted relative to what happened later as the full story of MIT's exposure to Fleischmann and Pons's data emerged (*see* Chapter 15). His conclusion made the three points that epitomised most informed observers' views at the time: '(1) The scientific community is excited about the possibility of a significant advance in the area of fusion energy research, but (2) is sceptical of results that have not been verified to this point, and (3) is very frustrated at the methods by which the discovery has been handled both in the scientific and non-scientific community.'

The day's testimonies had started off very positively: test-tube fusion was the greatest discovery since fire and held potentially unlimited promise for the future. The University of Utah's back-up team had built on this euphoria even though 'the science isn't yet proven'. But as the day wore on the tide began to turn. Not only was the science 'not yet proven' but, it began to emerge, had not really reached a level of maturity to warrant some of the claims being made. And finally there was the news from the major labs, with their wide expertise and quality instrumentation more sensitive and sophisticated than those in Utah and who saw nothing, and even were exposing flaws in the Fleischmann and Pons data. Although the media would have a field day attacking test-tube fusion at the Baltimore meeting of the American Physical Society on 1–2 May, it was the later part of the Congressional hearings that was the end of the beginning for Fleischmann and Pons. It was the first serious public questioning of the claims for 'test-tube fusion'.

The day had been productive in suggesting ways forward. Furth had outlined the necessary proofs for test-tube fusion and advertised the importance of hydrogen as well as deuterium. Ballinger made the administrative suggestion that seed money be available to enable small scale efforts to get under way, as distinct from a large lump going into one pot in Utah: verification first. Saltmarsh suggested that the major labs collaborate with Utah and with Brigham Young University by 'bringing a range of different diagnostic equipment to bear on an already working experiment'.

Committee members and the media took up Saltmarsh's suggestion and asked for reactions.[1] 'We're doing just that,' Pons replied, adding that they had set out to conduct a collaborative experiment with Los Alamos, the DOE-funded lab in New Mexico. 'They will come up and make the experiments they want to make on our system, bring their electrochemists and let them go through our methods of measuring the thermal output. And when they are satisfied with what they see then they will take that experiment away.'

This kept the credibility alive through the spring until 14 June when Sid Hecker, Director of Los Alamos, announced that the laboratory had finally given up. They had tried to set up this simple matter of working with Fleischmann and Pons, but months of inactivity on the university's part caused the attempt to collapse. James Brophy of the university admitted 'I can see how Los Alamos will be a little peeved. The scientists want to tell everything but the patent attorneys tell us to say absolutely nothing.'

How did neutral observers perceive it the congressional hearings? The sharpest summary of impressions was perhaps that in the leader article of the *New York Times* which, following Ira Magaziner's advice that the Congress should gamble tax payer's money on the Utah experiments, gave its own suggestion. Given the present state of evidence, rather than put money behind test-tube fusion, Congress 'might be better advised to put it on a horse'.

# 11
# THE CALTECH STORY

The heat in Utah had been measured by chemists. Jones had seen no heat but had detected some neutrons—he is a nuclear physicist, not a chemist. The Indiana team (see Chapter 8) consisted of physicists and they were applying their expertise in the natural way: 'If this is fusion there must be neutrons. Jones sees some but not enough to fit with the heat that the chemists claimed. But are there neutrons even at the Jones level? We are nuclear physicists so we shall look for neutrons.'

In early April few details of the experiments were known and no one was sure if the two Utah groups were seeing the same phenomenon or not. To make definitive experiments, ideally one wanted a team of people with various talents—an electrochemist measuring heat alongside nuclear experts looking for neutrons and other tell tale signs of fusion such as tritium or helium production. At Caltech such a group set to work within hours of the news from Utah and had been working '24 hours a day for a month' when Doug Beck, a nuclear physicist colleague, spoke to me during the weekend get-together of 23–24 April. The group included Charlie Barnes, with many years experience doing experiments on nucleosynthesis—essentially the production of elements as in stars, Nathan Lewis, an electrochemist who would later be the star performer at the American Physical Society meeting on 1 May, and nuclear theorist Steven Koonin, who later became a member of the DOE committee charged with examining the data of all groups claiming to see cold fusion phenomena.

Caltech is sited in Pasadena, between the Los Angeles smog and the foothills of the San Gabriel mountains. It has several Nobel Laureates on its faculty, one of whom, the great Richard Feynman, had died only weeks prior to the news from Utah. He was arguably the greatest theoretical physicist of the latter part of the century and is best known to the public for his lectures and books and, above all, for his penetrating observations during the enquiry into the Challenger

Space Shuttle disaster, which illustrated his ability to get right to the central point of a complex problem. In the case of Challenger, he performed a simple experiment in front of the television cameras; by dropping a rubber ring into a glass of iced water he showed dramatically that the rubber went brittle and would not instantly respond to sudden temperature changes. With this he undermined NASA's claims that the rubber 'O-rings' were suitable for use in a space craft where sudden temperature changes would occur. Thus he identified the crucial component that had failed and released gas with explosive consequences. Had he lived a few more months, many people suggested, this present saga would have been close to his heart and he would have rapidly pinpointed the essential ingredients.

Among the green lawns and fountains surrounding the yellow sandstone walls and red tiled Mexican style roofs one would not immediately have thought that there was a laboratory of nuclear physics—"the Kellogg lab", with a research tradition spanning 60 years.

Every Friday afternoon, to end the week, there is a seminar with refeshments and wine to encourage attendance and 'stimulate discussion'. This is among the more pleasant settings for such occasions in my experience, the location being a cosy room whose walls are lined with bookshelves and which is adjacent to a larger library. The audience typically numbers up to a couple of dozen, or at least that is how it was when I gave a talk there in the fall of 1988. But on 21 April the topic was Caltech's investigations into test-tube fusion and the venue was changed; 1000 people came to the main auditorium to hear Barnes, Lewis and Koonin speak.

Lewis reported that he had tried, unsuccessfully, to get information from Pons on the details of the experimental set-up. The Caltech team had to resort to subterfuge, using photos of Pons and his cell as displayed in the media, and from these building as near a replica as possible.

Here for the first time was a hint of the phenomenon that would become more and more apparent during the following months: the pursuit of test-tube fusion was not going to be a normal scientific enquiry.

The Caltech story divides into two parts. The first is the period up to the seminar of 21 April, of which I learned during 23–24th from Doug Beck and which has been documented in detail by Douglas Smith in the *Caltech Engineering and Science Magazine*[1]. At this stage the Caltech scientists believed that test-tube fusion had been discovered in Utah and set out to replicate it for themselves as the first stage in deepening understanding and pushing forward into a new field of science. The second half of the story concerns the

193

subsequent ten days. By then they were becoming convinced that the claims for test-tube fusion had been oversold and were very probably wrong. The appearance of the Utah group in Washington, and the optimistic claims being made for use of public money, stimulated Lewis and Koonin into a dramatic appearance at Baltimore on 1 May where they made a stinging attack on Fleischmann and Pons.

## *24 March to 21 April*

On Friday morning, 24 March, hundreds of scientists at the several universities and research institutions in southern California learned of test-tube fusion from that morning's paper, and of these Reginald Penner and Michael Sailor, two young post-doctoral chemists at Caltech in the Los Angeles suburb of Pasadena, decided to try the experiment for themselves. The news seemed to imply that the two chemists in Utah had made the greatest discovery since fire in an experiment straight out of a freshman chemistry class and with materials that Penner and Sailor could most probably find easily in their own lab. Who knew where it might led? The urge to try and see test-tube fusion for themselves, the day after it had been discovered, was irresistible.

They found pieces of palladium wire in the lab rather easily: finding the heavy water proved harder. Eventually they went to the chemistry stockroom to see if they could obtain any there. They received good news and bad news: the good news was that they could buy some; the bad news was that a cupful cost over one hundred dollars. They hesitated for about ten seconds and then decided to gamble that their supervisor, associate professor of chemistry Nathan Lewis, would be able to charge it to his research contract and surely would approve the expenditure when he learned the news. So the pair bought the heavy water and began to build a fusion cell in a corner of the lab.

Unknown to them, Lewis had heard about fusion the night before on CNN and from talking with Chuck Martin, an electrochemist and colleague from Texas A & M University, who had called Lewis at home desperately wondering what was going on. Martin and Lewis agreed to share any information that they gained.

On Friday morning when Lewis set off for work, he had decided to wait until a scientific paper was published in order to get details before setting out to reproduce the phenomenon. However he found his post-docs hard at work in the lab and he asked them what was going on. They told him about the test-tube fusion idea and how they were going to try it for themselves. Lewis seemed sceptical but said 'OK. You've got a day', and with that remark set himself on course, in the immortal words of Andy Warhol, to become 'famous for fifteen minutes'.

Just as Fleischmann and Pons had not forseen the consuming reaction to their own research and to the press conference, so when Lewis said 'you've got a day' no one expected that six weeks later they would still be at it, part of a multidisciplinary effort involving some twenty people, and that Lewis would be on television and making the front page headlines of the *New York Times* telling of their work.

Penner and Sailor made several experiments that first day, including not only heavy water but also plain water within the cell.

The plain water was a 'control experiment'. Penner and Sailor were looking for neutrons coming from the fusion of deuterium nuclei in the cell's palladium which, when captured in the surrounding water bath (containing $H_2O$) would emit 2.22 MeV $\gamma$-rays (the same idea as in the Fleischmann–Pons experiments). If there were indeed *dd* fusion taking place then there should be such $\gamma$-rays seen in experiments using $D_2O$ in the cell, but not from experiments using light water. However, if both cells generated gammas they couldn't be coming from deuterium fusion.

That they immediately realised the importance of doing control experiments with plain water is significant. The essential need to do this was obvious to any competent scientist, the numerous times this question was asked of Fleischmann and Pons shows how obvious it was, and two young postdoctoral scientists realised it on the very day that they set up their first experiments. Yet Fleischmann and Pons, it later transpired, who were claiming to have been investigating cold fusion for up to five years, seemed to give no clear answer as to whether they had made such a control or what the outcome had been (*see* Chapter 16). Hence the forcefulness of comments that Steve Koonin, professor of theoretical physics at Caltech and a major player in the cold fusion investigations, made later (page 213).

But these doubts about the Fleischmann and Pons experiments only emerged later. On that first day the positive nature of the press conference led no one to suspect that test-tube fusion would prove difficult or impossible to verify or that there would be any reason to doubt. If there were heat coming from the cells as clearly as Fleischmann and Pons seemed to be claiming, and if, as they also claimed, it were due to fusion, then gamma rays from that fusion would be easy to detect—indeed so easy that Penner and Sailor set up their initial cells over Polaroid films in the expectation that the gamma rays would fog them.

However, when they looked a few hours later they found that the

Polaroids were unaffected by either the heavy or the plain water cells. Clearly the fusion was not going to be so easy to see after all; indeed, Fleischmann and Pons were reported to have been at it for years. Also, Fleischmann and Pons were still around to tell their story and had not seemed to take any precautions against the radiation, so the radiation levels must have been lower than one would initially have guessed from the accumulated knowledge of fusion. Penner and Sailor expressed relief as it had not immediately occurred to them to take any precautions against a cell that could have been producing a large blast of neutrons and gamma radiation.

Lewis had not been idle either. Having told his postdoctoral assistants to try the experiment he had been called by Steve Koonin who was on leave from Caltech at the Institute for Theoretical Physics in Santa Barbara. Koonin was already aware of the story having been called up by Ryoichi Seki, an associate who had heard it on a Japanese language broadcast cannibalised from the London news scoop. Koonin advised Lewis that he should take it seriously and suggested that he work with Charlie Barnes, professor of nuclear physics at Caltech's Kellogg Nuclear Laboratory who had an outstanding neutron detector, thereby making good radiation measurements to complement Lewis' electrochemistry. When Lewis called Barnes he learned that Barnes too had been aware of the news already for a day, courtesy of Seki, and, what's more, had been about to call up the chemistry department for advice on the electrochemistry! Thus when Lewis called him up, the large collaboration was born.

Ralph Kavanagh, physics professor, and research fellow Stephen Kellogg (no relation to the cereal magnate who funded the lab's construction) had built a neutron detector that is 100 000 times more sensitive than the University of Utah's detector. According to the news reports, the Utah group was seeing neutrons coming from their fusion cell, so Caltech's sophisticated detector promised to capture them with no trouble.

The Caltech detector in total weighed a tonne, and is a cube slightly smaller than the passenger space in a car. The operative piece—that detects neutrons—is small, a mere 30 cm diameter cylinder containing a dozen 'proportional counter' tubes filled with helium (technically, helium–3, the lighter isotope of the element). A ten centimetre square borehole in the centre is where you place the sample whose neutron output you want to monitor. Neutrons hit the helium and undergo the reaction

$$n + {}^3He = {}^3H + p$$

with the effect that, analogous to the case of the $BF_3$ counter, the neutral neutron has converted into electrically charged particles which ionise the surroundings. Electrons are thus ejected, and wires trail to a computer which records their signals and, by inference, neutrons.

If this was the whole story, detecting neutrons would be easy. The problem is that there are neutrons around the lab that have nothing to do with the sample sitting inside the detector and as you're not interested in these you have to shield the detector from them. We are continuously being bombarded by cosmic rays from outer space which smash into atoms in the atmosphere and, among other things, produce neutrons. These cosmic rays might even hit the detector casing and produce a neutron at the last moment. There is natural radioactivity all around us and this could also fool an unprotected detector. So to eliminate this unwanted 'noise', the small cylindrical detector is buried within a tonne of paraffin wax, polyethylene plastic and graphite blocks.

These make quite an effective shield against spurious neutrons and as such are the first line in protection. However, even this isn't perfect and so there is a second line of defence against a particularly penetrating component of the cosmic rays—electrically charged muons (the heavy siblings of electrons, *see* Chapter 4). To alert the detector to these 'false' signals the experimenters surrounded the central detector with inch thick sheets of special plastic which emit flashes of light when charged particles—such as muons—pass through them. News of these flashes travels instantly to the computer that monitors the detector. The computer's program then performs the following logic. 'A flash means that a cosmic ray is passing through. Check to see if I recorded a neutron at the same time. If I did then it's probably related to the cosmic ray and not from the sample inside the detector. Throw it away from the neutron collection.' Of course, you lose a few genuine neutron signals this way because sometimes a real neutron comes through the cell at the same instant as a cosmic shower passes through. However, you can correct for these because the computer keeps account of what percentage of the time it was 'switched off' or 'dead', so by knowing how many 'real' neutrons were collected in the rest of the time it can estimate how many were missed in the 'dead' time.

The computer program also runs checks in reverse. When a neutron is detected, the computer verifies that no flashes occurred at the same time ('same time' really means within about a millionth of a second) and only counts the neutron as 'real' if there is no coincident flash. In the jargon this is called 'anti-coincidence counting' and the choice of what is 'same time' is called 'gating'—the gate is either open or closed.

Pons and Fleischmann, we now know, had no such check against background noise, and neither had Jones, details of whose work had yet to reach Caltech. At this stage no one knew that Harwell in Britain had been doing experiments for some weeks with a neutron detector comparable to Caltech's and, as would turn out also to be the case at Caltech, saw no 'real' neutrons. As such Harwell and Caltech were, in their respective continents, making the first truly definitive quality neutron searches.

Having taken Koonin's advice and talked with Barnes in the nuclear lab, Lewis' group brought a cell over to Barnes immediately after lunch on the Friday. Barnes' group (consisting of Steve Kellogg, T.R. Wang and Bruce Vogelaar, a graduate student about to complete his PhD) and the Lewis trio were all present, naturally. But word was getting around and large numbers of spectators were gathering to witness what they thought would be an historic moment. The cell, manufactured by Lewis' team to be as near as they could guess to the Fleischmann and Pons specification, was gently put into the nuclear group's neutron detector. If the cell was indeed a fusion reactor, the detector would burst into life. Within moments they would know.

Nothing happened. The detector carried on as if nothing had changed; there were no neutrons.

Meanwhile in Santa Barbara Steve Koonin, the theorist, had been thinking about how test-tube fusion might work. He had rushed off to the library to see what was known about hydrogen in palladium and, like many from the nuclear world to whom palladium had been just another element until the events of 23 March threw it to their attention, was amazed by how much had been written about it, even going back 50 years to the papers of Paneth and Peters (see Chapter 2).

He had no idea of how easy or hard it would be for test-tube fusion to be real, and so with Mike Naunberg, a professor from the University of California in Santa Cruz who was also visiting Santa Barbara with Koonin, he made a simple calculation to see how hard you had to squeeze the deuterium nuclei together. They found that there were some errors in the old literature but even their new calculation showed that the correct rate was still incredibly slow. Koonin later described it this way: 'A mass of cold deuterium the size of the Sun would undergo one fusion per year.' They also discovered that proton–deuterium fusion takes place 100 million times faster than deuterium–deuterium fusion—a result that motivated some later experiments—but even that did not help.

Theory seemed to say that test-tube fusion was impossible unless something could enhance the natural expectation by 57 orders of magnitude; if Fleischmann and Pons were right then something utterly new was involved.

Lewis spoke to Koonin again that weekend and became enthused.

He spoke with Pons in Utah on the evening of 24 March and also sent a message to Pons via *Bitnet*—a worldwide computer network linking universities and research institutions.

By Monday afternoon information was coming in by phone, fax and *Bitnet*. A new genre in scientific communication was being born. During the next months news of seminars, discussions on how to build cells, any titbit of information was disseminated on the computer bulletin boards. You logged in at any time and read the latest gossip or hard news and sent in any insights you had gathered yourself. Koonin and Naunberg wrote a summary of their calculation on test-tube fusion on Thursday 30 March, less than a week after the news broke, and sent it out to colleagues on *Bitnet*. These in turn passed it on to their contacts—simply by inserting the command 'forward Koonin mail to . . .'. The 'paper'—really a set of electrical signals propagating through the air—flashed round the world like an electronic chain letter.

Electronic chain letters were already beginning that first weekend and the first important news that Lewis gleaned was that Fleischmann and Pons's electrolyte was an alkali containing lithium in some form. About a week after they had begun, the crucial details of the Fleischmann and Pons experiment arrived—once more by electronic means, this time by fax.

Faxes of Fleischmann and Pons' paper soon populated the globe. The 'Adam and Eve' were five copies of the original paper that Pons had given to people in Utah and on the cover of which was 'Confidential—do not duplicate'. One of these was copied to a colleague in Texas and a fax went to *The Wall Street Journal. The Wall Street Journal* became a source for several people; you could call up the science desk who would transmit a copy.

Lewis' copy arrived by a more direct route. Pons had faxed a copy to J. Bockris at Texas A & M, from whom Chuck Martin had received it. Martin called up Lewis and such was the excitement that he dictated the entire paper to Lewis over the phone. They discussed some of the important and obvious points, and also found lots of questions that neither of them could answer. Martin faxed the paper to Lewis about an hour later; it was a messy copy and Lewis got an original from Pons three days later in the mail. Lewis and Martin also agreed that they would share anything that they found as this was 'too big to go without independent double checking and verification'. They also agreed that if either saw anything they would swap samples and check each other.

From the paper Lewis learned that whereas Caltech were using thin palladium wire, Fleischmann and Pons had been using rods up to 4 mm thick, and that the fusion heat effect increased with rod size.

Lewis' team now hunted around for more supplies of palladium. They found some in the physics lab where someone dismantled an old gas, purification apparatus and pulled out some palladium thimbles. (Palladium is used in catalytic converters—it is like a sponge for gas which is in effect the property that had attracted Fleischmann and Pons's attention in the first place). The thimbles were filthy, covered with dust and grease, but when cleaned up turned out to be 99.9 per cent pure palladium.

The Caltech group was desperately trying to get hold of thicker palladium wire and Lewis was calling people all over the United States, finding out information and getting raw materials. It was becoming hard to find palladium—one experimenter told me 'you could almost "feel" the surge of cold fusion experiments going on as palladium became unavailable and the market price rose'. Lewis drew so many blanks that when at last he found some, and not far away—at the David H. Fell Company in Long Beach California—he panicked that it might slip away if he did not get there fast. Lewis pulled out the keys to his own sports car and tossed them to Sailor, for whom this was too good an opportunity to miss. He ran out to the car, stopping at the cash dispenser on the way and taking out 100 dollars.

Long Beach is some 40 miles from Pasadena across the Los Angeles freeway system. 'Rush hour' is a misnomer as 'rushing' is usually all but impossible, and even in the relatively slack periods of the day a single accident can cause severe snarl-ups. Luckily there were no hold-ups but, even so, by the time he had reached the company offices the price of palladium had gone up: a rod, not quite 50 cm long and the width of coat hanger wire, now cost 120 dollars, twenty dollars more than he had on him. Fortunately the company trusted him and told him to send a cheque for the balance. In fact, Fell was so interested in the Caltech work that he let them have palladium at cost price and later on even helped to arrange casting of samples for them.

The experiments with bigger rods needed more time. This is because the deuterium first has to enter and saturate the lattice of palladium atoms. There's been extensive research on the speed at which hydrogen diffuses into palladium and so, knowing how big the rod is, you can work out this 'charging time'—the time it takes to saturate the rod. For the small rods twenty minutes had been enough; for the palladium thimble they calculated that if fusion were to happen, at least ten hours of charging would be needed. But after three days of looking for neutrons and gamma rays, nothing had shown up.

That same weekend, 1 April, they discovered a further necessary ingredient in the Fleischmann and Pons set-up. A graduate student, Pat Santangelo, had been in the library reading Fleischmann's old papers—an essential part of the research process—which described how he had used palladium cells, similar to the new fusion cells, for separating hydrogen and deuterium. In the small print Santangelo discovered important news: Fleischmann baked the palladium in a vacuum for several hours in order to drive out any gases that were already absorbed in the rod. Then Fleischmann would start the cell up by running the electric current in reverse—the palladium being the *anode*—which attracts oxygen instead of hydrogen (oxygen to anode, hydrogen to cathode). This oxidises the surface, making it porous. Only then would Fleischmann switch the currount round to its normal mode—palladium as *cathode*—which drives off the oxide leving a porous palladium metal. The effect is that the surface area for absorption is increased, and Caltech re-discovered a 100-year-old startling fact: normally electrolysis of heavy water evolves gas at both electrodes, oxygen at the anode and deuterium at the cathode; however, after the above pretreatment *no* gas came off at the cathode—*all* the deuterium was going *into* the palladium cathode.

Caltech decided this must be the key—'overloading' with deuterium thereby increasing the chance of fusion.

So in the first week of April they built an experiment with the materials, electrolyte and currents identical to those used by Fleischmann and Pons. As they did not know whether or not there were special conditions necessary, they built many cells with small variations among them.

Rumours kept arriving all the time which had to be followed up to find the fire behind all the smoke. One was that cast palladium was important.

Casting involves pouring molten metal into a mould where it cools and solidifies. The latest rumours suggested that the palladium had to be cast, and to cast metal you first have to melt it. This isn't as easy as it sounds. Palladium melts at 1550 degrees which is higher than the melting point of glass, so glass containers are no use. Graphite containers are no use either: they stay solid but palladium dissolves carbon. To do the job successfully you need to use a special high temperature furnace. Lewis' team learned that the professor of geology, Ed Stolper, had one, so they called him up and got permission to use it—that very night.

Their first attempts were disastrous. A plug in the bottom of the mould melted and the palladium ran out and was lost, but eventually they succeeded and completed the casts at about three in the morning. Late to bed and early to rise—the scientists were in the lab six hours later. Thus the Caltech group had cast palladium rods nearly a week ahead of other groups.

Later they learned that Fleischmann and Pons had been using

regular extruded wire and had not used special cast palladium at all. In any event, Caltech still saw no excess heat.

In mid-April Martin's team at Texas A&M University reported seeing excess heat and sent Caltech some of their palladium as previously agreed—an example of scientific co-operation: 'I see something. Please help to prove me right.' Replication is confirmation—except that it transpired that Caltech saw nothing even with the Texas cells. (Later it transpired that Texas had been mistaken and had not seen heat either.)

The fact that they saw nothing, plus the persistent rumours of magic ingredients in the cell, made them think that they must be doing something wrong. The nuclear physicist Charlie Barnes commented that he really believed that the different cells the chemists were making, with their subtle variations between one and the next, would eventually produce one that 'worked'. Although his colleague T.R. Wang's knowledge of nuclear physics couldn't make sense of the Fleischmann and Pons fusion claim, nonetheless they really hoped that Fleischmann and Pons were correct at some level, and that the Caltech group would eventually be able to replicate the effect. As nuclear physicists, experts in studying nuclear reactions and their products, Barnes and Wang were worried that there must be something that they had overlooked, but could not identify what it was.

All across America various groups were trying and some were claiming to see things. This news tended to appear at press conferences (such as the announcements by Georgia Tech, Washington and Texas A&M accompanied by television pictures showing the experiment and garish flashing lights) or by gossip on electronic mail *Bitnet*. Every rumour of success was chased by Barnes and Lewis who would engage in marathon phone calls, talking through the experiments step by step to find out exactly what was being done. They hoped that this would enable Caltech to generate the effect for themselves, but the conversations had a different outcome: they led to an accumulation of wisdom on the subtle mistakes that can be made.

With this wisdom, Caltech became troubleshooters, being able to identify the source of error in some of the claims being made. This resulted in some of the optimistic claims of heat or radiation being reconsidered or even withdrawn. An example of this is the case of the Utah group who had claimed to detect helium—a possible product of fusion. Caltech showed that the levels they were seeing were essentially the same as you would expect to find in the air of a typical science lab: liquid helium is used as a coolant in labs and evaporates, and glass can absorb measurable amounts of helium from the air. Thus ended the helium detection claim.

Other groups claimed to see tritium being produced, which would be clear evidence for fusion. Caltech showed what was wrong here too. Tritium is radioactive and, as such, can make certain substances glow or 'scintillate' (airport runway lights and some EXIT signs glow due to a tritium source within them). To detect tritium you mix up a special 'cocktail' and any tritium will make it scintillate; the more tritium, the more scintillation. The problem, as Caltech graduate student Bruce Tufts discovered, was that the electrolyte in the cell also gave scintillations when mixed with the cocktail as a result of well known chemical processes that had nothing to do with tritium. 'Are you sure that there are absolutely *no* traces of the electrolyte contaminating the cocktail?' Caltech would ask, and thus a claim to have found tritium would be undermined.

With Caltech continuing to see nothing—no heat, no radiation, no fusion products—and finding flaws in more and more of the positive claims coming from elsewhere, the team's expectation gradually changed. They had set out convinced that they would see something. When they failed they had decided that they must be doing something wrong, but increasingly it began to seem that it was *others* who were wrong, not them. Slowly the suspicion dawned that there was little or *no* fusion.

It was at this stage that they gave their first internal report on their work, in the seminar at 4 p.m. on 21 April.

In the seminar Lewis made a criticism which he later repeated to the audience at Baltimore. He stressed that it is difficult to do calorimetry (heat measurements) and balance the books in an open system—one where the gases can escape. He was worried about closing the system because of the danger that the evolving hydrogen and oxygen gases might recombine explosively. He said that he would be interested to hear of the results of anyone who did calorimetry with a closed cell but warned 'Kids, do not do this at home'. In summary, he was already rather sceptical about the Utah chemists' measurements and about the possible heating from the electrical current and whether that had been taken into account, and the bottom line in his own experiments was that there was no evidence for excess heat, certainly nothing like the Utah chemists had claimed.

Barnes spoke next and gave a detailed discussion of the nuclear physics aspects of the experiments. He pointed out that Jones' signal was very feeble. In order to assess its significance Jones had subtracted it from what he believed to be background noise. At Caltech the group had measured the background and could tell that there was no significant difference between gross neutron rates with the cells or without them. If there was any real signal it was so near to the background level as to be indistinguishable.

203

Barnes also criticised Pons and Fleischmann's claims to have detected neutrons, in particular that they had measured the background in an unreliable way by taking the detector 50 m away from the cell and measuring the neutrons there. Neutrons can come from a whole variety of sources, not least cosmic rays hitting material in the lab. A lead block could produce a lot of neutrons in one region of the lab which would not be seen 50 m away; conversely some other effect in the local environment at the remote site might be generating neutrons which would not be seen where the cell was. The only reliable way of measuring the background at the cell was to remove the cell and measure neutrons, or better, to measure neutrons when the cell was not working and to compare with those seen when it was working. The chemists had not done this and so it was hard to assess the significance of their claim. Notice that the same criticisms are being made independently by nuclear physicists in Britain as in the US mid-west and on the west coast. Standard practices in nuclear measurements were being violated, but chemists are not nuclear physicists; here is another example where research often requires taking advice from other experts.

Koonin reviewed the theory and made his remark that in the normal situation, the normal fusion rate at room temperature was such that if you had a cold mass of deuterium equal to that of the Sun, then one pair of nuclei would fuse in a year. Could one believe that by putting them inside palladium metal where, in fact, they would be even further apart and thus even less likely to fuse than when in a molecule of heavy water, one could suddenly emerge with a commercial reactor?

He then commented upon some of the theoretical ideas that were already trying to come to terms with the experimental possibility.

Although fusion of deuterium in the presence of electrons is slow, it is well known that when the electron is replaced by the more massive muon, this can encourge the deuterium nuclei to encroach and fuse more easily. This, after all, is the science of muon catalysed fusion that Jones had been pursuing before he entered this present line of research. The enhanced fusion rate is a result of the muon being more massive than the electron, so Koonin asked the question: 'How much more massive would the electron have to be in order for it to boost the fusion rate to the sort of level that Jones or even the chemists need'? The answer was tantalising and seductive. It transpired that if you went to the equations and changed the value of the electron mass while leaving everything else unchanged, the resulting prediction for fusion shot up rapidly. You only had to increase the electron's mass fivefold for the natural fusion rate to be that claimed by Jones. Increase it by a further factor of two and the fusion could even attain the levels needed to explain the heat claimed by Fleischmann and Pons.

The catch is that you cannot simply go and change the mass of the electron in an equation and still believe that you are dealing with the real world. It is true that in studies of electrons in solid materials the electron acts as if it has a larger mass (and several people wrote papers pointing out that maybe the electron's 'effective mass' in solids was a key to the issue). However, Koonin pointed out that this was a shorthand for describing the electron's motion in the solid and had little or nothing to do with its role when helping two nuclei to fuse. This 'easy solution' only appealed to people who did not understand where the concept of 'effective mass' in solids came from.

He also discussed a further problem: how do you get rid of all the energy? If it did not appear as neutrons, nor as tritium, then where was it? Not only was the amplification of the fusion rate 55 orders of magnitude (that is 1 followed by 55 zeroes, not merely a factor of 55 as some media reports seemed to imply) which is beyond imagination, but one also had to believe that it was a totally unknown type of fusion overlooked in four decades of study. One might possibly have missed it, but it would be something additional to the normal; you could not turn the normal pathways off. There had to be neutrons.

There were no members of the press in the audience, but there were several non-scientists. Koonin told me, 'I have a vivid recollection of overhearing someone in the audience saying on the way out that Fleischmann and Pons were really just holding some secret back and we all were not as smart as we thought we were'!

It was news of this meeting that Doug Beck brought to me on the Sunday, 23 April, and it was then that I decided to start following this up seriously. Clearly something weird was going on. On the one hand there were claims for saving the world that were being publicised ever more vociferously in the media by the day, while on the other hand there were already the murmurings in the community of physicists at least, and more widely by some accounts, that there was nothing there. I decided to start talking with people at Oak Ridge, to understand how much was physics, how much chemistry and how much metallurgy. But first I decided to go to the American Physical Society meeting in Baltimore to see what had become of the informal remarks that I had been hearing.

A week is a long time in politics, said Harold Wilson, a former British Prime Minister; the same was even more true of test-tube fusion in the last week of April. The Utah lobbyists came to the Capitol in

Washington while on the west coast the Caltech team completed their investigations to the point where they were convinced that Fleischmann and Pons had made important errors.

## 21 April to 1 May

When Penner and Sailor and Lewis, then Barnes and Koonin and a growing band of Caltech scientists had begun their quest, their original hope had been to see fusion, confirm Fleischmann and Pons' claim and thereby help establish a real natural phenomenon in the ledger of knowledge. But now their strategy changed. Suspecting that there might be no large scale fusion at all, they now prepared to set careful upper limits on how much fusion there could be. In the last week of April Lewis and postdoctoral assistant Gordon Miskelly built an exact replica of one of Fleischmann and Pons's cell to see if any peculiarity in its design would produce anomalous heat measurements. They did so by using spies—the media, indeed the same media that Fleischmann and Pons had used to tell their own story.

The *Los Angeles Times* had photos of Pons holding one of his cells, and the picture was clear enough to show how the cell was made. There was a little nipple on one side where the glass blower had sealed it off which showed that there were two glass walls, probably evacuated like a thermos flask. Pons' finger set the scale and so Lewis' team used their own fingers as a measure to gauge the size.

Fleischmann and Pons had been filmed in their lab by the local TV station. Caltech obtained the videotapes and could see everything they needed to know about the cells, and even how Fleischmann and Pons were doing the measurements. They could see where Fleischmann and Pons had put the electrodes and where the temperature was being measured, and could even copy down the readings from the thermometers. In short, they could tell how the whole experiment was built up and what was going on.

One thing Caltech couldn't tell from this was how much of a vacuum there was in the Fleischmann and Pons 'thermos' cell. So to cover all possibilities Miskelly designed the Caltech cell with a small valve so that they could pump the space between the walls to any amount of vacuum.

Miskelly made a further refinement: he drilled several small holes in the cell's lid so that he could put thermistors in different locations to look for hot spots in the cell. Fleischmann and Pons' design relied on the bubbling gas to distribute the heat and to keep the temperature

206

uniform, but Miskelly believed this to be inadequate. This was an important test because the heat balance was very sensitive to accurate temperature measurement throughout the cell. A small difference between the real and measured temperatures of the cell could translate into a large change in the calculated heat output.

Miskelly was proved right. There were variations in temperature throughout the cell; the bubbling gas was not enough to even it out. When news of this became public it helped to form much of the media's perception that the work by Fleischmann and Pons was flawed. However on this particular point the Caltech group was probably wrong. Although this cell would need stirring it had little to do with the cells that the Utah chemists had used. According to Fleischmann and Pons, Caltech had made their measurements using smll electrodes in a reconstruction of a large cell which Fleischmann and Pons had built for work on 2 cm diameter rod electrodes that had not been used in their work at that time.

Like many other readers of the Fleischmann–Pons paper, the Caltech group had difficulty interpreting the meaning of some of the numbers quoted (*see* the appendix 'Excess Heat in Calorimetry' in Chapter 5). The 'excess heat' was quoted as a number without specifying 'relative to what'. The excess as a percentage of different measures of heat or power was also listed, though as we show on page 351, these numbers were rounded off to complete integers and prevented easy translation into consistent figues for absolute power flows.

Lewis described his problems thus: 'In my conversations with Chuck Martin and with numerous other scientists here and at other places it became clear that nobody could reconstruct exactly the raw data that Fleischmann and Pons had observed.[2] What were the excesses defined relative to?'

'Miskelly and I tried various ideas and eventually concluded that there were two options in which most (but not all) of the table could be calculated consistently—the excess powers were either defined relative to the total input power, or relative to the heating power that Fleischmann and Pons assumed in the absence of recombination. The latter was very dicey, the former was more conservative. We did not understand the third column (using the 0.5 volt value) at all, but these were the highest numbers of excess powers which were being quoted to the news media in support of the excess heat in the process.'

Eventually Lewis realised that Fleischmann and Pons had 'depolarised the anode' so that it converts hydrogen into water instead of evolving oxygen. In this scheme no price need be paid to liberate fuel and so for a given power production the excess would be greater relative to the input power. The Caltech group then got together one evening and discussed how they could replicate this themselves. However they could not do it and so they called up Chuck Martin and asked him to try. Martin also failed to reduce the voltage that low.

But by now Caltech had begun to realise that there was a flaw; that it was not possible to run a cell at such a low voltage. Fleischmann and Pons claimed that charging the palladium rod with $D_2$ involved a potential of 0.8 volts and if they had indeed done this, then it would be impossible to maintain this state with only 0.5 volts. Lewis thus began to suspect that Fleischmann and Pons had 'not actually measured these large excess powers but had merely calculated them.

'But these were the 4 to 1 and 10 to 1 numbers that were always being quoted in support of the large excess powers! We asked Martin to ask Pons about this, to which Pons replied to the effect that he "measured everything".' Lewis was most confused.

Two days later Fleischmann and Pons were at the US Capitol in Washington DC testifying before the House Committee on Science, Space and Technology. The television channel C-SPAN showed the hearings and Lewis taped it. He was utterly distraught with what he saw.

In the congressional testimony Fleischmann pointed out for the first time in passing that, as Lewis suspected, column C was 'projected, as is commonly done in fusion research'. Lewis recalls that 'This was mentioned in passing and it went right by most people but I caught it, replayed the video five times to make sure, and knew that we were on the right track. Why had they not said so earlier? Why were they not saying that they had actually measured ten to twenty per cent excesses and had projected the others from some hypothetical calculations? This is when I decided that we had to go to the APS meeting the next week.'

By this point the Caltech team knew for certain that the popular impression that test-tube fusion was the holy grail for the world's energy problems was in fact utterly misplaced. They knew that some of the claims to have confirmed test-tube fusion were flawed, and that some of the original claims of Fleischmann and Pons were not borne out by their data. It began to look to them as if the whole test-tube fusion story might be little more than a house of cards. Yet meanwhile the Congressional committee was being advised by Chase Petersen, president of the University of Utah, to donate 25 million dollars to set up a cold-fusion centre in Salt Lake City. The promise of cheap unlimited energy was so seductive that Lewis was concerned lest the negative comments of other scientists seemed to be losing out. If the Caltech results were correct, the taxpayers would be throwing their money down the drain, having been led to believe in something for which there was no foundation, and which was based on 'evidence' which was incorrect. In particular, Fleischmann and Pons showed what they claimed was evidence for fusion radiation—gamma rays coming from neutrons—which a group at MIT were in the process of showing was completely false.

## Caltech decides to go public

Up to the fateful 26 April the Caltech scientists had intended to follow the time-honoured route and write a paper which would be peer-reviewed and published in the scientific literature.

Lewis told me: 'As a group we had decided not to talk to the press and keep a low profile. We were only going to publish our results when they were ready. Numerous phone calls from *The LA Times* and other major media outlets went unreturned. Barnes, I and both of our groups had agreed that we did not want to circumvent peer review at *any* cost.' The press conference style of the University of Utah had raised expectations of the whole world and, while scientists were still carefully checking the claims, the Utah administrators were keeping up the momentum, outflanking opposition by selling to Congress what the private world of science was beginning to undermine. Many felt that not just *scientific* ethics were on the line.

Lewis was upset that Fleischmann and Pons had 'distorted many of their findings and had not been totally scientifically critical or forthcoming with their raw data to their peers', and felt that someone should present this to the scientific community. He knew that almost no one else had got to the bottom of the excess-heat calculations, nor had information on the likelihood of the helium and tritium being artefacts, and that the Caltech team might be in a unique position with crucial information that the scientific community and the public should be aware of.

James Brophy of Utah had attempted to justify Fleischmann and Pons going public as follows. He argued that the work would have a major impact on society and 'sometimes you have more responsibility to the public than to the scientific community'. But from 26 April when he spoke to Congress, there began to emerge the possibility that Utah's desires could have a significant call on the taxpayer. It was at this point that Caltech felt that *they* now had 'more responsibility to the public' and that to maintain silence would be irresponsible.

The American Physical Society, APS, was due to meet in Baltimore five days later on 1 May, and this was where Caltech decided to go public and expose what was going on.

Lewis spoke to both the Provost and the President of Caltech about going to the meeting. He decided to go and present their work publicly there only because it would be a scientific meeting where he could get feedback. Such a presentation does not circumvent peer review, indeed it is potentially an occasion for the most severe kind of peer review, for if someone points out your error, it is not in a private letter but in the open in front of hundreds of your colleagues. Following any feedback they would (and did) submit their results for publication. A presentation to scientific peers followed by submission to a journal is a common procedure in the scientific community; such a scientific technical presentation is essentially different from a press conference.

To go public meant that they would have to be very sure of their facts. To tighten their case the Caltech team were up late every night in the hectic few days remaining before 1 May, the date of the APS convention. They had reams of data, all negative to test-tube fusion, which they double-checked.

The special session on test-tube fusion had been hastily arranged at the last moment and Lewis had managed to get onto the programme thanks to Fleischmann and Pons not taking up their invitation. Steve Koonin, who was already scheduled to speak, relayed the importance of the contribution to Jo Redish, chair of the organising committee. Thus Lewis was listed to speak at 8.45 p.m. Eastern Standard Time (three hours ahead of California). He had flown over from the West Coast to Baltimore while the experiment was still being wound up in California. He took blank sheets of plastic for use on the overhead projector for his talk, with the intention of filling them in with information only at the last moment when his group faxed him the data. He would then copy the data onto his transparencies while still at his hotel.

The team in Pasadena completed it all on the morning of 1 May, that is about 1 p.m. Eastern Time. Their collaboration had been conceived through electronic communications, news by fax, details by *Bitnet*, cross-examinations of other groups by telephone and spying by TV video tapes. There were less than six hours before Lewis was due to speak 3000 miles away when they fed the data into the fax, where it turned into electronic impulses which were instantly decoded at Lewis' hotel on the opposite side of the continent.

Thus Lewis prepared his damning testimony and set out to face the 2000 delegates in the convention hall. By 1.30 p.m. out west their immediate job was completed. By nightfall scientists would be typing summaries of the meeting into their computers for transmission to colleagues back home and CBS, ABC, and all the electronic media would be beaming pictures of Lewis around the globe.

# 12
## FROM SPRING TO FALL

Throughout April groups all over America had been trying to produce the test-tube fusion phenomenon for themselves. There were rumours passing around the electronic mail networks but no clear information on what they were finding, so the imminent approach of the American Physical Society (APS) meeting, due to be held in Baltimore during 1–4 May, began to take on a special importance. Here for the first time the physicists would get together *en masse* and discuss progress.

Most of us did not yet know how the results were turning out; during my discussions with several physicists over the weekend of 23 April (page 157) I had been struck by a feeling that nothing was being seen but it was still too soon to be sure, and by a fear that something was being done wrong—that 'the Utah people are very confident so there must be something that we are overlooking.'

The extensive work at Caltech, and their detailed contacts with other groups, had put them into an advanced position in the race to confirm or refute the phenomenon. Lewis had identified several potential errors in the calorimetry, they were seeing neither neutrons nor gamma-rays, and they were aware of work at MIT that showed the chemists' gamma-ray data to be suspect. Serious doubts were beginning to emerge and Caltech were confident.

Steve Koonin recalls: 'The whole business was being acted out very publicly with Fleischmann and Pons making unsubstantiated claims, they weren't answering questions and we were starting to disbelieve the whole business. So we decided to hit hard at Baltimore.'

So many abstracts of papers were submitted that the APS organisation arranged a special session, which eventually went beyond midnight and continued the following day.

The delegates converged on Baltimore like crowds arriving for the Superbowl and with a similar height of nervous excitement. There was a cloudburst an hour before the special session began but people preferred a soaking rather than risk losing a ringside seat. I arrived at the Baltimore Convention Center with 45 minutes to spare. Lewis was being interviewed by a television crew and two kibbitzing reporters. Moshe Gai, who was making a more sensitive measurement of neutrons than had Jones and was carefully monitoring against neutrons from cosmic rays, told me of interest in high places. This was the first hint that I had that test-tube fusion research was being followed with interest in the White House. My excitement and anticipation mounted.[1]

There was no trouble finding which hall the action was due to take place in as the noise from the developing crowd identified it at two floors and fifty metres range. Well over a thousand people were already there; media took care of the right hand side of the hall and the front five rows were roped off for speakers and officials of the APS. This was unlike any physics conference that I had experienced in twenty years; it was more reminiscent of a political convention. Until five minutes before the start none of the main speakers had appeared as they were entrapped in the wings of the hall by the media. Edward (Jo) Redish, a nuclear physicist from the University of Maryland who was the chair for the evening, attempted to call the meeting to order and the hubbub gradually stilled except for one radio interviewer with boom microphone who continued, isolated like one who continued with 'Hallelujah' during the pause in Handel's famous chorus.

Redish began with a brief address for the benefit of the media which explained how events had led up to this night, what the scientists were trying to achieve and how they would eventually come to a consensus.

'You are serving us as interpreters to the general public. It is important to understand how certain aspects of the current debate illustrate the normal processes of science. We're searching here for answers to the truth of how the world works. We do not accept a new idea easily because it's so easy to get it wrong. We challenge every new idea very intensely to make sure that it's real before we incorporate it into our scientific map of the world. We have to do this. Given human nature it is easy for us to replace truth by wishful thinking. It's the process of challenge, it's our demand for repeatability and our insistence on a detailed consistency that makes science work and gives it its power. Usually preliminary results are challenged internally by the working group, then by peer review and then often again in the open scientific literature by other groups before science goes public. What you are seeing here is a part of the way we work. I see it as a plus that the general community is seeing how science functions, but it's important that they not misinterpret what is going on. There is always a range of results in the scientific community. Some are well understood and firmly believed; others are in process and taken as working hypotheses under challenge. The most intense

212

challenge occurs when a new result appears at first to contradict our expectations which are built up from what we know. Tonight we will discuss such a case. Whether the deuterium–metal interaction can lead to fusion or not, I do not yet know. But I can promise you that the standard workings through of science will find the answer before long.'

Neither Fleischmann nor Pons was present; the star of the evening was expected to be Steve Jones, who gave a confident and modest account of the group's work, including sight of the notarised notebook, but seemed somewhat reluctant to get to the real point, and spent a lot of time showing entertaining cartoons and telling anecdotes rather than concentrating on a detailed and thorough presentation of the data. He stressed that his group found only small excesses of neutrons that appeared to be produced by fusion, but at levels so low that no excess heat could be measured. The heat that would correspond to his level of neutrons, compared to that claimed by Fleischmann and Pons, was like 'a dollar bill compared to the national debt'. Small though this was, even this trifling amount was criticised by Douglas Morrison, a physicist from CERN in Geneva. Morrison had come to the meeting to review the situation in Europe, and after Jones' talk reminded him that his analysis of his data had been criticised in CERN, where it had been said that Jones had overestimated its significance by failing to take proper account of fluctuations in the neutrons arriving in the cosmic ray background.

In the minutes before the meeting began, the rumour went around the hall that Lewis had the data that would destroy the chemists. Before him though was Koonin, who reviewed some of the theory behind fusion as well as the overall situation, namely that test-tube fusion should be impossible by many orders of magnitude according to theory, and that experimentally no one was convincingly able to reproduce the heat and radiation. He concluded with the stinging accusation that we were suffering from the 'delusion' and 'incompetence of Messrs Fleischmann and Pons'. There was significant applause and also a boo or two. Even in this excited atmosphere there were gasps of astonishment; delegates were, after all, at the Spring meeting of the APS and this was not the sort of thing that one normally heard on such occasions. He added, to justify this remark, 'I know what Nathan Lewis is going to tell you.'

Then followed Lewis who introduced himself, to laughter, as 'an electrochemist—but also a concerned scientist'. He described in detail the basic ideas of the process, how his group had tried but failed to get any good details or answers to questions from Pons, and how they had eventually been able to reconstruct the apparatus by means of photographs of Pons in his lab with a cell. He had a lot to say and said much of it very rapidly. Few delegates absorbed all the details there and then, but the impression came over that a serious, careful and exhaustive programme had been carried through. Most impressive was Lewis' documentation of how his group had been able to replicate Fleischmann and Pons' 'wrong' calorimetric data by purposely making bad experiments, in particular by not stirring the liquid. There was no doubt that he was telling the physicists what they wanted to hear, and the applause at the end of his speech persisted for over a minute before intervention from the chair terminated it.

Although the stirring criticism received much publicity, it was technically not the most damning item and was probably flawed, as the photo of Pons with the cell that Caltech had used as a guide was of a large cell that had not been used in experiments; such a cell would need agitation whereas smaller ones, as used in the experiments in Utah, were adequately 'stirred' by the agitation of the bubbles of evolving gas.

Much of the audience, who had come with sceptical but open minds, decided that Lewis had closed the subject for good and the hall emptied. Speakers came and went for another three hours with a common message: 'We looked at this, we looked at that and we saw nothing.'

Back at my hotel the meeting was already making headlines on CNN, the continuous news television channel. For weeks the story had been about the amazing new world that beckoned, courtesy of test-tube fusion; of the messianic appearances in talks to chemistry conventions and universities and of the appearance of the Utah team in Congress. In one evening, and primarily due to Lewis' talk and Koonin's outspoken criticism of the two chemists, suddenly the emphasis had swung the other way. Now it was the sceptics who were in the leading role.

Test-tube fusion had reached its zenith and, that night at least, it looked to be all over bar the shouting. I had to leave and return to Oak Ridge National Lab in Tennessee where I was working. Lewis and Koonin may have killed test-tube fusion as far as the media was concerned, but Oak Ridge and the other DOE labs still had some way to go before they could make a complete and final assessment for the Department of Energy.


## Post-Baltimore

On 3 May at the Oak Ridge National Lab I had my first sight of Don Hutchinson's heavy water fusion cell bubbling away. The beakers were no larger than small picnic flasks, and wires and tubes emerged from lids on top as if from a patient in intensive care. The bubbling liquid looked as if it were boiling water being prepared for a cup of tea; it was hard to believe that these homely pots could really be a fusion reactor.

They were using state of the art equipment, far superior to that used at Utah, and it showed that there were no neutrons coming from the cell. There were four other experiments then in progress at Oak Ridge, each working round the clock to find the answer to this most urgent of questions. The Department of Energy, the Administration, even President Bush wanted to know, without delay, if the claims to have harnessed the Sun's power on a bench top were true or not. Billions of dollars could depend on the answer.

News of Nathan Lewis' 'stirring' performance at the Baltimore APS meeting was by now common knowledge. If you hadn't already seen it splashed across the *New York Times*, 'Fusion Claim is Greeted with Scorn by Physicists', or if you missed it on the Public Broadcasting Service Radio News, you only had to flick onto the CNN and there would be Lewis, or Steve Koonin the other instant star, with their unequivocal criticisms.

Even so, not everyone was quite so sure as Lewis, nor as swayed as the 1800 attendees had been, carried along on the wave of excitement in the crowded Baltimore Convention Hall. Cooler heads wondered to what extent Koonin and Lewis were using the same tactics that Fleischmann and Pons were being accused of. And 500 miles away from the heat, reflection made one cautious.

I had taped parts of Lewis' talk and relayed its contents to Hutchinson's team. They were immediately full of questions—'How can he be sure of that? Where did he get those numbers from? How can we be sure until we have seen his written report?' Was it all over? Might there yet be some hidden loopholes?'

In science you have to find things out ultimately by direct experience, making use of the work of others; it is not sufficient to accept facts on mere hearsay. Lewis, as any other reputable scientist, would expect to be called up and questioned about his talk, discuss its details, defend his claims and help to expand the collective wisdom, both of the questioner and of himself as the two exchange information. Eventually you reach a question which no answer satisfies. *Then* you have elucidated the boundary between certainty and confusion and identified where to direct your research effort most effectively.

And that is what was indeed happening. In the days following the APS talk, Lewis was on the phone twelve to fourteen hours each day answering questions, explaining in detail how they knew the various things that he had announced, discussing details of the heat calculations and faxing copies of his transparencies so that people could mull over them at leisure and fire back further questions later. Hutchinson was one of many such callers and Lewis gave him the details that my inadequate summary had missed.

The interaction by phone, electronic mail, and fax between Lewis and other APS speakers was completely open during the early weeks of May. This contrasted with the paucity of detail coming from the experiment that had started the furore. The difference, and it was this that was rankling those who were name-calling, was that it seemed to be impossible to get consistent, clear-cut, useful answers to all but a few questions from the Utah chemists and that, a month after the first media event, no useful detailed information on what were the essential features of the cell could be gleaned.

Apart from the irritation, there were more serious financial concerns. Taxpayers' money was now being used in the national labs, on official demand, and at several universities prompted by natural scientific curiosity, to run experiments built around gossip and rumour; the central characters were perceived to be away garnering further money and support from Congress, or meeting the President's staff in the White House instead of appearing at the APS meeting (less than 50 miles away from Washington) to explain. 'Science is built on replicability', Peter Bond had observed, 'So why don't they help us replicate their discovery?'

What people were erroneously assuming was that the chemists had spent five whole years of concentrated work on the fusion enterprise and were holding back information for machiavellian reasons, led on by an urge for patents. However, no one realised that the research had been spread rather more thinly, much of it having been in a rush during the final weeks, and the reason it was so hard to get information from them was because they did not yet have all the answers as to what conditions were the best, whether it was necessary to have cast palladium or whether one had to make special preparations—these were uncertain even to Fleischmann and Pons. And since March they had been unable to reproduce some of their earlier phenomena and were temporarily, at least, in a similar state of confusion to many others.

It was in the aftermath of this turbulent atmosphere that I was trying to understand more precisely what Lewis had said and to explain it to the experimentalists at Oak Ridge.

I am a theoretical physicist, most at home with nuclear particles. Here, in the questions from Hutchinson's team, I saw at first hand what was plaguing the whole business. The jargon of electrochemistry—'base, cathode, IR heating'—was deep in my memory from school chemistry, many years past. In trying to drag it up from the depths I would lose the thread of their agitated staccato argument.

In turn, comments from me on nuclear or particle physics required many repetitions for minds whose expertise lay in materials, lasers and electrochemistry. The areas of maths that one person was at home with were unfamiliar to another. Science is a vast detailed range of subjects in which one may spend a lifetime and become expert in but a small part of one discipline. This research into test-tube fusion involved many disciplines, each with its own jargon and values. Images in the mind of one person did not necessarily, through words, evoke the same images in the recipient.

In the course of time many people will look back on the spring and summer of 1989 and realise how much they learned of areas of science that had been foreign to them, and think of the new friends that they made. Faces that had previously been anonymous when seen in the cafeteria now became real people: Micha who is an electrochemist, Ken who is a metallurgist. But during those early weeks, when the rush to understand was most urgent, it was frustrating to be confronted repeatedly by one's lack of knowledge in many different areas of physics, chemistry and engineering.

And there was real urgency. At Wednesday noon each week from April until who knew when, all the people involved in the test-tube fusion experiments, together with experts in the various areas of science and engineering that these questions overlapped with, would meet with Mike Saltmarsh, head of the lab's 'regular' fusion programme, to share news of progress, any tricks they had learned and details they had discovered, and to answer questions that had come up or that arose during the lunchtime discussion session. Following the meeting each group had to send in a report, deadline Thursday noon, from which Saltmarsh could then prepare and submit his weekly report on progress to the Secretary for Energy in Washington.

This was all a result of the command from Washington (page 144) that an answer had to be found urgently to the test-tube fusion question and that efforts had to be stepped up at the national labs to this end. The meeting on 3 May at Oak Ridge alone involved 45 PhD scientists and engineers and was being acted out in similar fashion in the other laboratories around the nation. Hundreds of people's precious time was being taken up, and research programmes in lasers, nuclear physics, chemistry and many other areas were being interrupted in order to respond to the command.

There were two main items on the agenda. One was to plan the reports to be presented at a 'Cold Fusion Workshop', a meeting of up to 2000 participants arranged for 23–25 May in Santa Fe New Mexico, and the other was to hear reports on the APS meeting, to put scientific dressing on the skeleton stories that had appeared in the media. The tone of the media reports had worried Saltmarsh, who was becoming concerned that many institutions could look bad after all the dust had settled, 'whichever way it turns out', and urged that we be careful not to get dragged into the nastiness.

The report on the APS meeting was the main item. The news reports had been unequivocal. The *Wall Street Journal* and *New York Times* had initially been relatively muted: 'Groups of Physicists releasing Reams of Data Dispute Claims of Cold Fusion' and 'Physicists Challenge Cold Fusion Claims' had been the headlines on the inner pages on the day following. But these had been prepared from the press releases and preliminary brief announcements and the abstracts of the talks. The *Wall Street Journal* caught the temperature of the acrimony early and by the morning of 3 May the *New York Times* ran the developing story as a front page lead.

Making the front page of the *New York Times* is one thing; where on the page you make it is another. In the USA it is common to see papers on sale in special stands, glass fronted boxes into which you put your 50 cents or whatever the price is, and then you can pull down the front glass panel and pick out a paper. The glass panel is the size of a folded newspaper, so through the panel you can see displayed only the top half of the front page. The upper half of the front page is in effect an advertisement for the paper. The exciting stories are there, and you can begin to read them without buying a copy, but then they continue on page 15 or go onto the lower half of the front page which is out of sight, folded back. A friend once explained the importance of this to me when a story of interest to her 'had made the front page of the *NYT*!' That was the good news. However, it was 'below the fold'. The reports from Baltimore were very much 'above the fold'; in fact they were the lead item at the top left corner.

It transpired that the morning that I had left to return to Oak Ridge National Laboratory a group of key speakers from the session of the night before had been assembled at Baltimore to discuss the Fleischmann and Pons' results. Eight of the nine members had dismissed them; only one—who was associated with the BYU experiments—was prepared to hedge his bets. 'Fusion Claim is Greeted with Scorn by Physicists' the *New York Times* thundered, and it carried on to report that 'top physicists directed angry attacks at Dr Pons and Dr Fleischmann, calling them incompetent, reciting sarcastic verses about their claims and complaining that they had refused to provide the details needed for follow-up experiments'.

The demolition job by Nathan Lewis had given everyone the key that they had been awaiting—the information on how the experiment seemed to have blundered. Accepting his criticisms together with the MIT group's criticism of the neutron capture gamma-ray peak and the general set of null results on radiation from many high quality groups, they had launched into the attack. His comments about failure to stir the liquid, with the result that the thermometer had probed hot spots and not recorded a true even temperature, was parodied by rhyme from Walter Myerhof, one of the panellists;

> *tens of millions of dollars at stake, dear brother*
> *because some scientists put a thermometer*
> *at one place and not another*

Such was the general flavour of the occasion. It was all highly coloured and many were disturbed at what they perceived to be the witch hunt atmosphere pervading a gathering of the intelligentsia. I put this to one of the leading participants some months later who commented 'hardly a witch hunt. More of a gentlemanly roasting wouldn't you say?' It was a release of pent-up frustration and the speed with which people jumped in, adding their voices to the throng, was akin to the swell of feeling in Romania once Ceaușescu was overthrown; doubts spoken privately one day could now be shouted in public as everyone turned out to believe the same thing.

Yet there was still the niggling possibility that something somewhere could have been overlooked. Was there some fluky condition in the palladium rod, for example, induced in the industrial preparation that might induce unusual phenomena, mimicking fusion if only transiently, that did not show up in other experiments whose rods were prepared differently? This was why it was so important to be able to bring good detectors to bear on the Utah cells which were claimed to produce heat. If 'the eye of newt is in those cells', as Peter Bond had said to me, then all that we were doing was proving that fusion did not occur in cells made with 'our' types of palladium. If there were a magic ingredient Fleischmann and Pons were not admitting to it.

An example of a 'special ingredient' is in the rumour of the cast rods.

In the middle of April some experiments at General Motors appeared to show excess heat with recast palladium, though normal rods gave no excess heat. (Co-operation with a team from the University of British Columbia, who acted as troubleshooters and came to GM to check, later showed that this result was an artefact due to overloading a potentiostat.) This news about cast rods was communicated to Pons and as the media were reporting that major universities and national laboratories were, in most cases, not getting positive results, Pons suggested that maybe casting of the palladium was essential. In his weekly news conference (the main regular source of news) on 24 April he said that cast rods rather than extruded ones were more likely to work. So far, he said, about three out of every five extruded palladium rods had failed to give heat whereas 'only one out of fifteen or sixteen cast rods has been "faulty".' He repeated this at Los Alamos on 28 April and added that the vaporised block—the phenomenon that Fleischmann and Pons claimed had made such an impact—'was cast by the way'. He added that when they were writing their paper they had not known that cast versus extruded was a crucial matter.

Materials experts at several institutions were surprised by this claim, and in particular Ken Farrel at Oak Ridge and the Caltech team tried to follow it up.

The feeling was that normally metal suppliers would not cast such small diameter palladium rods unless it was by special order; and that if it was so important that special orders had been made during the 'five years of experiments', then this would have been specified in the Fleischmann–Pons manuscript.

Calls to Johnson Matthey in England and in the US revealed that no cast rods had been made for anyone in the past several months. Further suspicion of the casting rumour could be gained from the fact that early claims of positive results from Texas A & M, among others, had been endorsed by Pons with no question about casting. (Indeed Appelby's group at Texas specifically stated that annealed, not cast, rods were used in his excess heat results.) Later it turned out that Pons' rods were not cast but were normal annealed rods, essentially the same as those being supplied to everyone else.

With the media so excitable about the test-tube fusion story at that time and with the Santa Fe meeting soon to take place, Saltmarsh rightly stressed that all should be cautious about describing results of the ongoing experimenters until they were ready to publish and defend their results. Many experimenters from around North America had reported at the APS meeting that they were seeing nothing—no excess heat, no neutrons—and were confident that they could quantify the 'nothing' for the neutrons as being far less that the magnitudes that Fleischmann and Pons had been claiming and, in some cases, maybe less than the amounts that the Brigham Young university team were claiming to have seen.

'We are under great pressure from the Congress and Administration in Washington to bring this matter to a conclusion, one way or the other' Saltmarsh reminded us. He also asked whether the experiments in the laboratory were now sufficiently advanced that we could state in that week's report that there is no fusion happening; that test tube fusion offered nothing as a world energy source.

The discussion was fairly brief. All that could be said scientifically was to report that nothing was being seen under the conditions studied so far[2] in the cells that had been made, even though the theoretical prejudice was that test-tube fusion was exceedingly unlikely. If we knew more details about the cell that Fleischmann and Pons had used then stronger statements could be made. But of course, those details we did not have, only pictures of the mock-up shown by the media.

Suspicion was not far beneath the surface, even in this restrained gathering.

'Has anyone actually seen if there is a cell?', someone asked, to be assured that its existence had been verified, that it was in the basement of the University of Utah chemistry building and that a team of Los Alamos scientists were eagerly waiting to turn their own detectors loose on it. The problem delaying them, apparently, was that it was under lock and key while legal aspects of the patent applications were being sorted out. So everyone once more was prepared to believe that the normal workings of science were on course and it was just a matter of time before the Utah cells would receive the usual scrutiny, and be incorporated in a larger scale experiment collaborating with nuclear experts from a government lab. As we shall see, this hope was ill-founded.

This readiness to accept that, even now, there could be some unread fine print, some loophole, shows how open minded scientists have to be if truth is eventually to emerge; 'the standard working through of science' that Redish referred to in his Baltimore address still had some way to go before we could state with certainty that we understood everything that needed to be understood.

The media had picked up Lewis and Koonin's exposé of the bad science but relatively little had been made of disturbing news from MIT. Fleischmann and Pons had claimed heat and radiation as their evidence for fusion; Lewis had been primary in demolishing the heat and had shown how badly the radiation and products evidence had been gathered. MIT had done more; they showed that the radiation data were flawed to such an extent that there was even concern as to the integrity of the claims by Fleischmann and Pons.

People close to this now began to question the motives of the Utah group and news of what they had said in presentations began to be looked at carefully. This information was disseminated on the electronic mail bulletin boards. People at talks by Fleischmann or Pons would send in reports of the talk; they were such stars that their talks were recorded on videotape and could be re-run at leisure. Gradually a disturbing pattern began to emerge. Inconsistencies began to emerge in their story; evidence claimed in one talk would be different in a later talk, sometimes contradicting parts of what had gone before. Some thought that this proved that some of the work claimed to have been done had not been done; others that it had been done but at such a rush, and with a graduate student's help, that Fleischmann and Pons were suffering communications breakdown in the hectic atmosphere.

The student was Marvin Hawkins. The first clues to the fact that Fleischmann and Pons's work may have been rushed came with an astonishing erratum to their paper. It changed some of the numbers that had appeared in the original version and regretted the 'inadvertent omission of M. Hawkins' name from the list of authors'. This occasionally happens in a large collaboration in physics where over 100 names may be involved, but when only two names appear one assumes that they should have been aware that someone was missing. The existence of this erratum seems to have been overlooked by many, as many of the references to the paper in the scientific literature still omit to mention Hawkins' name. It was not until a year later, when I talked with Fleischmann and Hawkins, that I learned that Fleischmann and *Hawkins* had done the bulk of the work in the last hectic months, and Pons had been primarily the spokesman during April while Fleischmann had been in Europe.

---

### SOME MAJOR MEETINGS

| | | |
|---|---|---|
| 1988 | 12 April | Erice, Sicily |
| | 12 April | American Chemical Society, Dallas, TX |
| | 26 April | US Congressional hearings, Washington, DC |
| | | Materials Research Society, San Diego, CA |
| | 1 May | American Physical Society, Baltimore, MD |
| | 8 May | Electrochemical Society, Los Angeles, CA |
| | 23–5 May | Cold Fusion Workshop, Santa Fe, NM |
| 1990 | March | First Annual Cold Fusion Conference, Salt Lake City, UT |

This is a reference list to meetings mentioned in the text. There have been other meetings such as that hosted by the Electric Power Research Institute in October 1989.

---

## From Baltimore to Los Angeles

In the aftermath of the American Physical Society meeting refuting Fleischmann and Pons' claims, the Utah team's credibility was severely dented.

The APS who had sponsored the Baltimore meeting had no doubts about who had won that round. Their weekly bulletin, *What's New*, sent out by electronic mail, *Bitnet*, to paid-up members had the following headline: 'The Corpse of Cold Fusion will probably continue

222

to twitch for a while even after two nights of assault at the APS Baltimore meeting.'

A disinterested observer might have commented 'but they would say that wouldn't they'. Even so, fortunes did seem to be taking a turn. The non-appearance of Fleischmann and Pons at the APS meeting had been reported to be because they had been too busy preparing for a Congressional visit to go and speak. However, the APS Executive had picked up a rumour that Pons was in Washington to meet with John Sununu, the chief of staff at the White House.[3] To check if this was so, a member of the APS Executive had called Sununu's office and asked 'Is Governor Sununu to meet with Prof. Pons?' The voice at the far end of the line replied in quintessential Washington fashion—giving the information while not giving the information: 'I cannot confirm that since the meeting is private.' The APS news concluded: 'But when the time came, Sununu stood Pons up. Fortunes change.'

On 8 May the media ran major stories about test-tube fusion. *Newsweek*, *Time* and *Business Week* each gave it their front cover and in the articles had managed to tie together the developing stories. With access to the many different groups working on test-tube fusion and having had the benefit of reporters attending the several press conferences and official presentations, the media were able here to present a coherent picture of the state of play. For many scientists this was the first good opportunity to read an overview of the new field. The reporters had investigated well and talked to many people at the heart of the action, and their stories tended to reinforce the worries that had begun to be aired in Baltimore.

Not just in the APS but across the globe the tide seemed to have turned. There was an interesting insight into the different ways that the various national media sold their stories to their different readerships and, in Britain, an indication of how important a news item test-tube fusion had become.

The political news lead story was that Mrs Thatcher's ten-year dominance of the polls was ending; her Conservative party was four points adrift and had just lost a 'safe' seat in a national by-election. Emotional news was also in high profile; 95 football supporters had been crushed to death in the semi-final of the national competition and the game, abandoned, had just been replayed with the usually rowdy crowd standing in silent tribute. These stories would normally have been expected to dominate the news in the serious and tabloid press. But they were displaced in the *Sunday Times* on 7 May; its unequivocal banner headline read 'Harwell Scientists Pour Cold Water on Cold Fusion'. While the US media kept trumpeting Stanley Pons, an American, with Fleischmann the senior man sometimes absent, sometimes as an appendage, the British paper patriotically announced that scientists at Harwell, 'the World's leading nuclear laboratory' had finally buried the hopes that Fleischmann and 'his former student' had found the key to solving the world's energy problems.

David Williams, the Leader of the Harwell efforts, had made some interim and limited statements to the press to the effect that his team had 'spotted neither heat nor radiation'. Harwell was now added to the growing list of world-class laboratories that were previewing how their research was panning out—itself an unusual step, but one regarded as necessary to combat the impression that Fleischmann and Pons must be correct because no one was contradicting them, a dangerous impression to allow to grow when major financial decisions by states, governments and private industry were being weighed on incomplete information.

Williams had expressed his concerns to Douglas Morrison, the physicist at CERN who had been editing the electronic newsletter on test-tube fusion for the seriously interested professional scientists. In particular Williams was worried by rumours that Harwell had been seeing neutrons radiated from the palladium rods, thus 'proving fusion', when in fact his group had seen none at all. Here, as Caltech had been, he was faced with a conflict between scientific scruples and social responsibility: his researches were incomplete yet there were external pressures on him to speak out and correct wrong impressions. Remembering previous fusion claims, in particular the ZETA episode at Harwell (page 37), he was reluctant to make detailed statements until he was absolutely certain, but Morrison argued that if he said nothing then it would perhaps encourage the incorrect rumours to become 'factoids' and be used by interested parties to support claims for funding based on the 'reality' of cold fusion. Morrison proposed that Williams make some statement to prevent the rumours, without compromising the research process. Thus Williams announced that Harwell had duplicated the chemists' experiments but with much more sensitive equipment. 'Given the numbers that we had been told to expect, we should have seen something leap out at us by now. This has not happened'.

A British scientist copied the article onto his computer and sent it via electronic mail to the bulletin board, and within 24 hours it was common knowledge all over North America. During the week subsequent to the APS meeting there had also been discussion of the gamma-ray signal that Fleischmann and Pons had cited in their paper as evidence for fusion, and people at large were beginning to understand how its shape and size were all wrong for it to be what the pair were claiming. Opinion among scientists was moving against test-tube fusion, or at least the Fleischmann and Pons version of it, very fast at this stage.

Yet following hard on the Sunday article about Harwell's news, the next day in the London *Times* Pearce Wright, the distinguished science editor, headed a special article with 'Proponents stage last stand'. Given what had happened at Baltimore, with the credibility of their data at stake, and added to that the news from Harwell, Wright's article was disturbing and added to the unreal atmosphere surrounding what should have been a clear cut investigation: the techniques were sophisticated but the questions to be answered and the methodology were not abstruse—'is there excess heat and fusion products to match'. The answers were increasingly seeming to be 'NO', yet Wright reported that Fleischmann and Pons would be attempting to buy time for more research when discussing their work at the Los Angeles convention on May 8, and that they remained adamant in their claims and 'are asking the US Government for several million dollars to confirm the explanation of their findings.'

Upping the stakes was James Brophy, the vice-president of research at the University of Utah. He insisted that their work was correct and said that 'people are trying to push science too fast. Pons and Fleischmann . . .' (the American order of names) '. . . worked for five years on this. People are trying to criticise it in four and a half weeks.' Although Pons and Fleischmann had set out five years before, the sum total time that they had spent on performing the fusion experiments together was probably nearer to the four and a half weeks than to five years, and in addition Fleischmann and Pons were a single pair; the four and a half weeks cited by Brophy was a month of intense work by hundreds of people with lots of experience and much better equipment.

The meeting was sponsored by the Electrochemical Society and in their advertising for the occasion they had invited people 'with positive results' to speak; reports from experiments that were showing no signs of fusion were not encouraged—an attitude that is unheard of at normal scientific meetings. Such talks were eventually allowed when strong lobbying stressed to the organisers that what they seemed to dismiss due to 'absence of evidence' was increasingly beginning to look like 'evidence of absence'.

Two thousand people were at the Bonaventure Hotel: last week the east coast in the rain, this week the west coast in the sunshine. In Baltimore the proceedings had been totally open, in Los Angeles no cameras or recorders were allowed. In Baltimore there was neither Fleischmann nor Pons nor sight of test-tube fusion; in Los Angeles they spoke for 40 minutes and Jones for 25 minutes, and were followed by a phalanx of supporters: Huggins from Stanford, Landau from Case Western, a speaker from Seattle on the sighting of tritium (which was already being discredited, page 155) and two groups from Texas

A & M University each with ten minutes apiece. Lewis from Caltech provided a lone voice for the opposition with a ten-minute presentation.

One of the Texas A & M presentations was by J. O'M. Bockris, a longtime colleague and friend of Martin Fleischmann and who would be one of the strongest supporters of test-tube fusion in the coming months. He gave a preliminary announcement in which he claimed to have cells producing heat and also large amounts of tritium, as much as ten thousand times that in the original electrolyte.

Throughout the year as other groups had trouble replicating the phenomenon, or were even withdrawing earlier claims, Bockris continued to see tritium—not regularly but on occasions, which cynics noticed tended to coincide with visits to his laboratory by administrators associated with the funding of his group's research. The unique ability in the whole of North America to perform such remarkable feats made the majority of the scientific community suspicious, convinced that the tritium was a contamination. Some of the gossip was even that someone was deliberately tampering with the experiment by introducing tritium into the cell to give the impression that it was being created by fusion.[5]

These sensational allegations were still a year in the future when, on 8 May 1989, people gathered for the Electrochemical Society meeting. Pons was certainly looking forward to a convention that he believed would be a great triumph, and was assuring colleagues that all the sceptics would be laid to waste.

Fleischmann and Pons claimed that now they were achieving even more dramatic effects than before, with more than 50 times as much energy coming out than they were putting in, and that a cell had generated a power of one watt for two whole days. Fleischmann said that there were not hot spots and that excess heat was being measured accurately. A titbit for the crowd was a suggestive, but not conclusive, attempt to respond to Lewis' damning claims that they had failed to stir the pot, leaving hot spots in the cell. Fleischmann and Pons showed a brief video of one of their cells bubbling. They had dropped some purple dye in the top and it diffused through the cell in less than 30 seconds, and so, they claimed, their cell was well mixed with no hot or cold spots. However, they did not show the real proof which would have been the temperatures from top to bottom, even though they said they had the numbers. It was certainly good theatre if dubious science as Fleischmann said to laughter, 'This argument of ineffective heat mixing really doesn't hold water.'

They said nothing definite about their tritium or helium measurements, the presence of which would have been proof of fusion products. Groups from MIT and Sandia offered to analyse small pieces of the Utah cathodes in less than three days to see if there were any fusion products—if there is fusion then what the deuterium had been converted into must be there somewhere; the chemists had seen the smoke, or so they thought, but where was the gun? Fleischmann and Pons responded that they were in the process of experiments that involved charging up the electrodes for a long time period and could not release them. Moreover, they said, they had made 'other arrangements' for such an analysis. However, no one revealed who was going to do it, nor how long it would take. The question of fusion products—were they in the Utah cell or not—remained as unanswerable as ever. (The analysis of the cathode to find if there was any helium was supposed to be completed by July and the results released, but a year later had not been.)

Was this proof or not? Were they being evasive? They certainly were in the opinion of Koonin and in the other reports from several physicists who had attended the meeting and had sent out summaries on the electronic mail networks. They made no direct response to many of the criticisms raised at Baltimore. Koonin reported 'prevarication and evasion' in the responses to questions.

Fleischmann and Pons insisted that their calorimetry was right, and Fleischmann remarked that he could not respond to criticisms 'until he saw full details published in a refereed journal'. As one of the major complaints from the physicists was that Fleischmann and Pons' submission to *Nature* had failed the refereeing process, that their 'preliminary note' in *Journal of Electroanalytical Chemistry* gave few details and contained material that they acknowledged that night to be in error, this was felt to be rather rich. On top of this, the veracity of their claims rested on hard data which had not been published and none of which, once more, were shown to the crowd.

In the press conference they did admit that there were 'lots of omissions and mistakes' in the paper—which had been pointed out in Baltimore but which were not addressed in the talk. There was intense questioning about the gamma-ray evidence for fusion which had been strongly criticised by many people for weeks, culminating in the refutation from an MIT group at Baltimore. This questioning continued into the press conference afterwards, led by Stan Lockhardt of MIT. Pons was trying to defend the gamma peak in an inconclusive way when Fleischmann said emphatically that 'the gamma peak is wrong'. Here was the first admission of error in the part of the work which dealt with the physics.

Also in the press conference Lewis, sitting next to Fleischmann, wanted to ask about the status of the search for helium—a smoking gun for aneutronic fusion—so that he could air his suspicions that the helium claimed by Pons in April was really not coming from fusion. In addition he wanted to clarify the status of the chemists' tritium measurements.

The status of the published tritium data at that time was rather confusing. In the Fleischmann–Pons paper was a 'figure 1B' showing an energy spectrum of the radiation emitted in tritium decay, but with an incomplete energy scale with its axis showing three equal spaced marks, two labelled 15 keV and 3 keV and one unmarked. As the energy had to be positive and a linear scale would imply that the missing label should read 'minus 9 keV' one needed instruction as to what the scale was. None was offered.[6]

There was another worry about the tritium results. The chemists had added a potassium compound to neutralise the electrolyte before determining the tritium content. This sounds harmless but a problem here is that potassium is naturally radioactive. One of its decay modes is to argon-40, which takes place by electron capture from the K-shell of potassium. This is accompanied by emission of radiation from argon which happens to be at 3 keV, precisely at the peak shown. Without knowing how much potassium was added it was not possible to tell whether the spectrum presented as evidence for tritium suffered from contamination from the natural radioactivity of the added potassium. Here was another question that people wanted answers to.

Finally the paper gave no information on how much tritium was present to begin with (there is some tritium in all water and relatively more in heavy water). When the chemists circulated a list of errata early in April there was still no mention of this.

Lewis questioned the tritium data, noting that there was a lack of detail in the paper—there was no quantitative information on the background. He then tried to ask about the status of their search for helium, but Fleischmann interjected: 'the tritium measurements are in the paper'.

Lewis responded, 'the foreground [signal] not the background', i.e. without the background it was impossible to decide how significant the numbers were.

Fleischmann reacted sharply and with emphasis said: 'the background is in the list of corrections to the paper', referring not to the original paper, nor to the erratum sheet that had been informally circulating, but apparently to a *second* list of corrections that most people, including clearly Lewis, one of the central characters, had not seen. 'That may be . . .' replied Lewis, trying to press on with his question about the helium. 'Then please don't rubbish it', responded Fleischmann, but by now the chairman had stepped in and the interaction stopped with the helium question unanswered.

Lewis told me later: 'I was shocked when Fleischmann loudly

228

asserted that the tritium background was in the errata because it was indeed not in the errata that were available at that time. Fleischmann was mistaken that it was in the published errata, but he asserted it with such confidence that I did not wish to challenge this point with the TV cameras rolling. So I backed down and said ''That may be . . .'', and wanted politely to ask him ''could you then please remind me of the number'' when he interrupted (with ''don't rubbish it'') and the chairman stopped the conversation without me getting either the tritium or the helium results out into the open.

'I felt rather badly about this since the film clip was run on many network stations and it made it look as if the Caltech group was not familiar with the data, so obviously we could not reproduce the phenomenon!'

Next day Lewis called Jim Brophy, the vice president for research at the University of Utah, to tell him that he had checked the erratum list and that the numbers were not there, and asked if Brophy could tell him the numbers in order to aid analysing the data. Lewis also requested that Utah did not encourage use of that clip anymore 'knowing that it was factually inaccurate'. On the next Friday night ABC ran it as part of a nationally televised prime-time segment on its news documentary '20/20'. Lewis feels that there was deliberate manipulation of the media in this case, whereas Utah argue that this was television footage of an actual event and that it is TV companies that decide what is to be shown. In any event Lewis was caught wrong-footed and had to grin and bear it.

Mention of the tritium background appeared in the final errata published in the *Journal of Electroanalytical Chemistry* (volume 263, page 187) but was not in the widely circulated errata sheet initially put out by Fleischmann and Pons. Finally in June there appeared a list of errata which consisted of the errata put out by Fleischmann and Pons in April plus some others.[7] Included among these new errata there appeared a comment about the tritium: 'Initial count rate 41 dpm; steady-state count rate 141 dpm' (dpm is 'decays per millilitre per minute'), a difference between signal and background that is consistent with natural enrichment during electrolysis (the extra bulk of the tritium relative to deuterium causes the tritium already present in the water to stay there while the deuterium gases off, hence there is a natural 'growing in' of tritium).

So much gossip had been disseminated by electronic mail that one wished that the accumulating errata could also have been transmitted so that everyone would have more efficiently learned of the state of play. Whether there were good data or not, whether there was evidence for fusion products, whether control experiments using light water had been done and with what results, was increasingly hazy and confused. A huge edifice was being set up. We were asked to enter the high rise of the twentieth century and assume that the foundations were solid; the builders said that it was all right, but no building inspector had been able to check.

At the weekly meeting at Oak Ridge on 10 May the summary of the Electrochemical Society Meeting in Los Angeles was that it had been 70% negative, and the rising feeling among scientists was that the claims for fusion were simply wrong. There was news that the Los Alamos authorities were having difficulties completing an arrangement to collaborate with Utah and that chemical assays were underway to look for helium in the palladium rods. The Utah chemists had failed to produce convincing answers to the questions about nuclear effects and had even admitted that some of their neutron data were wrong. The DOE labs had by now devoted thousands of man-hours to the task, and the strategy now was to plan to finish the experiment in an orderly manner.

There was to be a three-day conference in Sante Fe, New Mexico on 23–25 May, attended by representatives of all groups working on cold fusion in the USA (though neither Fleischmann nor Pons attended, which was widely regarded as a major question mark about their enterprise). The hope was that the laboratory experimentalists would write up reports within two weeks of the Sante Fe meeting so that the lab could issue a final report on the DOE-directed effort.

During the discussions there were some examples of how easy it can be to be misled in detecting neutrons. Francis Perey's group had been using a $BF_3$ counter associated with one of the experiments and noticed on two separate occasions that the background neutron count rate doubled for several hours. This was so surprising that they could easily have thought that it was activity coming from the cell, but on the second occasion they were present and were able to take other data and ascertain that it was indeed background. They were in a building with 'negative pressure', meaning that air and any other gases got sucked in from the outside. Although they were sure these data had nothing to do with the cold fusion cell, they had no idea where they came from. However, others present knew of similar experiences, one recalling how at the National Bureau of Standards someone had an intense neutron source 400 m away and a $BF_3$ counter had detected this. The moral is that neutron counters can pick up signals due to remote happenings, and that rises in background need not be signalling activity in electrolytic cells. Monitoring with several detectors (a point stressed also at Harwell)

230

is a necessary insurance. Detectors can also respond to 'noise': the discussion noted that turning on a sodium lamp had made electrical noise in the detector as if 700 neutrons had hit, while a nearby bolt of lightning induced signals similar to those that tens of thousands of neutrons would have given.

Hutchinson's group at Oak Ridge had by then been operating their cells for several weeks.

Three of their cells contained annealed palladium rods, purchased from Johnson Matthey, which were 10 cm long and 6 mm in diameter. The other was of cast palladium with twice the diameter and was fabricated from stocks at the lab. They had begun by monitoring for neutrons, which was nearest to their speciality, and found no excess above the (continuously monitored) background level. They then began to focus attention on the reports of excess heat.

They had started these electrolysis experiments, measuring heat, on the Friday evening 5 May and were inputting slightly more than 30 watts of power and were measuring output of about 26 watts; they calculated that the power used up in electrolysing about 2 cm$^3$ of water per hour could account for most of the difference. It had been operating for around 100 hours by the time of the 10 May meeting. However during the following weekend one of the 6 mm diameter rods[8] started to produce a positive excess of 2 watts (when power used in electrolysis was included in the sum).

By the time of the Sante Fe conference they had been monitoring excess heat (in this one cell only, for two weeks) but did not know how good their calorimetry was. The heat continued like this for about three weeks in all, up to the end of the month, at which point they decided to lower the temperature of the bath surrounding the cell, by 5°C, in an attempt to study the effect of temperature change on the imbalance. Previously they had raised the temperature and the heat imbalance persisted, but now the excess power disappeared and repeated changes of temperature and current were unable to reproduce the effect.

They continued the experiment for a total of ten weeks until mid-July, but saw no recurrence of the phenomenon; the other cells remained in balance throughout. One check that they made was to interchange the 'active' cell and a null cell in their calorimeter sockets to verify that the monitoring apparatus was working properly, and found that the measured imbalance in the cell persisted.

While the cell was producing heat they monitored for neutrons and saw none (limit to be less than $10^{-24}$ neutrons/sec for each $D-D$ pair). They found no significant increase in tritium. A memo sent to the DOE panel on 22 June noted this result and also that Charles Scott's group at the lab had seen two short periods of apparent excess

231

but of less than twenty per cent and lasting only a few hours. The DOE report noted that these effects had not been reproducible and were continuing to be investigated. A group at Los Alamos was reporting isolated bursts of neutrons, as had one experiment at Livermore in California. Argonne scientists had also noticed 'false positives', where neutron backgrounds had risen occasionally, but these were isolated examples: the message across the rest of the spectrum of the DOE laboratories was of no signals. It was at this time that Harwell reported that its extensive range of experiments had seen nothing.

By the summer of 1989 the message from the world's major labs, and in the emerging official reports in Europe and the USA, was that there was no reason to believe that a new nuclear process, solid-state fusion, had been discovered, and that the heat phenomenon—if indeed there were any—did not warrant special attention; anyone wanting to research into it should compete for funding as any other area of science had to.

# 13

# INTERNATIONAL REACTIONS

Everyone with any experience of nuclear physics realised that if the chemists' claims for cold fusion were true then the story was more complex than the media were advertising. There were not just the obvious energy implications and their possible impact on geopolitics, but there were also more worrying concerns. Could this form of cold fusion have military applications, would the neutrons from it be useful for enriching uranium and hence have applications in conventional fission, and was it a source of tritium—the fuel of hydrogen bombs?

If test-tube fusion had indeed behaved as the headlines following 23 March 1989 had claimed, and been as easy to produce, then much of the hype might have been justified. Baroness Hooper made a statement in the British Parliament early in April speaking from an official briefing from government scientists which gives a sober assessment of the situation, while also recognising the excitement that everyone felt:

'Speculation about the possible implications of these reports has come before the scientific evidence of the basic experiments has been validated. . . . Before any sensible assessment can be made of the feasibility of scaling-up the experiment to the point at which larger power outputs can be obtained, a fundamental understanding of the process will be necessary to establish if it is possible and safe to do so. However, it would also be quite wrong to stand aside and ignore these claims since the potential gain is so large.'

The possibility of a pollution free energy source 'too cheap to meter' was evident to media, scientists and administrators alike. The commercial benefits and need to protect patents were also in the front of many people's minds, as Chase Petersen of the University of Utah had remarked: 'Being one week ahead of the pack in fast moving science projects is enough to keep people focused here . . .'. So on these grounds alone it was natural that government-sponsored laboratories worldwide set up experiments and reported to their paymasters.

While the attentions of the media and public focused almost exclusively on test-tube fusion as the solution to the world's energy problems, there were, as we have seen in earlier chapters, other hidden agendas and concerns that generated interest in government and other official quarters. These included the impact on budgets for the large scale hot fusion programmes, particularly in the USA and Europe; practicalities of turning a small-scale test-tube phenomenon into a commercial reactor and the impact this would have on world supplies and prices of palladium; the strategic military consequences of a readily available and cheap source of tritium and neutrons—the worry that Fleischmann had had all along; but first and foremost there was the question of whether the phenomenon was real and, if so, what its properties were.

The awareness of military applications and nuclear sensitivity was perhaps most acutely felt in India. However, in Europe and America these concerns seem to have been evaluated and seen as being less urgent than the impact on the major investments that were already in place for hot fusion. If test-tube fusion was verified, would it be an alternative, a replacement or an irrelevancy? It is one thing to have made a scientific discovery, but there is still a long way from there to achieving a commercially viable product. Even if test-tube fusion were a real phenomenon it was not immediately clear that it would ever have any commercial relevance.

On 22 March, before the Utah press conference, Bullough and Williams from Harwell met with Lomer, Director of Culham where much fusion research goes on, and began to discuss how test-tube fusion would affect official attitudes to the ongoing work of the Atomic Energy Authority and the European Fusion Programme at JET (Joint European Torus). That weekend there was a press statement issued jointly by Lomer of Culham, Bullough the UKAEA's Chief Scientist, Iredale the director of Harwell, and Gittus the UKAEA's Director of Communications. This began by summarising the achievements so far, noting that fusion is a major part of the UKAEA's work, centred on Culham, and that Britain is part of a fourteen nation consortium at JET. The fusion of deuterium fuel has been achieved at up to 250 million degrees at JET and fusion reactions regularly occur but not yet at a sufficient rate for a commercial fusion reactor. 'Our long association with fusion research naturally makes us interested. Therefore we are trying to reproduce [the Fleischmann and Pons experiments] at Harwell from information given us by Fleischmann—a consultant.' They stressed that this would take time and commented, 'If confirmed it will be the *first occasion* that chemical processes have produced nuclear reactions', a choice of words implicitly admitting the possibility that the test-tube fusion phenomenon is real.

The news naturally made the lead in the internal newsletters that the various laboratories sent to their staff. The JET newsletter of 28 March was based on the press release but a slight change of words gave a rather different opinion on the reality of test-tube fusion for the employees' consumption: '. . . it will be *extraordinary* if chemical processes have produced nuclear reactions. Hitherto it has been thought that nuclear and chemical domains are quite separate.'

While the reality of the Utah experiments was still an open question, on Tuesday, 28 March immediately the lab opened after Easter, Fleischmann came to Harwell and gave his first report of the experiments before an audience of critical scientists including four Fellows of the Royal Society and representatives from JET, CEGB (Central Electricity Generating Board) and Harwell—a collection of theorists, chemists, materials experts and nuclear physicists. This was the first intensive peer review as the Utah press conference had not been accompanied by a scientific seminar. Crucial questions were raised about the evidence and the controls to check against errors, and the answers began to make several present wonder if there was perhaps less to this than met the eye. Bullough prepared an official memorandum of the meeting which formed the basis of much of the subsequent British perception of the work.

Fleischmann was questioned in great detail about the validity of the calorimetry results, and 'gave a confident and careful justification of his results. As he was unable to explain his results through any classically acceptable chemical process, he was led to believe that the heat output must have a nuclear origin and hence invoked the concept of *dd* fusion occurring within the palladium electrode.'

Three major worries about the 'evidence' emerged right away. First there was the fact that control experiments with cells using ordinary water in place of heavy water, and electrolytes using sodium rather than lithium, had 'unfortunately and surprisingly not been performed'.[1] The second worry was that there were so few details about the measurements of tritium—a product of fusion and prima facie evidence, if correct. However, the experts noted that 'the shape of the spectrum presented was not as would have been expected given the description of the measurement.'

The third concern was the most serious in that not only did it cast more doubts on the claims for fusion, but it also raises questions about the data as presented in the published paper (Chapter 15).

'A sodium iodide detector had been used in an attempt to detect the gamma rays produced following any capture of fusion neutrons in the water bath surrounding the cell. *A peak at 2.5 MeV* was claimed.' However, [this peak] 'was at the wrong energy (it should have been at 2.2 MeV)'. The audience concluded that 'the calibration was therefore suspect and it was difficult to see how it could be so far wrong . . . The raw data were not provided and without such detail it was difficult to assess the reliability of the claims or the energy of the gamma rays involved.'

Reaction from the Culham fusion experts was that the work presented was scrappy but not amateurish '[Fleischmann] certainly believes that there is a considerable energy release . . . it is hard to see how a competent electrochemist could be mistaken on this scale. At times twice as much heat has been released as was put in.' Had the raw data with heat output as a function of time been shown, then some of this confidence might have been less.

The main conclusions of the meeting included:

(1) 'The central problem is that of unexpected heat generation at a level [that] could not be met by chemical processes.'
(2) 'If the heat production were due to conventionally accepted fusion processes, a factor of $10^8$ to $10^{10}$ more neutrons should have been emitted than were claimed to have been observed.'
(3) ' . . . not many other channels for fusion were felt to be available [for producing the heat].'
(4) 'The only explanations for the observations appeared to be either (a) an error in the calorimetry measurements, or (b) a nuclear reaction other than those considered to date was open, or (c) some entirely new and unexpected phenomenon had been uncovered.'
(5) 'As a result of parts (1) to (4) it appeared that a verification of the calorimetry measurements was vital.'

Following this meeting, news of Jones' work reached Britain and the rather scanty information led people to regard it as a confirmation. In particular, the report stated that he saw very low yields of *dd* fusion neutrons and so Harwell decided to improve the sensitivity of their neutron detection.

These memoranda were written on 29 March. Awareness of the possible implications of the fusion news grew rapidly.

On 30 March an internal memorandum noted that there could be fusion reactions involving the lithium in the electrolyte which produced energy, and hence heat, without neutrons. This reinforced the belief that heat measurements were important: 'It is tempting to speculate on [the phenomenon] as a power source but first we must confirm the heat.' On 31 March the Harwell authorities received a note that fusion may be raised for discussion in a select committee of the British Government in April. There was also comment about the possible impact on world metal markets, that a major demand might force up palladium prices, and the palladium was mined primarily in South Africa and USSR with their obvious political sensitivities. The writer provided the information that one third of palladium went into electrical manufacture, one quarter into dental and medical applications, and one fifth into vehicle exhaust catalysers which were already placing big demands on the metal. Taking Fleischmann's figures for the heat production per $cm^3$ of palladium and scaling up to that needed to produce a gigawatt of power—enough to illuminate ten million 100 watt light bulbs and not adequate for the needs of a large city—already cost 100 million dollars and commercial power would require a doubling of world production. The conclusion was that 'if the phenomenon is real, then practical *large-scale* realisation may therefore be difficult unless more common materials can be substituted [in place of the palladium] or higher energy densities achieved.' So even if test-tube fusion was real, energy utopia was further away than some of the media-led euphoria acknowledged.

There had been much concern in Britain about the future of nuclear power in Britain and about research and development in this area of power generation following the privatisation of the CEGB by the British Government. A special committee in the House of Lords—Britain's equivalent of the US Senate—dealing with science and technology was due to meet on 14 April and the Department of Energy requested a briefing on 'Test-Tube Fusion' for Baroness Hooper, the Parliamentary Under Secretary of State at the Department of Energy. It was in this that the cautionary remarks about premature speculation, cited at the start of this chapter, were made. And the CEGB, along with many other organisations, naturally wanted full details of developments, requesting a briefing from Harwell for their directors' meeting in April.

Cold fusion also was discussed in the House of Commons. It is a regular procedure for members of parliament to table questions to be answered by ministers in the house. The question is written in advance and answers prepared; however, the member may then ask a previously unannounced supplementary question.

Dudley Fishburn (Conservative MP for Kensington) tabled a question asking the Secretary of State for Energy what information his department had on 'low temperature nuclear fusion'. Civil servants prepared an answer for this, with the help of Harwell. However, the art was to guess what the supplementary question would be and to brief the minister for all contingencies. With the high interest in the subject, the civil servants working for the energy minister suspected that other MPs might take the opportunity to range more widely over nuclear issues and related topics. So this single question led to half a dozen people researching briefs on fusion, nuclear R&D, the future of the UK Atomic Energy Authority, the greenhouse effect and general environmental matters.

Throughout Europe the interest among scientists was as intense as that in the USA. On 14 April Glenn Seaborg and Admiral Watkins had met with Governor Sununu and President Bush in the White House. Seaborg had recommended that a special panel be formed. After the weekend, on the 18th, messages had gone out summoning the heads of fusion research at the US Department of Energy laboratories to an emergency meeting in Washington DC on 19 April (Chapter 8) and from this a national coordinated research effort began. In Europe 'Euratom' commissioned a similar programme.

Admiral Watkins then acted on Seaborg's advice and on 24 April instructed John Schoettler, chair of the Energy Research Advisory Board at the US Department of Energy, to:

1. 'Review the experiments and theory of the recent work on cold fusion.
2. Identify research that should be undertaken to determine, if possible, what physical, chemical, or other processes may be involved.
3. Finally, identify what R & D direction the DOE should pursue to fully understand these phenomena and develop the information that could lead to their practical application.'

He also requested that the Board should provide an interim report on the first item by 31 July and a final report on all items by 15 November 1989. The motivation for this was 'because of the potential benefits from practical fusion energy.'

In Europe, too, administrators were beginning to coordinate efforts early in April. Harwell's experiments were well under way when the initial approaches were made by the European Community. At this stage it created somewhat of a problem as they could well have been ahead in a race for an industrially significant development, and so wanted careful discussion on who would own what if the programme succeeded. On 14 April a meeting took place in Brussels involving scientists from Europe's various fusion and multidisciplinary laboratories. This was analogous in some ways to the American meeting that took place in Washington at the DOE at about the same time, and the resulting strategy was also similar.

There was debate on how this should be funded—how much from existing national programmes and how much from central funds. The priorities were 'through exceptional procedures' to coordinate European efforts and through labs associated with the European fusion programme to answer these questions:

(1) Is the Fleischmann–Pons phenomenon real and will it produce energy at a rate which might lead to a practical reactor?
(2) Is it fusion, and if so what is the relative value for scaling up against the existing 'hot' fusion programmes?

Representatives from around the continent made reports. As in the USA a large-scale effort was already underway in several independent centres. David Williams reported on the Harwell work, which was the most comprehensive of the European labs. Reports from Garching (Germany) and Frascati (Italy) also confirmed Harwell—that so far, at least, no one saw anything.

In the USA the plan was for all the laboratories to send in weekly reports to Washington. A committee of experts then produced and interim report in the summer and a final report in November on the status and likely exploitability of test-tube fusion. In Europe a similar situation occurred but without the committee. An interim report came out in August.

At the meeting in Washington and at that in Brussels, no one had any positive effects to report. Among those present in Brussels on 14 April were representatives from Frascati in Italy. However, that weekend, events took a dramatic new course.

## Italy and the nuclear referendum

Many people who felt uneasy about nuclear power but who felt themselves to be part of the 'silent majority' were stung into vociferous action after the Chernobyl disaster. Pressure groups sprang up in many countries urging governments to rethink their nuclear policies. In Italy there was a plan to develop nuclear power and this led to a national referendum under the rather dry heading of 'Repel the Energy Plan of the Electric Company'.

The intention had been for the company to build a number of nuclear plants and this became a central issue in the referendum. The Yes-No vote was rather ill-balanced and left a huge middle ground barren. In effect, 'Yes' implied agreement with the number of power plants that the power company proposed to build whereas 'No' meant that you were against nuclear power in general.

The 'No' votes won. This caused a restructuring in Italy's energy industry in the latter part of 1988.

Particularly vulnerable was ENEA (Energia Nucleare e Energie Alternative, not 'European Nuclear Energy Authority' as claimed in the *Guardian*. ENEA is an Italian equivalent to the UKAEA at Harwell in Britain). A good strategy for a healthy future at ENEA seemed to be to convert from fission, which was out following the referendum, to fusion. Fusion was perceived as 'clean' in Italy, and was politically acceptable in that it was not explicitly excluded by the nuclear referendum.

With all the interest that the referendum had generated, every journalist in Italy was instantly an energy expert; 'they knew their kilowatts', one Italian scientist said to me. They knew the amount of energy that Italy requires day to day and as soon as the news of cold fusion broke, all the Italian journalists had an opinion. Science in general and physics in particular has had high profile in Italy, especially since the Italian Carlo Rubbia won the Nobel Prize in 1984 and became Director General of CERN, Europe's centre for particle physics. This all increased the media excitement about test-tube fusion and put pressure on Italian scientists to be into test-tube fusion early. For ENEA, who were seeing fusion as a way out of their post-referendum difficulties, the news was a god-send.

Francesco Scaramuzzi is a low-temperature cryogenics expert at ENEA in Frascati near Rome. He had titanium to hand and knew how to pump deuterium into it. He used titanium chips to maximise the surface area, making it easier to force the deuterium in, then cooled it with liquid nitrogen and pressured it to 40 atmospheres with deuterium. This was unlike the Fleischmann and Pons experiments and only marginally like those of Jones, the common features being deuterium and titanium metal. Scaramuzzi was doing good science in that he was doing what he knew best and was testing to see how cold fusion depended on the conditions. He brought in some neutron experts from the tokamak division at the Frascati lab and during the first weeks of April looked to see if anything happened.

Whilst the materials were cooling down the scientists noticed that the detector fired, apparently recording a burst of neutrons. They then repeated the cooling cycle several times during the week in the hope that the phenomenon would happen again, but it did not and by Friday, 14 April (the same day that Euratom was discussing cold fusion in Brussels), they were ready to give up the attempt. They depressurised the cylinder containing the titanium chips, opened it to the vacuum pump, took away the liquid nitrogen coolant and went home.

The following morning the apparatus was warming up from liquid nitrogen temperature towards room temperature and they noticed that the neutron counter was counting at a relatively high rate and had been for some hours. The source appeared to be emitting 10 000 neutrons each second.

The senior management at ENEA knew that Scaramuzzi was doing experiments and work leaked out that something strange had happened. With ENEA's survival at stake and with the vast desire for fusion at the laboratory there was naturally much pressure to make a public announcement: in the new tradition of test-tube fusion they held a press conference on Tuesday 18 April and put in patent applications for this new process. However, unlike some other claims for test-tube fusion this one also involved an internal seminar and discussion within the laboratory. The news conference claimed success and this was hailed as further proof of test-tube fusion by many supporters, though it had little obvious direct connection either qualitatively or in amounts of radiation with what the prime movers, Fleischmann and Pons, had claimed as the effect.

As in many of the 'confirmations' this appears to have been another example of premature public announcement which the science did not convincingly warrant. After the initial excitement much more experimentation was done but the effect was not reproduced.[2]

News of this propagated by fax and electronic mail. A paper appeared with details of the experiment, and while interesting it was clear that this was an announcement of a result that was still a long way from being confirmed and that the experiment would be continuing. The programme at Frascati, as at other laboratories, continued carefully—which is to say that until completed there would be no statements of how things were progressing. Nonetheless, rumours sped around to the effect that more bursts were being seen, followed by other tales that there were no bursts. As with all rumours, often the story bore only passing similarities to the truth.

The power of rumour was illustrated in early April in Britain. Everyone in the scientific community knew that Harwell had a large team doing experiments that would probably be definitive, but no news leaked out as to what they were finding. Even so, rumours spread, and one which flared briefly was that they were seeing tritium. On checking this it was traced to someone having been asked to look for tritium in heavy water for the experiment—and that they had found some! But further enquiry revealed that the experiment in question had not yet begun—the measurement had been to determine how much tritium was in the heavy water *before* the experiment began, against which later measurements, during and after the experiment, could be compared.

This rumour about the tritium was quickly killed and did not propagate far. However, a rumour that Harwell was seeing neutrons, also without foundation, spread further and even affected the British Ministry of Defence.

Forty years earlier, the original British patent[3] for a fusion device, a forerunner to modern tokamaks, had been classified secret—not because of the commercial jealousy of a potential energy source, but because the authorities were afraid that *dd* fusion would supply a copious source of neutrons with military applications. Hot fusion technology is today a large-scale venture, and the few countries who have developed it are quite capable of producing neutrons by more efficient, though high technology, means. But if small cold fusion cells were producing watts of power, and in consequence billions of neutrons each second, the strategic implications of the phenomenon could be very worrying. Rumours that Harwell were seeing neutrons disturbed people at the Ministry of Defence who were relieved to learn that the rumour was not true.

It was the possible military potential of the products from test-tube fusion which worried Fleischmann intensely and which was commented upon privately by several scientists, and, had the phenomenon produced nuclear radiation in amounts commensurate with the watts of heat that were claimed, then governments would have paid very close attention to the implications. It was awareness of this, and suspicion that the developed nations would classify results on test-tube fusion, that influenced the early research programme in India.

## Indian reactions

India has a large scientific base, statistics putting it third in the national rankings of the number of working scientists. It has a total of four million working professionals, though many of them are administrators rather than genuine research and development scientists. India is strong in mathematics and theoretical physics. It has an active nuclear research programme spanning the whole spectrum of that field; there are scientists working in high energy particle physics who visit CERN, the European Nuclear Research Centre in Geneva; the Tata Institute for Fundamental Research in Bombay is a world class institution. There is a considerable nuclear physics programme including a cyclotron at Calcutta which studies nuclear structure and makes isotopes for use in medicine.

Indian nuclear reactors that produce power have a controversial history and there is widespread speculation that some of these installations may have been used to make the essential ingredients for atomic weapons. India bought fuel rods for its Taranpur nuclear plant from the USA. Now the USA wants its rods back as it is worried that India is extracting plutonium from them for use in nuclear weapons, but India refuses to return the rods as it has already paid for them.

242

The Pokran nuclear test in 1974 showed that India has the capability to make atomic weapons. The US, USSR, Britain, France and China have both atomic and hydrogen bombs; India has the atomic bomb only, and there is speculation that Israel, South Africa and Pakistan may have too. The early 1970s was a period of high tension between India and Pakistan and also China, and it was in this period that India exploded its atomic bomb. It is unclear what further developments have taken place in India since then. The atomic bomb is the first step to a hydrogen bomb; an atomic device is the trigger for a hydrogen bomb and the missing ingredient is tritium.

Following the Pokran nuclear test in 1974 both the USA and Canada severed nuclear ties with India, visits by Indian scientists to US laboratories were restricted and more general trade links were cut. One casualty of this was the shipment of helium to India, hence the interest in the hot springs as a helium source north of Calcutta (see Chapter 2).

India's nuclear programme has been based on a Canadian design of reactor (CANDU) that uses locally available uranium and heavy water. In 1988 India had made an agreement with the USSR to import two Soviet light water reactors. These will use enriched uranium, the Soviets supplying the fuel under an agreement that the spent fuel, which is a source of plutonium, be returned to the USSR. The reactors will be built under safeguards agreed with the IAEA (International Atomic Energy Authority) such that there is no weapons capability. Some people within the Indian Department of Atomic Energy are aggrieved that the government has agreed to these condition, seeing it as an erosion of India's nuclear self-reliance and sovereignty.

In these circumstances, the application of test-tube fusion as a source of neutrons for uranium enrichment and also as a source of tritium was not lost on the Indian authorities. Neutrons added to $^{238}U$ make $^{239}U$ which decays to make fissile $^{239}Pu$ (plutonium). India also has large reserves of thorium $^{232}Th$ and the products of test-tube fusion could convert this to fissionable $^{233}U$.

The Indian authorities also, naturally, saw the benefits that might accrue for energy production from test-tube fusion, and the possibility that these might revolutionise India's energy programme.

Generating electricity to satisfy the needs of a growing population is a continual problem which is especially marked in some areas. In Calcutta for example, after sundown the lights come on, and within minutes the power fails. Petrol driven motors generate electricity in the big hotels and offices and smoke and grime in the air. Wood burning in the shanties and villages adds to the extensive pollution and is so excessive that it is seriously depleting trees and devastating the environment.

These problems of energy production and curbing pollution were in many people's minds when test-tube fusion was first announced, and were already part of the stimulus behind an initiative involving the Prime Minister, Rajiv Gandhi, and the scientific community. At the end of 1988 Gandhi had asked his Scientific Advisory Council to suggest strategies for incorporating science and technology into India's national plans up to 2001 and for integrating these ideas into the next five-year economic plan to begin in 1990.

So when test-tube fusion was announced, the news reached Gandhi who contacted the heads of the national research institutes personally and encouraged them to pursue the phenomenon. Test-tube fusion from a press conference in Utah with its attendant publicity had influenced the leader of one-tenth of the world's population.

Test-tube fusion touched several needs, including energy and a nuclear substitute. It also set a psychological challenge to scientists and administrators; test-tube fusion had been discovered in the USA, so the maxim expressed to me by some Indian scientists was 'Show that it's true in India. Indian science is as good as that in the USA. Help give self-confidence to Indian science.' Such desires can lead to a lack of critical controls and play on human nature, namely the desire of an institution or group that is not regarded as 'First Division' to prove that it is. This was present in Utah too where some expressed their pleasure to be proving themselves to be as good as the 'east coast establishment'.

Even for those scientists who pursued test-tube fusion for its energy applications, or even initially simply for scientific curiosity, the military significance was not lost. Several Indian scientists who had been involved in test-tube fusion expressed to me the feeling that if it had turned out to be real then other countries would pursue it for their own ends and that it might in consequence become classified. 'It was important for us to get in at the start', Yogi Viyogi told me, his remarks being similar to those of Ron Bullough at Harwell who had said, after learning of test-tube fusion, that it 'could have major ramifications and here was a chance for Britain to get in on the ground floor.'

Viyogi was one in a group of nuclear physicists at the Variable Energy Cyclotron Centre (VECC) in Calcutta who set up a test-tube fusion experiment under the direction of Bikash Sinha. VECC is a small cyclotron—an accelerator source of protons, neutrons and ions for use in nuclear research. The scientists' professional lives dealt with nuclear physics and detecting radiation, and so they concentrated on looking for radiation products as well as doing rudimentary heat measurements and electrochemistry.

They started within a few days of hearing about Fleischmann and Pons' claims. The Calcutta cells were rather smaller than those in Utah.

244

They attached them to the electrical supply and after charging up for seven hours the temperature rose from 25°C to 53°C—from air temperature to that of a hot bath. These sudden temperature changes were due to changes in the crystalline structure of the palladium cathode as it responds to the stress and strain of being loaded with deuterium. To those who were not aware of this property—which was most non-specialists at the time—this heat output was very dramatic and was the cause of some of the erroneous claims to be duplicating the Fleischmann and Pons heat effect. But the VECC group were more surprised to find that as the temperature went up so the neutron counter registered a burst of neutrons with an intensity three times that of the usual background, and this continued for about ten minutes. Another neutron burst happened three hours later.

The VECC group made another experiment, similar to the first but that now they replaced the palladium with titanium. Here after fourteen hours there was a neutron burst and the temperature rose, though this time only to 40°C. In both experiments the neutron bursts were about three times the background level and this convinced them that the neutrons were indeed coming from the cell. As they saw a sudden heat rise too, they decided to call a press conference.

The announcement on 12 May made as big an impact in India as the original press conference of Fleischmann and Pons had done in the USA. The Indian announcement was only four days after the Electrochemical Society meeting in Los Angeles at which Fleischmann and Pons had been forced to make a retreat and withdraw some of their claims. The Indian reporters took heed of this saying that 'Indian scientists are therefore playing it extremely safe before claiming any major success. But beneath Dr Sinha's carefully guarded official statements one can detect the excitement that has built up in the last few weeks over the issue. It is an excitement which clearly conveys that the Indian scientific community is on the verge of a breakthrough'.

The *Telegraph* (of India) picked up the theme of national pride: 'Indian laboratories are on a course followed by the greatest labs in the world'. The front cover of the colour magazine on 28 May carried the headline 'The frontiers of fusion: India's dramatic surge ahead' and a letter to that paper on 21 May anticipated this with 'As a young scientist one felt proud of our senior colleagues for being on a par with western researchers.' But the letter had a sting in the tail: 'Why were the Indian scientists so long [in reproducing the phenomenon]? Could they not publish their findings sooner? And now that the two western researchers have admitted errors in their experiments, what will our scientists say? Will they stick to their claim?'

To be fair to Sinha and his colleagues they had said at the press conference that they were now carrying on the experiment under more controlled conditions to satisfy themselves if their results were real, and said also that 'while one cannot outright rule out the increase as due to nuclear fusion as claimed by the two foreign scientists, the increase could also be attributed to such outside sources as cosmic rays.'

When a sense of reality as to the significance of the Calcutta test-tube fusion statement was brought in by the US Science Counsellor in India, Mr P. Heydemann, it was perceived as part of the West's plan to keep test-tube fusion off the front pages. The *Telegraph* quoted him as saying that local news reports claimed an Indian breakthrough but 'How can this be a breakthrough? India is only replicating what some scientists outside have done.' The paper defended this by noting that 'The activities in the Indian labs however prove that the country is not merely in the business but is quite ready to move ahead.' And then it came to the core of the situation 'Given the immense implications of test-tube fusion, it is certain that India's attempts will not be liked by the West.'

India was convinced that if test-tube fusion was real the West would classify it as secret, and that there was already an attempt in the West to divert attention from test-tube fusion by mounting a disinformation campaign. The paranoia that Peter Bond had experienced in the USA: 'You are seeing test-tube fusion but are keeping quiet because the big oil companies have bought you out' was now seen as an official conspiracy against Fleischmann and Pons.

Journalists asked Sinha to comment on *Nature*'s editorial which had stated that Fleischmann and Pons's claim was 'unsupported by the facts'. Sinha replied: 'There are reasons more than meet the eye for all this.' The *Telegraph* and other papers had no doubt what this meant: 'Indian scientists are of the view that it is the defence implications that have led to the mauling of the Utah team.' As the *Telegraph* reported in May and as Viyogi and Srivastara repeated to me in December, 'The stakes are really high. Forget about fusion's potential to turn the tables on oil-rich Arabs; the making of H-bombs could change overnight. H-bombs are fusion devices but they need a fission bomb to detonate first to provide the thrust to squeeze together the fusionable atoms. With cold fusion these bombs could be cheaper.'

The head of the national Bhabha Atomic Research Centre in Bombay, Dr P. Iyengar, was quoted to have said that test-tube fusion could be the 'easiest neutron source ever invented and if the strength could be multiplied then it could pave the way for new fissionable materials.'

These news reports clearly illustrate the three driving forces behind the research: a wish to prove that Indian science is as good as that in the West; an awareness of the military potential; a belief that the

Western Powers were conspiring to suppress test-tube fusion if real.

The strategic implications were on everyone's minds on 18 May when the Indian Department of Atomic Energy called a meeting of physicists from atomic laboratories throughout India. Journalists weren't allowed in, Iyengar stressing that the scientists needed to do more experiments to be sure and that he didn't want to be hauled up by the scientific community like Fleischmann and Pons had been for rushing to the press before doing enough work. However, the media decided that the meeting was 'Top Secret'. The *Telegraph* said: 'The outcome could alter the calculations of our planners including the ones who allocate funds for the country's defence . . . they were indirectly pondering the possibilities of making bomb production manuals all over the world obsolete all of a sudden.'

Were military implications really an impetus? The media may have perceived it this way and the scientists felt it so. Iyengar however preferred to emphasise the immediate application of the neutron availability as a good source of neutron radiography in medicare. All of these implications were apparent to anyone who appreciated the role of nuclear phenomena in India's strategy. And the desire to show that India could do it was also natural. The original motivation for Sinha though had been to test whether the heat and gases around hot springs could be due to fusion.

However, there was a hidden agenda for the VECC scientists in their press conference, namely to get additional media attention for plans to develop VECC.

In the Calcutta *Statesman* of 13 May next to the story about test-tube fusion at VECC there was a headline 'Super Cyclotron at Salt Lake soon'. This bizarre conjunction did not refer to Salt Lake City Utah but to the suburb of the same name in Calcutta, home of the VECC laboratory.

VECC scientists wanted to build a more powerful machine at a cost of 75 million rupees. In normal circumstances this would be much larger than its predecessor, but the VECC scientists planned to use the increasingly popular superconducting technology to make its size similar to the present cyclotron at half the cost. There were only four machines of this type operating in the world, and one in India would give a prestige boost to Indian nuclear physics. Conventional magnets were replaced with a liquid-helium cooled magnet coil made of superconducting material. This material, a niobium–titanium alloy, is commercially available in the USA and other advanced western countries. The new machine will need 42 kilometres length of wire made of this metal and the VECC authorities were concerned that the US Administration may block its export for political reasons, claiming that this high-tech material might find its way to the USSR, even in exchange for nuclear technology. There are other sources, for example Europe or Japan, but it needs scientific and political support in India if this project was to have any chance of going ahead. On the grounds that all publicity is good publicity they were aware that test-tube fusion could get the VECC superconducting project national attention.

Calcutta science correspondents are all junior reporters who happen to have a science education. 'Science' has low priority in the papers. But make this a national achievement and it will be a big story, good for the reporter, and with a good chance of getting on the front page. So to some degree the test-tube fusion press conference was bait.

The work at VECC continued but their early results were not reproduced. As they became better acquainted with their apparatus and extended and improved it they realised that the early 'signals' were probably artefacts. After several weeks and no further signals, they gradually wound up their effort and returned to the research that they had been doing before the cold fusion furore began.

I asked them why they had not published their null results as a scientific contribution to help balance the well publicised spurious claims from elsewhere. Their response was akin to that that I heard in other places: it is a lot of work to quantify how near to zero an 'absence' of signal is. They had only entered this research in the belief that the effect would be easy to reproduce and, once reproduced, they could then move on with the new important field. However, when it gradually transpired that success—major energy, national and international prestige—was unlikely, then the reasons for taking time out to pursue it also died.

In Bombay, however, at BARC (Bhabha Atomic Research Centre) a national programme of research continued and on 18 May a special meeting was held.

During the next six months several groups at BARC reported measuring heat and neutrons and even tritium. The first burst of neutrons was sighted on 21 April but workers elsewhere believe that this was a faulty detector—some neutron detectors are very sensitive to shock, electrical impulses or damp and can give signals as if neutrons were around. The amounts of tritium claimed were quite large, far in excess of upper limits deduced in many other labs. Wherever this came from it was not from $dd$ fusion, as tritium produced that way has enough momentum that neutrons will be disrupted from the surrounding material, and BARC saw no neutrons in association with the tritium. Whether the tritium is real or an artefact, and if real how it is being introduced, are still unanswered questions.

From the brief descriptions in the report it was hard to gauge a well-formed opinion on these experiments. A clue may lie in the introduction which admits that 'many cells which were tested did not give any positive results [and] are not included in the list'. Here again one has the non-reproducibility aspect, the failure of an essential precondition for a sure, scientifically acceptable, result.

I was in Calcutta for only a few days and tried unsuccessfully to make contact with BARC in Bombay. Strikes and engineering problems with the planes made air travel impossible, the train was impractical and the phone defeated me: even placing a local call in Calcutta can strain the system; a call to Bombay seemed to be beyond the limits of the telecommunications infrastructure, even though in cold fusion India was apparently leading the world.

Ed Wrenn, a physicist from Utah and member of Salamon's team (see Chapter 17), visited BARC in February 1990 and identified some of the strengths and weaknesses of the experiments. By February BARC was claiming to see tritium and neutrons. Wrenn's analysis is rather perceptive. He found the experiments to be of high quality as expected at a national laboratory with much nuclear expertise. The neutron measurements were being made with a $BF_3$ counter and as noted by many scientists, this is susceptible to subtle contamination. A problem is that powerful electromagnetic fields or inductive currents in the vicinity of the apparatus can be easily picked up by cables as noise and confuse the detector, which responds as if recording a genuine neutron. This was a problem encountered widely and Wrenn noted that in the Bombay lab there was a 3 MHz (megahertz) induction heater being used which could induce electrical activity in the detector much as a radio signal is picked up by a receiver. There were also very large electrolysis units with currents exceeding 40 amps. With all this electrical activity around there was, in Wrenn's opinion, the possibility that the neutron detector was responding to the noise. This is a suitable point to recall comments that David Williams, the electrochemist from Harwell, made about his experiences during Harwell's experiments. 'Like many chemists I thought that you just go out and get a neutron counter from the stores, point it at what you're interested in and register the counts. But it is not like that at all. It was impressive to me the rigour with which the experimental physicists went about it. It turned out that there are lots of reasons why you get signals on neutron counters that have nothing to do with the presence of neutrons.'

To guard against spurious readings it is useful to have several detectors directed at a cell, as at Harwell where over 40 detectors were used in concert. Spurious activity in a single detector can then be isolated and ignored. A problem at BARC, as with some experiments finding tritium at other labs, such as Los Alamos, is that experts on tritium tend to work at places where there are large amounts of it and hence ample opportunity for contamination of samples with it. BARC has a uranium reactor which uses a lot of heavy water and neutrons from the reactor produce tritium in the heavy water, so it is easy to find yourself in a 'dirty' environment. Indeed one of the experiments at BARC was conducted in a laboratory adjacent to a heavy water facility where sources of tritium are clearly available.

Other reasons for scepticism about the data's relevance for cold fusion are that the experiments are not reproducible, that the amount of tritium does not match with the neutrons—if tritium were produced by *dd* fusion it would have enough energy to knock out neutrons from the surrounding deuterated material; the lack of such neutrons implies that over 95 per cent of the tritium had origin other than *dd* fusion and the heavy water plant is the favourite culprit.

Wrenn's summary was that although the experiments were of 'technically high calibre, they are not definitive in showing the radiation spectra expected from test-tube fusion.'


## *Official conclusions in Europe and the USA*

In the USA official reactions to the test-tube fusion furore had initially been played in private with DOE laboratories working to verify the phenomenon at the urgings of the headquarters in Washington. But reactions were soon out in the open with the Congressional hearings on 26 April, which are outlined in Chapter 10.

The British Embassy in Washington rushed news of the proceedings to the Cabinet Office and Department of Energy in London. It summarised that the hearings left people 'no clearer as to the validity of the Pons/Fleischmann results.' The full written testimonies were sent to London by Diplomatic Bag. The writer had assimilated the events and their implications remarkably well, noting that Huggins' heat measurements lent some support but that he had not checked for radiation, and also emphasising that none of the US government laboratories had yet managed to replicate the effect.

The University of Utah's agenda for significant federal funding was reported, as was the US Energy Secretary's direction to all the DOE laboratories to intensify their research efforts and the formation of an inter-disciplinary panel to conduct an independent review.

There was little hard evidence that the effect was real but suddenly several institutions were in the hunt. The US authorities were interested, the British Embassy reported this to the Cabinet Office in London, and Harwell and then Euratom responded! Everyone was chasing because no one could afford to be last if the race turned out to be real.

The diplomatic telex concluded that the promised collaboration between Pons and Los Alamos National Laboratory could prove seminal and that a workshop, due to take place during 23-25 May, could produce more clarity. So, uniformly, cool, well-informed heads were able to continue to exercise scepticism even in face of the media hype and rumour. But everything was going ahead under the assumption that the normal procedures of science were at work: that Pons would collaborate with Los Alamos such that science could work through to a clear conclusion; that a whole range of probes bearing on a working cell would soon reveal what was responsible. Once the source was known, the way would be open for development.

But this was not, it transpired, normal science. The collaboration between Pons and Los Alamos never took place, and foot-dragging by a university whose lawyers were worried about protection of patent rights was blamed. And when a team from the university physics department was admitted to the Pons' laboratory at the urging of the university administration, they found *no* sign of fusion radiation at all (see Chapter 17). In the normal course of science that would settle the issue, but Pons responded by claiming that they had been looking at dead cells. Far from the passage of time leading to a clear answer, it raised more and more questions about the conduct of the experiment.

The story had begun in ways abnormal for science and so it continued. In a single week in April Pons and Fleischmann were the lead stories in *Time, Newsweek* and *Business Week*. President Bush was so impressed that he wanted to learn about it from experts at first hand and received a personal briefing from Glenn Seaborg—a nuclear chemist from the University of California and former head of the US Atomic Energy Commission.

At the White house meeting with Bush, Seaborg said that he was puzzled by the claims and urged Admiral Watkins to set up a committee of eminent physicists and chemists to help clear up the affair by comparing the various claims and counter claims and sorting out inconsistencies. Watkins took up the suggestion and appointed a panel of twenty co-chaired by chemist John Huizenga of the University of Rochester and Norman Ramsey, the Harvard physicist who won a Nobel Prize later that year. Various members of the panel visited six labs, including those at Brigham Young and the University of Utah, and also attended a three day conference on cold fusion at Santa Fe in May.

The panel estimated that 30 million dollars had been spent in the USA in universities, national labs and industrial laboratories; the global figure probably exceeding 40 million dollars. And what did this panel conclude had become of this outlay?

' . . . the experimental results on excess energy from calorimetric cells reported to date do not present convincing evidence that useful sources of energy will result from the phenomena attributed to cold fusion. In addition, the panel concludes that experiments reported to date do not present convincing evidence to associate the reported anomalous heat with a nuclear process.'

'Neutrons near background levels have been reported in some $D_2O$ electrolysis and pressurized $D_2O$ gas experiments, but at levels $10^{12}$ below the amounts required to explain the experiments claiming excess heat. Although these experiments have no apparent application to the production of useful energy, they would be of scientific interest, if confirmed. Recent experiments, some employing more sophisticated counter arrangements and improved backgrounds, found no fusion products and placed upper limits on the fusion probability for these experiments, at levels well below the initial positive results. Hence, the panel concludes that the present evidence for the discovery of a new nuclear process termed cold fusion is not persuasive.'

The panel recommended that there be no special programmes established nor research centres devoted to cold fusion. Any research projects should compete for funds as should any other area of science.

The report contained a preamble that summarised the scientific requirements for a discovery to be accepted, and contrasted this with the cold fusion status.

'Ordinarily new scientific discoveries are claimed to be consistent if reproducible; as a result, if the experiments are not too complicated the discovery can usually be confirmed or disproved in a few months. The claims of cold fusion, however, are unusual in that even the strongest proponents of cold fusion assert that the experiments, for unknown reasons, are not consistent and reproducible at the present time . . . Consequently, with the many contradictory existing claims, it is not possible at this time to state categorically that all the claims for cold fusion have been convincingly either proved or disproved.'

These conclusions paralleled those in the Euratom investigations. Their interim report published in August, soon after the US team's own mid-term report, included the following:

'The positive, erratic cold fusion findings are still considered with widespread caution and are contrasted by an overwhelming majority of consistently unsuccessful attempts to detect any signs of cold fusion from electrolytic cells. Most of the laboratories attribute the discrepancies to inaccuracies in measurements, others to non-reproducibility of a new and not understood process.

'The positive findings include: one excess heat burst, measurements

252

ruling out cosmic-ray effects on cold fusion signals, detection of neutrons and recently of many more tritium atoms than neutrons. There is no doubt that irreproducibility is the bane of these phenomena, and that reaching a sufficient degree of reproducibility is a prerequisite for any possible serious development.'

As the report itself remarked:

'The general views and recommendations of the US Panel . . . could equally well be applied to the European effort.'

One European lab, Harwell, had been very confident, on 15 June announcing that it had decide to end their research into cold fusion. Ron Bullough said:

'This work demonstrated our capacity to mount a thorough programme in basic science at short notice and the ability to put together a sophisticated cocktail of scientific expertise and equipment is the unique attribute of the Atomic Energy Authority,' but concluded that, 'The potential benefit and scientific interest in test-tube fusion, together with Government's need for information and advice meant that the subject had to be investigated. However, results to date have been disappointing and we can no longer justify devoting further resources in this area.' David Williams was on record asserting the original claims for cold fusion to have been 'a mad idea'.

To those who want to believe that test-tube fusion is real, notwithstanding this range of effort, the two continental reports are claimed not to be independent. The leading European experimenters and the US panel met at meetings and discussed data and impressions. Thus, say some, they mutually missed 'the truth'. But this misses the point of the scientific method: open discussion, comparison, extension of experience, elimination of errors, expansion of intelligence. The European and American efforts were open and collaborative, contrary to the picture of secret work painted by Ira Magaziner in the US Congress and in contrast to some of the lack of information surrounding the Utah lab.

Some 40 million dollars had been spent pursuing a phenomenon in the good faith that the evidence presented was genuinely obtained, the question being whether or not it was correct. But there was a counterpoint: while at one extreme the chemists were stars and the centre of attention all the way from magazines to presidents, in parts of the scientific community there were the rumblings of discontent. However in Utah test-tube fusion was supported positively despite the contrary indications from other scientists. When *Nature* did not publish Fleischmann and Pons' paper, the state governor, Norm Bangerter, combatively asserted that 'We are not going to allow some English magazine to decide how state money is handled'[4].

# 14

# TEST-TUBE FUSION:
# SCIENCE OR NON-SCIENCE?

## *A magic ingredient*

By the end of the spring 1989 it was clear that there were problems. Of the two signs of fusion—heat and nuclear products—some saw one, some saw the other, most saw neither and very few saw both. The measurements required sensitive instruments which were liable to error. Even those groups who reported results acknowledged that they were irreproducible; sometimes they worked, at other times they did not. This began to convince many people that there was something unscientific about the claimed phenomenon.

Douglas Morrison, a physicist at CERN in Geneva, noticed another unscientific property about test-tube fusion: whether or not you saw it seemed to depend upon where you did the experiment. Morrison commented in the summer of 1989 that reports originating in northern and western Europe always reported 'No fusion'; southern and eastern Europe reported 'Yes.' He divided the USA into two groups—the major laboratories and a region which Morrison referred to as 'in the greater *New York Times* readership' was one and reported a resounding 'No'; the rest, such as the south and north western mountains reported 'Yes.' This is not how science is supposed to work, and although you might accuse Morrison of having selected his definitions of areas after the event, the origins of the continuing claims and counterclaims agreed to a high degree with his thesis. He concluded that here we were witnessing a classical example of 'pathological science', well-known to students of 'N-rays' and the canals on Mars, as well as many other effects periodically claimed at the borderline of detectability that are visible to believers but not to others. The usual cycle in such cases, he noted, is that interest suddenly erupts, there are a few sudden verifications followed later by many careful refutations. The claimed effects are not reproducible regularly and are at the borderline between significance and background noise. The phenomenon then separates the scientists into two camps, believers and sceptics. Interest dies as only a small band of believers is able to 'produce the phenomenon', and finally it is lost to science, sometimes disproved by the development of more sophisticated measuring instruments, such as improved telescopes or, in one extreme case, actually landing on the planet Mars. However, even in the face of overwhelming evidence to the contrary, the original practitioners may continue to believe in it for the rest of their careers.

In April, in the US Congress, Fleischmann had spoken positively of many confirmations from 'far afield'. But as news of negative results, some from major research centres, began to appear, reporters began to harass the two chemists. Fleischmann and Pons fought back, the former repeating to reporters the litany:

'We have worked for five years on this and if people think they can have a quick dabble and go and find the result, then fine, let them do that but that is not our style. Our style is to do definitive experiments over long periods of time and we disapprove of the way in which some experiments are being done at the moment.'

Pons added that in many cases we were seeing 'Experiments done in the morning, papers written at lunch and published that evening.'

Fleischmann reiterated: 'We have sat for five years on this and I think that people doing the work should respect that.'

One feature of test-tube fusion that everyone agrees on is that it is not reproducible. Reproducibility of a phenomenon is a central criterion for its scientific acceptance and so it is this aspect more than any other single one that causes the greatest scepticism. Douglas Morrison labels it 'pathological science' accordingly; others vent even stronger opinions asserting that this proves that there is nothing to test-tube fusion and that people who insist in believing in it are deluded. Herein lies the source of the religious fervour — it is less an issue *within* science, rather it is a question of whether there *is* science. Does irreproducibility show that there is no real phenomenon (hence unworthy of the label 'scientific'), or is it because there is some unknown condition, present in 'successful' experiments and absent in others?

Serious scientists who persist in studying test-tube fusion adhere to the latter view. They have considerable leeway as many experiments, whether 'positive' or 'negative', do not address all the issues nor report in sufficient detail to be tested for reproducibility.

As examples, most experiments have followed the initial examples of Fleischmann and Pons and/or Jones, using palladium cathodes for electrolytic charging and titanium or its alloys for gas charging experiments. Most electrolytic experiments used platinum anodes. Now, one group at Texas A & M claimed to have measured tritium produced in experiments using *nickel* anodes; for some people this suggested that maybe nickel was a special ingredient. However others have used nickel and seen no tritium; also there have been claims in India for tritium without using nickel. The moment one identified a potential 'eye of newt', it could be checked and eliminated.

There was still much leeway though. Even with palladium cathodes and platinum anodes there had been almost no characterisation of their compositions, microstructure or homogeneity; these played little part in attempts to understand test-tube fusion. There were persistent and fluctuating rumours that palladium when cast in moulds would give heat in the experiments whereas extruded palladium does not. There were some who insisted that 'cast' palladium is a key; yet there were also negative results with the former and claims for heat from the latter. You can buy rods from several manufacturers and if one of these turned out to give heat, for whatever reason, it could have great commercial advantage. Fleischmann and Pons' rods were supplied by Johnson Matthey and some have wondered if this company has a secret production process; however other groups using Johnson Matthey palladium (in particular Harwell) saw no heat.

If the eye of newt is in the preparation of the rods then nobody knows that it is, except, perhaps, in one case of the tritium seen at Texas A & M and possibly at Los Alamos too, where it turned out to have been in the rods from the outset. This is one example where a 'phenomenon' depended upon certain details (which in this case had nothing to do with test-tube fusion occurring in the rods). Keeping track of all the fine details would have been easier if more complete reporting of some experiments had been made. Regrettably there was much poorly documented science.

The idea has been to load palladium with deuterium. Is there some optimal concentration, amount of deuterium per palladium atom, that will 'induce' the process? Various experimenters report that they have managed to load between seven and eleven deuterium atoms for every ten palladium atoms, but some report heat while others see none. Most experimenters don't report details of their procedures so it is difficult to assess how accurately the deuterium amount has been measured. The measurements were often made by weighing the cathode before and after loading, however some experimenters have noticed deposits on the surface of the cathode, caused either by lithium from the electrolyte or trace elements in the liquid, and the amounts of these need to be known and taken into account before the weight gain due to deuterium alone can be assessed. In the majority of cases it is not possible to tell from the published papers whether all of the necessary procedures have been carried out and thereby to know with confidence what the true deuterium content was. Some experimenters have attempted to force the deuterium into the cathode by pressure instead of electrolysis. However, exposure to the gas does not ensure that any significant amounts of the gas are absorbed and in several of this class of experiments there is no evidence that the metal contained any deuterium at all! Such doubts can be seized on by test-tube fusion aficionados to suggest that failure to see heat was because the required conditions had not been correctly set up.

But what are the 'required conditions'? To this there are no good answers and several suggestions have ignored the fact that the behaviour of deuterium and hydrogen in palladium metal is not a dark mystery; in fact a great deal is known to metallurgists, physicists and chemists already. In particular they know how long it takes for deuterium to diffuse in and saturate the palladium, and also how it behaves once it is in place.

The original idea that fusion might occur was based on a misconception: that when deuterium is packed into palladium at pressure, the chance of the deuterons coming together and fusing would be significant. Fleischmann and Pon's original paper claimed that the effective pressure induced by the electrical potential would approach $10^{27}$ times atmospheric, however, this huge number has very little to do with the real pressures that deuterium atoms achieve in the metal. The ratio of deuterium atoms to palladium is well known as a function both of temperature and pressure. To achieve a ratio of one to one, which is typically what is measured by the test-tube fusion experiments, requires at room temperature a pressure of only some 10 to 20 000 atmospheres—which may sound large but is nothing unusual, certainly nothing like the billions of billions suggested in Fleischmann and Pons' paper.

The picture that these large numbers create in many people's minds is of deuterium atoms being squeezed so close together that they have a chance to fuse. However the reality is very different. Palladium has a close packed crystalline structure known as face-centred cubic. Deuterium takes up vacant sites among the regularly spaced palladium atoms, overall forming the corners of an octahedron. Studies of the palladium – deuterium system can be made by scattering X-rays (X-ray crystallography), and a lot is known about the way that the palladium dissolves the deuterium. The closest approach between the neighbouring deuterium nuclei when inside palladium is 0.17 nanometers, more than double the bond distance in deuterium gas molecules which is 0.07 nanometers. Deuterium gas does not spontaneously fuse, and the situation within palladium is even *less* advantageous, quite opposite to what many people were led to believe by reports in the media.

Another important feature is the way that gases diffuse into, and out of, palladium. The starting hypothesis of Fleischmann and Pons was that you must load with deuterium and, when the palladium was saturated with it, fusion would occur. When at the Baltimore APS meeting in May 1989 several experiments reported no heat, the excuse given by Fleischmann and Pons's supporters was that it takes many hundreds of hours of charging with deuterium if an experiment is to be successful. However, Fleischmann at Erice on 12 April said that one to two days was adequate for a 1 mm diameter rod and that the time grows as the square of the diameter such that one week should be allowed for a 2 mm diameter rod and one month for a 4 mm one.[1]

A group at Texas A & M reported seeing dendrites — small tree-like growths on the surface of the palladium — in their experiment in which they also reported tritium. It is quite common to find such dendrites during electrolysis (such as in electroplating) so it is not clear why they should have anything to do with test-tube fusion. A suggestion was that there might be significant electric fields at the tip of the dendrite which accelerate the deuterium nuclei so that they have a better chance of colliding and fusing. However, this idea fails as soon as quantitative estimates are made: voltage drops in the cell are no more than in a small battery and so cannot accelerate deuterium to more than a few eV of energy, whereas fusion requires much higher energies (the temperature at the centre of the Sun corresponds to about 1000 eV, or 1 keV). And after these theoretical concerns there is experimental refutation too: the Harwell experimenters noted dendrites but found no heat and no tritium.

## Calorimetry

The story began with Fleischmann and Pons claiming to produce up to 10 watts of excess power per $cm^3$ of palladium, maintaining this for 120 hours giving a total of 4 megajoules of energy. The heat claimed by Martin's group at Texas was subsequently withdrawn and on 1 May the Caltech group announced at Baltimore APS meeting that they saw no excess heat. However, on 9 May at the Electrochemical Society meeting in Los Angeles three groups reported heat (though one subsequently withdrew its claim) and there was the first announcement of an energy 'burst', producing 4 megajoules in two days (fifty times the electrical input).

The first major evaluation of test-tube fusion took place during 23–25 May at Santa Fe, New Mexico. There was a mixture of reports, some claiming to see heat, others finding none. There were no confirmations of the 'bursts' of energy that had been reported at the Electrochemical Society. One of the more significant, though unreported, happenings at Santa Fe was that the first meeting took place of the Department of Energy Panel that had been formed in response to the 24 April letter from Admiral Watkins, the Secretary of Energy, in which he called for a review of the test-tube fusion claims. This panel included physicists, chemists, nuclear engineers, geologists and materials scientists. During June and July members of the panel visited the University of Utah, Brigham Young University, the groups at Texas A & M who claimed heat, and some experiments at Caltech and Stanford. The most important aspect of their work was that they saw the various pieces of experimental apparatus and evaluated them on the spot. As such they provided the only extensive 'behind the scenes' investigation, and their report raised several disturbing questions. Anyone intending to make a major investment into test-tube fusion would do well first to study their report before deciding how much to gamble[2]. The quotations in what follows are taken from it. One should be stated right away: 'In none of our visits to the different sites did we see an operating cell that was claimed to be producing excess heat at that time.'

In much of this book I have concentrated on the evidence for fusion, or rather, the lack of it. Yet there is still, a year later, the tantalising possibility that the heat (or more precisely the power) input from the electricity supply and that measured from the cell do not balance— some groups insist that there is 'excess' power.

Although the basic electrolysis experiment sounds simple, the careful accounting of the power is rather subtle and requires considerable care. As was the case with the misleading neutron counters in the example of nuclear measurements, so can the unwary be led into error in the calorimetry.[3]

First, what is happening. As outlined earlier, as electrical current passing through heavy water, deuterium oxide, the liquid splits into constituent deuterium ($D_2$) and oxygen ($O_2$). Energy is used up in doing this and could be recovered if the $D_2$ and $O_2$ gases recombined to form again heavy water. If you electrolysed heavy water and closed the cell so that the gases totally recombined within, then (if there is no 'new' heat source such as fusion) the total electrical power in and heat power out should balance; the electrical current will heat up the system until it reaches a temperature where the heat leaking out balances that going in. Nearly all such experiments have reported no excess heat; the 'heat producers' are almost all using 'open' cells, where the $D_2$ and $O_2$ gases can exit the cell and are assumed by the experimentalists to have done so without any recombination. When drawing up the power accounts ledger the experiments have included how much power can be gained back by recombining all of the evolved gases; but if some of the gas have already done so, then you will be counting it twice and end up with an apparent excess.

Many groups took precautions against this error by measuring the volumes of the emitted gases and compared them with the volumes expected from the amount of electrical charge (current multiplied by time it has been flowing) that has passed through the cell. Fleischmann and Pons in their paper of March 1990 claim that the volumes of gases evolved from their cells agreed with the amounts expected and so they are not subject to this error. However, the DOE panel noticed that in some other experiments these measurements were 'not made at all', and in other cases were 'not made with sufficient precision and accuracy to assure the absence of recombination at a level that could account for thermal excess.' There were even more astonishing errors uncovered by the DOE panel; 'In a surprising number of instances, cells were operated with platinum or palladium surfaces exposed . . . above the solution.' The venting gases could well come into contact on these surfaces, breaking the assumption that there was no recombination. Deficiencies such as these are not mentioned in written papers; they only became known by informed visitors to the relevant laboratories.[4]

A related problem is the possibility that deuterium produced at the cathode may bubble across to the anode and combine there with oxygen, converting back to heavy water within the cell. Such an occurrence would affect the calculations about how much gas should be being vented. 'In virtually all cells the anode and cathode operate without a diaphragm separating them,' and so some recombination will occur and contaminate the conclusions.

Fleischmann and Pons claim that any reaction between deuterium and oxygen in their cells is incompatible with 'the marked differences between blank experiments and electrolysis of $D_2O$ using most palladium cathodes.' This alludes to the fact that in experiments with *platinum* electrodes their heat in and out balanced, as well as the fact that in some experiments with large (0.8 cm diameter) palladium rods in heavy water the heat had also balanced. These latter 'should' have been 'fusion cells', and later on heat generation with 0.8 cm rods was seen when the rods were prepared by 'modified procedures'. If a large batch of cells of one type always gave heat while a large number of another never did, then one would have a convincing proof; at present it is still a matter of preference whether you wish to concentrate on the 0.8 cm rods that give heat, or those that do not do so, as being the 'anomalous' ones. Not until there are published details of all rods can one form a statistically significant sample.

In addition to these experimental problems there were several examples where the numerical evaluation of the data and assessment of errors were incorrectly or badly done or, in some cases, not done at all.

The input power is the product of the amount of current and of the voltage. The practice is to hold one of these fixed (say the current) and then to measure the other (in this case the voltage). The relation between current and voltage depends on the total electrical resistance of the cell and this varies during the experiment as liquid levels drop during electrolysis and rise on topping up, and also vary when the temperature changes. The current may be fixed very precisely, to one part in a thousand, but the measured voltage can vary instantaneously by several per cent so it is the measurement of the latter that determines the accuracy.

One reason why proper care to eliminate these errors can turn out to be critical is because the amount of excess power claimed is comparable to the amount of the electrical power input, and rarely exceeds the amount *assumed* to be available if the vented gases could be recombined. The point at issue here is that the quoted total heat output already assumes that this heat of recombination is still available and so is added in. However, if some already recombined, you are illegally 'topping up' the sum. Thus incorrect practice could account for the supposed excess heat.

The DOE panel commented that there had been a noticeable lack of attention to the statistical assessment of errors, and that in some cases, where heat was being claimed, 'a group's claim of excess heat is not supported with results of sufficient precision to allow such a conclusion. More usually it is not possible to assess precision from reported results because the result is reported from a single run and no error bars are provided for the measured parameters.' When I saw Martin Fleischmann in February 1990 I remarked that failure to quote errors was one of the reasons why some people had commented adversely on their claims. He insisted that the errors had been quoted and, on checking his own paper, expressed astonishment when he realised that indeed there were none. The DOE panel noted: 'Conclusions in this area simply cannot be accepted without a thorough assessment of the measurement errors. In its visits and conversations the members of the panel were struck repeatedly by the absence of critical assessments of this kind.'

They cited three specific criticisms: (a) lack of sufficient controls; (b) failure to assess errors; and (c) incorrect use of 'significant figures'.

The lack of sufficient controls was one of the initial criticisms levelled at Fleischmann and Pons. A 'test' cell involves deuterated palladium cathodes in heavy water; 'control' cells feature either platinum cathodes in heavy water, or palladium in plain water. If test cells regularly produced heat then a single demonstration that a control cell balanced would be definitive. However, everyone agrees that excess heat is, at best, erratic and that many cells with palladium and heavy water give no heat. Consequently many test cells and many control cells are needed; to be statistically meaningful similar numbers of both types of cells need to be studied.

When Fleischmann and Pons produced a more extensive set of calorimetric data in 1990 they included several cells with light water for which the heat balanced, and cells with heavy water for which there was excess heat. Yet their statements through the year had also implied that there were several heavy water cells that gave no heat and light water cells that did (though no mention of the latter appears in the paper).[5] Until a complete set of data is available, one cannot safely conclude that a case has been made for a reproducible phenomenon.

With regard to the failure to assess errors the DOE panel noted that measurement precision was often assessed by one or two rather than repeated measurements, or that the accuracy in reading a voltameter was assumed to be given by the ability to locate the position of an indicator on the scale without account of fluctuations in cell voltage or other small experimental matters. When a set of data points with some amount of scatter was fitted by a straight line, whose slope could be a critical calibration in the final evaluation of excess heat, it was often done so without taking account of the possible deviation in the slope from the cited value. Thus even when errors were quoted, their true value was frequently under estimated.

This spills over into violation of simple rules concerning figures. Input and output power are measured to accuracies no better than two or three significant figures, and yet are tabulated to four or more. This gives the impression that measurements are more precise than in fact they are. 'Many investigators do not say in public reports exactly how the measurements are made; only by visiting the laboratories does it become evident that tabulations of quantities with so many significant figures is wholly unjustified. This practice is insidious in the absence of real assessments of precision and it could explain why some groups see positive heat effects only erratically.'

Among other concerns there is the question of how one determines

262

the power output from the measurements of temperatures and electrical quantities. This is the problem of calibrating the cell.

Most groups that reported excess heat had calibrated their cells by temporarily adding power, for example by putting some electrically resistive element in the cell, and seeing how the steady temperature changed. This gives a measure of the 'differential heat transfer', $k$, that is a correlation between power and temperature changes. You do this while the cell is running and can then calculate the total power balance by multiplying $k$ (the 'power per degree') by the number of degrees difference between the cell and the external sink.

The cell might be at 50°C and the external water bath at 20°C. The small heat input via the resistance gives the measure of $k$ at or about 50°C. If the magnitude of $k$ is the same for all temperatures down to 20°C then you can infer the power in the cell. Many experiments assume that $k$ is constant like this.

However, in practice $k$ is not so simple and has a complicated dependence on temperature and power. You need to know $k$ over a range of these parameters and suitably average them. This complicated dependence has an unfortunate consequence in that it causes this calibration method to overestimate the evolved heat power. An error of a few percent in the assumption that the magnitude of $k$ is a constant 'would invalidate nearly all reports of excess heat'.

These are but a few of the several technical concerns raised by the DOE panel. They noted that these experimental shortcomings were widely encountered. 'No single study was compromised by all of them but no positive report that we were able to study in detail was free of such problems.' And there was a significant comment directly about the experiments: 'In none of our visits to the different sites did we see an operating cell that was claimed to be producing excess heat at that time.'

There is one possible example of heat production that could be real and awaits explanation. There have been reports of abrupt changes in cell temperature which appear sporadically usually after some operating condition has been changed, such as a change in the current. Often these are small and are subject to the same concerns about accuracy as outlined above. However, there have been reports of large bursts, with temperature rises of tens of degrees—even boiling the electrolyte. These seem to be too big to be dismissed as calibration errors. However they had been mentioned only in press reports or in talks and detailed written reports were not available for assessment at the end of 1989 when the DOE completed their report. They described them as 'mysterious and perplexing'.

\*　\*　\*

The heat excesses seen sporadically by Hutchinson and by Scott at Oak Ridge were among these 'mysterious and perplexing' phenomena.

When we left them, on page 231, Hutchinson was planning new experiments to begin in September. These were only just started when I went to the discussion meeting at Oak Ridge on 4 October 1989. The contrast with the hectic days of April highlighted the whole community's feeling about test-tube fusion. In April through June the meetings had been weekly, with as many as 50 people attending. By October they were down to twice a month and on 4 October there were only five people present, one of whom, Misha Petek, an electrochemist working in Charles Scott's group, reported that their cell had been producing 2 to 3 watts excess heat, some 37 watts going in from the electricity supply and from the temperature rise they computed a net 40 watts of heat was going into the water, an 8 per cent excess over what they were supplying.

The water in the bath surrounding the cell flowed continuously, entering cold and leaving slightly warmer. The electrolysis was splitting the $D_2O$ inside the cell into deuterium and oxygen but in the closed cell these recombined into $D_2O$; none was lost. Originally they had run the external water bath at 24°C and the heat account balanced, and stayed balanced as they varied the amount of electrical input. After 1300 hours, nearly two months, they reduced the temperature of the input water to 8°C and from this moment the excess heat began and had continued unabated for 200 hours. Their rod of 3 mm diameter, 8.5 cm long palladium and the current density, 0.6 or 85 mm amps/cm$^2$, were all similar to those of Fleischmann and Pons.

The power input and output were being monitored continuously and the computer printed out a report every five minutes. It showed 37.1 watts in and 40 watts out at 2 p.m., five minutes later a little more was going in, 37.6 watts and 39 watts out, by quarter past two it was 36 in and 43 out. For over 300 hours the record showed that this had been going on. Awaiting the printer became compulsive and each successive datum helped to build up the conviction that there was a real effect — though whether it was from the deuterated palladium or some other cause one could not say.

Hutchinson's group meanwhile had built four cells with the same characteristics as their single successful one: a smooth annealed palladium cathode and no precharging with deuterium. By October it was clear that, according to their calibrations, each of these cells was producing excess heat in amounts that rose in proportion to the electrical current so long as the current density exceeded about 50 mA/cm$^2$.

In one of the cells they replaced the $D_2O$ with $H_2O$ and the excess heat died away—the cell's heat came into balance. Hutchinson said 'We were proud of that. When it went into balance we were really thrilled. It verified a lot of things.' They replaced the $H_2O$ with $D_2O$ but it stayed in balance. The three other cells continued to generate excess heat.

They terminated the experiments in December because they ran out of money. By this time they had electrolysed 16 kg of heavy water, vaporised it into the atmosphere, at a cost of \$400 per kg. So here are two reasons why a closed cell where the gases recombine into $D_2O$ is best: the book keeping is more reliable and you don't gas away your money. Their plans for 1990 are to run a pair of cells, one $D_2O$ and one $H_2O$, in every experiment that takes place, to have internal recombination of the gases into $D_2O$ and to collaborate with other groups from DOE laboratories who have not seen heat. By this strategy one side or other should begin to focus in on the source of the phenomenon, or of the error.

## Tritium and helium

'This must be aneutronic fusion,' Martin Fleischmann remarked in 1990. As the first anniversary of the Utah press conference approached it was clear that neutrons, if produced at all, were at very low levels. If Fleischmann and Pons were right to claim that power is produced at rates of watts per $cm^3$ and that it is due to fusion, then there have to be of the order of $10^{12}$ particles produced each second. Whether these are neutrons, protons, gamma rays, tritium or helium, they must come at this rate: nuclear reactions liberate of the order of 1 MeV per fusion and you need $10^{12}$ each second to reach the required rate. At the very best neutrons fell short by factors of billions. So attention focused on other signs of fusion products, notably tritium and helium.

In $dd$ fusion there are only a very few possible outcomes. Each deuterium nucleus contains 1 proton and 1 neutron, so the reaction produces either nuclei of tritium ($t = 1$ proton and 2 neutrons) with the spare proton ejected, or helium (2 protons and either 1 neutron in $^3He$ or 2 neutrons in $^4He$). In the case of $^3He$ production, the spare neutron is ejected. These processes and the way the energy is shared among the products in 'normal' fusion is

$$d + d \rightarrow {}^3He(0.8) + n(2.5) \qquad 1.9 \times 10^{12}$$
$$d + d \rightarrow t(1) + p(3) \qquad 1.5 \times 10^{12}$$
$$d + d \rightarrow {}^4He + \gamma(24) \qquad 2.6 \times 10^{11}$$

where the numbers in the parentheses are the energy in MeV and the numbers on the right are the number of reactions each second for 1 watt output. There are small traces of tritium and also individual protons around in heavy water, so for completeness here are the other possible reactions of the hydrogen isotopes:

$$p+d \rightarrow {}^3He+\gamma(5.5) \qquad 1.1\times10^{12}$$
$$d+t \rightarrow {}^4He(3.5)+n(14) \qquad 3.5\times10^{11}$$
$$p+t \rightarrow {}^4He+\gamma(19.8) \qquad 3.1\times10^{11}$$

Each of these reactions requires the various nuclei to penetrate the electrical repulsion of their mutual positive charges. Fusion of the nuclei of heavier elements, such lithium and palladium, have even greater electrical forces to overcome due to their larger electrical charges, lithium being three and palladium 46 times more charged than $p$, $d$ or $t$. It is difficult enough to accept that the hydrogen isotopes can achieve fusion at room temperatures at the rate required to explain the heat, let alone when the problem is even more out of reach because of the larger charges of lithium and palladium. To bring this problem into context once more, remember that at room temperature a galactic mass of deuterium would produce roughly one fusion per second; watts of heat require $10^{12}$ such fusions in a small flask.

The fact that theoreticians cannot accept this does not of itself mean that it cannot happen, but there must be more evidence than just the heat if one is to convince others that nuclear fusion is taking place—the deuterium (or something—lithium, palladium?) must be changed in order to release the energy. And what it changes into is something that we should also be able to find. Moreover we know how much should be there. I could list all of the possible fusion channels of all the elements in the cell, palladium, electrolyte, nickel, but if nuclear reactions are producing the heat, then whichever process is responsible it must possess one feature which is already apparent from the list of reactions that I drew up above: *whatever the nuclear reaction is, at least one hundred billion ($10^{11}$) of them must occur each second to generate a watt.* This follows quite generally from the fact that the energy taken up in holding nuclei together—the 'binding energy'— is of the order of a few MeV (million electron volts). Now since 1 joule is approximately $6\times10^{18}$ eV then, very approximately, we can say that 1 watt (one joule per second) is approximately $10^{18}$ eV/sec or $10^{12}$ MeV/sec, in other words, one hundred billion releases of ten million electron volts of energy per second.

*The products must be present in amounts commensurate with the heat and, if you know what to look for, can be found.* Thus experiments have looked for the traces of helium and tritium that should be present.

Complementary to this, and in many ways better, is looking for evidence of fusion as it happens. While the tritium, or helium, or new isotopes of palladium (or whatever your preferred mode may be) is being produced, neutrons or protons are being rearranged. There must be some $10^{12}$ of these occurrences each second per watt, and their products are nuclear fragments carrying energies thousands of times greater than the electrons bound in the atoms of the material. These fragments will knock the electrons out of their parent atoms and can excite the nuclei of those atoms causing them to emit gamma rays of characteristic energies. When Salamon's group in Utah looked for signs of these radiations in Fleischmann and Pons' cells, it was doing more than eliminating just the *dd* fusion channels but was implicitly constraining, albeit qualitatively, the possibility of there being *any* significant exotic fusion processes.

The earliest piece of the fusion claim to be disproved was the process $d + d \rightarrow {}^3He + n$; the neutrons are easy to see in careful experiments and the debate is whether they are present at the level claimed by Jones. As this is orders of magnitude below that required to explain Fleischmann and Pons' results I have not pursued the fascinating story of Jones' work beyond 24 March 1989; it is the Fleischmann and Pons claims that drove the media interest and the appeals for funds. The implicit question was whether the small rate, or even absence, of neutrons could also implicitly rule out the other main channel, $d + d \rightarrow t + p$, as this is known to occur at essentially the same rate as ${}^3He + n$ over a wide range of energies.

This near equality of the two production rates is a consequence of the basic symmetry of the nuclear forces between proton and neutron. This symmetry is broken by the electrical forces which repel the protons. So some have suggested that the two deuterons prefer to come together oriented such that their neutrons make closer approach than their protons. In this case it is more likely that a spare remote proton breaks off, thereby favouring the $t + p$ channel over the ${}^3He + n$. This asymmetry would only become noticeable in low energy collisions, as in test-tube fusion, being masked in the high energy collisions that have been extensively studied by the nuclear physicists.

Supporters of test-tube fusion have appealed to this in the hope that it might excuse the absence of neutrons in favour of tritium and protons. There are even spurious claims that experiment supports this, for example, in his early talks on test-tube fusion, Martin Fleischmann claimed that the $t + p$ channel excess was known to high energy physics and he has referred to the original discovery of the $t + p$ process by Rutherford, Oliphant and Harteck in 1934. This paper states: '*Rough estimates* [my italics] of the number of neutrons suggest that the reaction (at 20 keV energy) which produces them is less frequent than that which produces the proton (and tritium).'[6]

However, this remark in 1934 was not quantified and was admitted to be 'rough', and more extensive data during the subsequent half-century have shown that the equality of the two channels is known to be preserved from high energy through 20 keV and down to about 5 keV. A reason that it is not as well known below this energy is because the individual rates are so low. However the rate is known at room temperature from muon catalysed fusion experiments. The neutron and tritium production rates are intrinsically equal (there is a *slight* favouring of the tritium channel due to the $t+p$ combination being slightly lighter and so easier to produce than $n+{}^3He$). As the temperature rises to around 200°C the *neutron* channel even dominates by up to 40% through a molecular resonance excitation. Not only does experiment show that $n+{}^3He$ and $t+p$ are essentially equal at room temperatures, but theory can even accommodate the subtle variations in the ratio at these low temperatures.

It is surprising that this popularity for tritium was ever credible, as experiments have limited this channel in the palladium electrolytic cell with even better precision than the well-publicised limits on neutrons. The idea has been to seek either the 3 MeV energy protons and/or the 1 MeV tritons, tritium nuclei, in the $d+d \rightarrow t+p$. Several experiments have looked for this and there are several that limit it with the similar precision with which Jones' experiment limits neutrons, namely to less than about 1 fusion per $10^{23}$ *dd* pairs per second, around a billion times below that corresponding to the heat that Fleischmann and Pons claimed.

Besides the direct detection of tritium or helium nuclei as they are produced, there have been searches made for their accumulation within 'heat-producing' palladium electrodes. This is more difficult than detecting them during production. The detection of accumulated tritium is by its beta-decay, where it emits an electron that can be identified. But tritium decays only at a rate of 5% each year, or put another way, out of every hundred million tritium atoms one will decay per minute. There is about 1 part tritium to $10^{18}$ of hydrogen in ordinary water; the normal manufacture of heavy water enriches the tritium and each millilitre of typical commercially available heavy water gives between 120 and 180 disintegrations per minute ('dpm/cm$^3$') from tritium decay. So it is important to measure the number of dpm/cm$^3$ in the heavy water before you begin the experiment. Experiments at all of the major laboratories did this, and found at the end of the electrolysis experiment that any increase in the tritium is due to electrolytic enrichment, and there is no evidence that any has come from fusion. The levels of tritium that

Fleischmann and Pons reported also appear to be consistent with this interpretation.

Experiments have been made in closed cells, and these find that the total excess tritium formed in the heavy water is less than $10^4$ tritium atoms per second and so can give an excess power of less than one hundredth of a millionth of a watt. M. Wadsworth of the University of Utah reported to the ERAB-DOE panel that in an open cell his group had observed a heat burst of 35 watts lasting for an hour and a half; however there was no excess tritium found after this burst, suggesting that the heat burst is not from the $d + d$ reaction.

Although there is no steady nuclear process going on that could produce microwatts, let alone the levels of power claimed by Fleischmann and Pons, nonetheless there were some anomalous observations of tritium that cold fusion believers insisted were proof of unexplained nuclear processes. Very occasionally a few experimenters reported (irreproducible) amounts of tritium in the heavy water samples drawn from their electrolytic cells after days of operation.

Groups from Texas A & M[7] and Los Alamos[8] and also from Bombay[9] have reported significant amounts of tritium at levels of $10^{12}$ to $10^{14}$ atoms per $cm^3$ of heavy water following several hours of electrolysis, and amounts far in excess of those that are possible purely by enrichment. These effects are not reproducible and are not found in all cells. A team led by Kevin Wolf at Texas A & M then looked for neutrons produced in similar cells and found that there was less than one per second, hence $10^9$ times below the tritium rates.

The Indian experiments also claim to find tritium levels comparable with those reported from Texas A & M. As in the Texas experiments, there is no sign of neutrons. The Indian abstract reports that the 'total quantity of tritium generated corresponds to about $10^{16}$ atoms suggesting a neutron to tritium branching ratio less than $10^{-8}$ in (test-tube) fusion.'

These levels of tritium, though only present in occasional cells, were individually too high to be dismissed as an error of measurement. However, it was also clear that the tritium was not being produced by $d + d$ fusion, for the following reason.

Any tritium nucleus ('triton') produced in $d + d$ reactions has an energy of 1 MeV which must be lost in the surrounding material. In an electrolytic cell this material contains a lot of deuterium, such as in the heavy water or in the saturated palladium electrode. The interaction of 1 MeV tritons with deuterium has been studied extensively by nuclear physicists and is known to be a rich source of neutrons, which emerge with an energy of 14 MeV. For every $10^5$

tritons produced by the $d+d$ fusion, their subsequent collisions with the surrounding material will produce one neutron; every $10^8$ tritons therefore will knock out a thousand neutrons. The Texas and the Indian experiments see none of these, so their tritium is being produced with no energy, which showed that it was not produced by $dd$ fusion nor by any other known nuclear reaction.

This tritium is almost certainly a contamination—it appears on isolated occasions and, at least in the Bockris group experiments at Texas, in quantities far exceeding anything claimed elsewhere (and bear in mind that the majority of the (few) experiments that claim any heat see no significant tritium at all). The claims from Bockris' group have been so remarkable that they have been controversial ever since Kevin Wolf first spoke about them at the Santa Fe workshop in late May 1989. There were clearly concerns about this in Texas as there was much gossip around the electronic mail networks and privately among many concerned scientists that the experiment had been doped. Bockris' paper specifically attempted to discount this, claiming that security around the laboratory was too good for this to happen. Yet there was also rumour that the security was rather lax and in *Science* on 15 June 1990 these concerns are well documented.

The first signs of official withdrawal of some of the claims to be producing tritium appeared in the *Wall Street Journal* on 6 June 1990. This article suggested that palladium rods used at Texas A & M and also at Los Alamos had been contaminated with tritium during their production process.[10]

This followed work done by Kevin Wolf, who had found traces of tritium in some palladium cathodes which had not been used in experiments. He also tested some of the liquid electrolyte from a fusion cell run in Bockris' lab that had shown a high tritium presence. The liquid had originally been $D_2O$ and now had tritium in it, but Wolf discovered that it contained some $H_2O$ as well. This fits with rumours that the source of the tritium had been a bottle of tritiated water that was known to exist in the lab. If someone had taken drops of tritiated water and added them to the heavy water then it would dramatically increase the tritium content of the heavy water, but also add plain water to it.

Was this where the tritium in Bockris' cells came from? Gary Taubes, in *Science*[10], certainly makes a strong case for this and time will tell whether such mischief has indeed taken place, helping to induce the wastage of millions of dollars and immense hours of effort in test-tube fusion research. Whether the tritium was introduced innocently or malevolently has serious implications for the ethics of the experiment, but whichever of these it is, for test-tube fusion the general message is destructive: the tritium emerging during the electrolysis has not been produced by a nuclear reaction within the experiment — there is no sign of any nuclear products ever having been produced at levels significantly above the ground.

As with helium, so now with tritium, Rutherford's edict of 1930 applied right to the end:

'The presence of an element has been mistaken for its creation.'

# III

# REVELATIONS

In Utah after 23 March 1989, and evaluations of what Fleischmann and Pons really did compared to what they were perceived to have done.

# 15
## THE SPY IN THE LAB

In Utah on 24 March 1989, the day after the press conference, the phones in the chemistry department were permanently occupied. Work on the test-tube fusion cells stopped as Pons, Fleischmann and Hawkins were overwhelmed by calls which came continuously from corporate America seeking details and from universities, labs and other institutions wanting talks and raising questions about details of the procedures. The fortissimo of interest had heard only a pianissimo of explanation.

Fleischmann was preparing to go to England and the plan was that Hawkins would drive him to the airport in expectation of meeting Jones there and sending off their papers. They waited at the prearranged place but nobody from BYU turned up. (Jones had faxed his paper to *Nature* the previous night following the news from the press conference that the chemists had already submitted a paper to an unnamed journal.) Eventually Fleischmann could wait no longer and said that he would have to catch his plane, so they express mailed the paper to *Nature*, Hawkins returning to the university and Fleischmann departing to San Francisco en route to Europe. For the next critical weeks, until the final week of April, Fleischmann was in Europe—Harwell, CERN, Erice in Sicily—and the focus of attention on Utah was increasingly handled by Pons. Hawkins, whose name had been omitted from the original paper, was transferred off the project at the end of March and took a vacation.

It was around this time that the first hints of problems arose.


Thursday 23 March had dawned with Fleischmann and Pons almost unknown outside their scientific circle; by the evening they had become celebrities, and were being fêted as having found the key to pollution free, abundant cheap energy, and portrayed as potential saviours of the human race. Rarely has serious talk of Nobel Prizes, fame and fortune followed so fast after a claimed scientific breakthrough. Rare is the student who, entering into a scientific career, does not dream that they may someday be considered for a Nobel Prize; for the two chemists that possibility suddenly seemed to be real. But there is also the nightmare that some disaster will befall, like Icarus who flew too near to the Sun, that in the moment of success something has been overlooked and that the foundations are flawed. For Pons and Fleischmann the nightmare threatened to become real when at his talk at Harwell on the Tuesday 28 March Fleischmann learned that a gamma-ray signal for neutrons, an important part of their evidence for fusion, should occur at an energy of around 2200 keV and not 2500 keV as in their data.

Their papers had already been sent out to the *Journal of Electroanalytical Chemistry* and to *Nature*. If the figure of these data appeared it would be a major embarrassment and create serious doubts about their claims to have evidence for neutrons. The paper had already been accepted by the chemical journal and the printing was imminent. A difficulty for Pons was that the data had originated with Hoffman and Hawkins and there was a desperate urgency to understand what had gone wrong, and to remedy it.

Meanwhile, in the physics department, people were bemused. Claims had been made in a press conference at their university to the effect that fusion had been achieved by chemists, who had detected nuclear radiation, and no one in the physics department, not even the nuclear specialists, knew about it.

Carlton de Tar, a professor of physics, tried to call Pons in the hope of arranging a technical talk about the fusion project, but Pons' phone was permanently busy. After several unsuccessful attempts de Tar decided that he had better go across the campus and see Pons directly in the chemistry department. He was greeted by a hectic scene: a member of the department was standing in the secretary's office trying to talk to Pons while Pons was handling a phone call, the secretary was meanwhile answering another phone, and the moment that either phone was put down it would immediately ring again. Through this de Tar was able to make a brief contact in which he learned that Pons was due to give a seminar on the subject in Indiana in about ten days time (4 April), and that the deluge of pressure that had descended on him made it unlikely that he would be able to fit in anything else very soon. De Tar then called James Brophy, the vice-president for research, and suggested that it would be nice if the first technical talk on the phenomenon were given at the home university rather than somewhere else, and as a result this was arranged for the following Friday, 31 March.

From all around the world physicists were calling up their colleagues at Utah assuming that the local physicists had given their seal of approval to the whole business. Indirectly the reputations of Utah physicists were on the line, and so they were very careful to tell people that not only did they not know about the Fleischmann–Pons experiment but that they had not even had an opportunity to see a written report ('preprint').

A number of people in the Utah physics department were very upset that the announcement had been made by the University without anyone having consulted them on things that they were experts in. Their only source of direct information had been the news conference and their only source of indirect information was the electronic mail from colleagues around the globe. The electronic network made everyone coincident in space and time; there was no longer any advantage in geographical location.

Physics professor Eugene Loh called James Brophy to explain their difficulties, and said that they wanted to talk with Pons about his claims to have seen gamma rays and neutrons, in general to substantiate things, and to see a copy of the paper that according to news reports had been sent to *Nature*. Brophy said that he had a copy and that Loh could come over to see it, but not to copy it nor to take it away, and that he could bring one or two colleagues but not more as he did not want a circus. The eventual party consisted of Craig Taylor, Orest Symko and Mike Salamon, the specialist in cosmic rays who only a few days earlier had had the chance meeting with university president Petersen on the plane, where he had been quizzed about fusion (see Chapter 6).

Salamon's work with cosmic rays had made him very familiar with ways of detecting nuclear particles, such as neutrons, and in recognising them. He also knew many of the pitfalls that the unwary can meet, the quirks of the electronic apparatus, the physical processes that take place when neutrons are captured in the detectors and the records that these leave behind. Here we see a group of experts who are not there to try and preserve some sort of 'physics status quo' but rather to satisfy themselves that the experimental measurements on nuclear radiation—their expertise—had been made properly. Salamon looked at Pons' paper, found its evidence for neutrons— one of the signs of fusion—and was immediately worried. He expressed his worries to Brophy and asked for an appointment to see Pons and discuss with him problems that were apparent to them all.

The paper mentioned two independent ways of having detected

neutrons. One used the Harwell neutron dosimeter, a $BF_3$-filled detector; the other used a sodium iodide crystal detector which records gamma rays. The $BF_3$ detector counts neutrons but gives no information about their energies; there is nothing to tell that they are from fusion as against high energy neutrons related to cosmic rays—this is where Jones had an edge over the chemists in that his detector also analysed the energy spectrum of the neutrons and was able to discriminate the crucial 2.45 MeV region of energy where the neutrons coming from $dd$ fusion should be. Their second method for detecting neutrons was to use a sodium iodide crystal which detects gamma rays and the signal is passed electronically to a visual display unit which shows a spectrum of the gamma ray energies. Most of these gamma rays come from natural radioactivity in the lab, but among these one is looking for a peak at energy 2.22 MeV (2220 keV) which originates from neutrons slowed down by collisions until 'thermalised' (moving at energies similar to the molecules at ambient room temperature) and which are then captured, producing deuterium and the gamma ray:

$$n + p \rightarrow d + \gamma (2.22 \text{ MeV})$$

When Salamon saw these two pieces of 'evidence'—from the $BF_3$ counter and the gamma ray peak—he noticed that each was flawed.

To ensure that any neutron signal was indeed coming from the fusion experiment one needed to show that there was a signal when a cell was giving heat and that there was none when the cell was inoperative or when a null-cell was present. Instead of doing this Fleischmann and Pons reported that the 'background' had been measured some distance away. All that this proved was that there were more neutrons at one part of the laboratory than another—a quite usual occurrence. If this difference had been large it could have been interesting, but the amounts at the two sites were so similar and the measurements involved such small numbers as to be insignificant.

The $BF_3$ counter recorded very few neutrons, by Fleischmann and Pons' estimate only one in a million, thus although it recorded only a few events [1] Fleischmann and Pons claimed that the true value at the source was around 40 000 neutrons emitted per second. This scale-up by a million is no problem for intense sources, but the significance of such low count rates and their susceptibility to random influences, even the noise in the circuits as the power supply is switched on and off, made the errors enormous. Extrapolating to 40 000 per second really included a margin of error that allowed the number to be zero.

Salamon regarded these data so un-seriously that he never even discussed them with Pons: they were at the level of 'maybe yes, maybe no'. What struck him more forcibly was that the gamma ray data measured by the sodium iodide crystal were definitely wrong, and it was these that he queried.

A well known feature of this gamma peak is its skewed shape, having a characteristic shoulder called the 'Compton edge' arising from the scattering of the gamma rays before detection (see page 362). Salamon told Pons that the smooth rise and fall of the peak shown in the paper had the wrong shape to have been caused by a gamma ray arising from neutron capture. When Pons asked what the problem was, Salamon explained the physics of the Compton edge to him. It might just have been possible for the peak to have been as smooth as this if Pons' had used a very large detector, and so they asked him how big it was. He indicated that it was about the size of his hand and at that moment they physicists knew that something was wrong and said that he should be cautious.

Salamon had also noticed another problem, the same one as had been pointed out at Harwell: the peak was at the wrong energy to be caused by neutron capture. 'I don't recall whether it was at 2.5 or between 2.4 and 2.5 MeV (we were reading the paper quickly), but it was *not* at 2.2 MeV where it should have been if it was a genuine signal for neutron capture. I told Stan and he said that he knew that and that it was a mistake—the figure had to be corrected. I asked him how did he know that the peak really was at 2.2; had he recalibrated his detector, and he said that he had and that the peak was now at 2.2 MeV.'[2]

And it was indeed at 2.2 MeV when the paper was presented to the world after 30 March, and was at 2.2 MeV in their published paper. Within a few days an erratum sheet was put out which included a further change—a new peak was exhibited, replacing the previous one but this too was at 2.2 MeV. Its relation to the previously claimed 2.2 MeV was not clear; there were no statements to show how this 2.2 MeV had been arrived at, no hint that their measurements had placed a peak at 2.5 MeV. But some people were already aware of this and a few, close to the centre of the action, became concerned about the nature of the evidence.

### The mobile peak

Hoffman, the radiologist, and Hawkins, the student, had measured gamma ray spectra. They found the familiar range of peaks from radioactive decays of potassium, thorium and the other elements that are present in the environment, and also saw a peak that appeared to be present only when they took measurements near the cell. Thus it seemed natural that this peak was associated with the cell and could be the required evidence for gamma rays produced by neutron capture. Interpolating between the potassium peak and other identified peaks of known energy implied that the putative signal peak was at 2.5 MeV.

Hoffman was used to measuring gamma-ray background radiation spectra but in rather different circumstances and with different

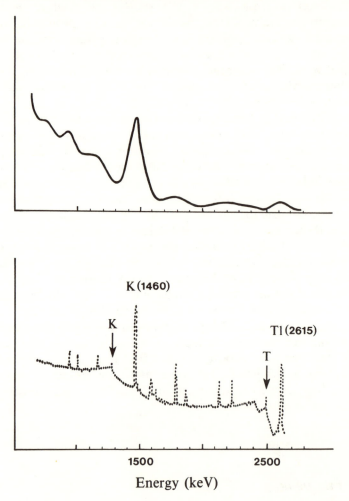

Figure 13. The top figure shows the background spectrum of gamma rays measured with a sodium iodide detector. The lower figure shows the same energy range measured with a high purity solid-state geranium detector (the vertical scale is logarithmic). Note the much finer resolution; the Compton edges of the K (1460) and T1 (2615) peaks are clearly seen (K and T arrows). Note also the sharp thalium peak (T1 2615) with this detector. Reproduced courtesy of R. Petrasso, MIT.

instruments from those on this occasion. His usual work was with solid state germanium detectors which have much better resolution than the sodium iodide detector that he was using here. With germanium detectors the peaks stand out like sharp lines instead of the rather broader humps that a sodium iodide detector reveals. In particular, this smearing makes it difficult to distinguish the neutron capture peak, which would be at 2224 keV, from the peak due to naturally occurring bismuth which occurs at 2200 keV.[3] The first reaction of many people was that the chemists had misidentified this nearby 2200 peak, not least because the peak was centred at that value and not 2224 keV!

Hoffman had obtained a spectrum which showed a clear potassium peak and at higher energies a series of bumps and, at the highest energies of all, two sharp peaks.

The data appeared on a screen as a series of peaks looking like a range of mountains, displayed against a quantity called 'channel number', and some work is required to translate from 'channel number' to 'gamma-ray energy'. These are shown in Figure 14; the prominent peak number 3 is Mount Potassium, which identifies the 1.46 MeV energy position on the bottom scale. Hoffman identified peak number 8 as thallium at 2.6 MeV and so the fusion peak should be to the left of *peak 8*, and they identified it as the pimple, number 7. What is more, this pimple was present in the spectrum taken over the cell and was absent in the 'background' measured remotely. Consequently they decided that this must be Mount Fusion.

Peak number 7 is nine per cent of the distance between their 1.46 and 2.6 MeV markers (peaks 3 and 8) and a linear interpolation between these locates the peak 7 as being at 2.5 MeV.

The flaw, though none of them knew this at the time, was that a true neutron capture gamma ray should occur at 2.2 MeV (where there was no special signal in their detector) not at 2.5 MeV. This 2.5 MeV peak had nothing to do with what they were looking for.

This erroneous 2.5 MeV peak formed part of their evidence when they submitted the manuscript to *Journal of Electroanalytical Chemistry* in mid-March, and this paper was accepted for publication without being examined by a specialist in nuclear physics who would have realised that the peak had the wrong features to be what was claimed. This was unfortunate for the two chemists as the acceptance of the paper cannot have dimmed their conviction that they had good evidence for fusion. When the press conference took place on 23 March they had not yet presented a report of their work to a scientifically critical audience of experts who could have pointed out this flaw. On 24 March, a paper was submitted to *Nature* with the 2.5 MeV peak in it; the referees included nuclear experts who knew where a genuine peak should have been.

But that was still in the future, as the paper took time to be processed by *Nature* and to be forwarded to referees.

The first public clues to the error came via Fleischmann. He had left Utah very soon after the press conference and returned to Britain for the Easter weekend. At this stage Fleischmann and Pons believed that the 2.5 MeV peak (Figure 15) was evidence for neutrons and, in his briefcase, Fleischmann had a copy of this figure among his notes for the lectures that he would give in Europe, starting with the presentation at Harwell on Tuesday 28 March. It was there, for the first time, that the problems began to emerge, as he was informed that a true neutron capture peak should be at 2.2 MeV. Thus at the beginning of April Pons sent out publicly their paper with an isolated peak centred on 2.2 MeV but with no discussion of how this had been measured there. In addition there was an error in that the wrong peak was drawn; the paper that was published in the *Journal of Electroanalytical Chemistry* showed Figure 16, which contains the same data points as in the original 2.5 MeV peak but the number of counts has been increased and the resulting curve looks similar to peak number 8 with some 20 000 events on the scale. Originally the peak had been at the wrong compass bearing; now it was being presented at the right compass bearing but was the wrong mountain.

Whatever this peak was it was not caused by a gamma ray from neutron capture, and within a few weeks an erratum to the original paper appeared. Not only had they omitted the name of their co-worker, Marvin Hawkins, which caused raised eyebrows, but they had changed the figure showing the 'fusion' peak. The shapes of the original and the replacement are compared as solid and dashed curves in Figure 17. Their erratum said: 'The y-axis [vertical axis] should be labelled 0 to 1000 . . . i.e. replace 0 . . . 25 000 by 0 . . . 1000 respectively, see below' and then showed the new figure. The peak is at essentially the same place but was only a tenth as large as they had previously claimed. But what was not obvious unless one actually went and compared with the original picture was that not just the numbers but *the shape* had changed. It was a *different* peak.

They did not explain how the change had come about, nor how two different peaks could have both been measured at essentially 2200 keV.

As became clear only later from Hoffman's spectrum (top entry in Figure 14; compare with the energy scale in bottom entry), there is no neutron capture peak at 2.22 MeV, no sign of neutrons, no evidence for fusion.

None of this was public knowledge in April. All that people had to go on was the single peak, asserted to be at 2.2 MeV and proof of fusion, even though it had the wrong shape. Even by the end of

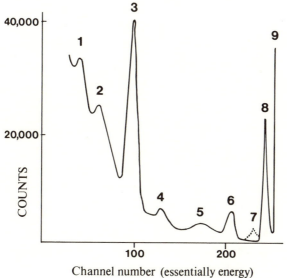

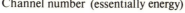

Channel number (essentially energy)

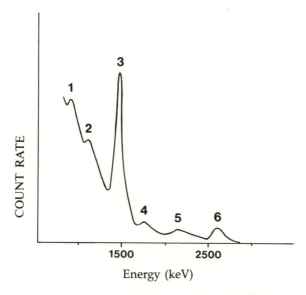

Energy (keV)

Figure 14. Gamma-ray spectrum from Utah (upper) and MIT (lower). *See* Figure 10.

May there had been no public display of the full mountain range measured in Pons' lab. It was only after some detective work by Petrasso's group that this was published, enabling the full story to emerge into the scientific literature.

283

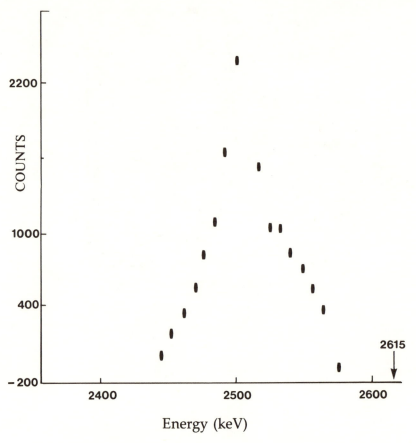

Figure 15. The data originally measured around 2500 keV.

The chemists had protrayed an isolated peak, and this alone was sufficient for the MIT team to be concerned. By going further and obtaining a spectrum from the TV station, the MIT group encouraged the Utah team to display, for the first time in print, their full spectrum. This was what people had wanted all along: for example, Richard Garwin had commented at Erice, 'It would be nice to see the whole spectrum.' And as soon as this whole spectrum was displayed, it showed to the world that there was no peak at 2.2 MeV and hence no evidence to support neutron capture.

The letter from the Utah group responding to Petrasso's paper began by denying that Hawkins had ever stated that the television picture was taken in their lab 'as it most certainly was not'. I asked Richard Petrasso of MIT about this and he resolutely confirmed that Hawkins *had* told him that and that any statement to the contrary was incorrect.

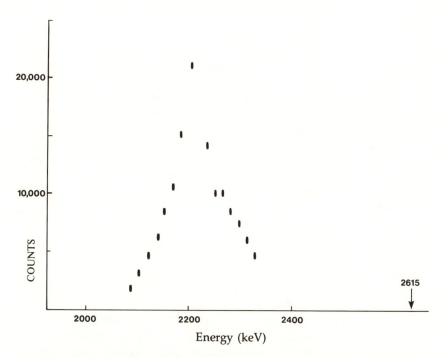

Figure 16. The signal peak data presented in the published paper. The position of the nearby 2615 keV peak has been added in this figure for reference. Compare with Figure 15.

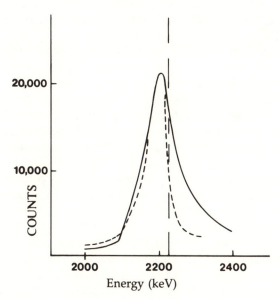

Figure 17. Original peak (solid) and the erratum version (dashed). *See* Figure 9.

285

The Utah group continued, wryly, that in view of 'this somewhat strange approach to the collection of scientific data' they would now display one of the complete sets of spectra that had been measured in their lab. This had its own irony in that the normal way to obtain others' scientific data is by reading it in the literature and that had the data, which they were now belatedly showing, been published where it should have been, namely in their original paper, there would have been no need for 'strange' approaches. And what's more, the data that they showed looked very much like those in the TV display! (Compare the photo and figures on page 169.)

They began by stating that 'we have repeatedly pointed out that we are well aware of the deficiencies of these spectra.' I have been unable to find any evidence to support that claim. When Pons showed the peak at Dallas, where he was enjoying idolisation as the prophet of the new age, he made no suggestion that there was any deficiency. Studying the videos and interviews with people who attended his seminars in Indiana, Utah and elsewhere show that some physicists in the audience were disturbed by the shape of the peak, but gained no concession from Pons nor any admission about its uncertain origins. An editorial in *Nature* on 6 July also comments 'there are no records of that reservation earlier than the meeting of the US Electrochemical Society at Los Angeles (8 May) by which time they had been sent (but may not have read) Petrasso's first draft of his complaint. This is a more serious retreat than they acknowledge.' (A statement announcing the retraction of the peak included the Orwellian remark that 'the data had been incorrectly homogenized'.)

Had they published in March what they were now revealing in their 29 June response to MIT's criticism, then it is unlikely that anyone would ever have taken them seriously.

They seemed to have retreated to where they came in—a 2.5 MeV peak as shown at Harwell and referred to in their original text, which they had then withdrawn and changed as it became clear that this had nothing to contribute as evidence for fusion neutrons. Now they were admitting that they had *no* 2.22 MeV peak, but instead of conceding that they had no evidence for fusion, they claimed (erroneously)[4] that this should not be taken as evidence against fusion because 'the search for this peak does not seem to be possible with detectors (of the type we are using).'

MIT made a reply which reminded them that it was not the TV picture but detailed impossibilities in the data that had been critical in showing that the 'evidence' was flawed. For scientists the withdrawal of the chemists' claim, for that is what it now amounted to, was clear if technical, and highlighted further problems with the history.

286

In their 29 June letter to *Nature* they claimed in their defence that their detector could not measure 2.2 MeV photons from fusion, yet they had previously shown a peak, which they *said* they had measured at 2.2 MeV and cited this as evidence for fusion. If, as they now claimed, their detector was blind then they had no justification for claiming to have detected fusion with it, and the 2.2 MeV peak was a fiction[5]; but if the detector was able to see, then the absence of the 2.2 MeV peak ruled out fusion.

One of the most important results of the exchange between MIT and the Utah group was that it proved that the latter saw no neutrons at all,[6] based on the chemists' *own* data in the vicinity of 2.2 MeV and with a cell that the chemists claimed was producing nearly 2 watts of power.

## *Postscript*

Meanwhile, following his hectic pressured measurements, Hoffman had set up some gamma sources in his own laboratory and measured these and the background spectrum carefully. This showed him that there were some problems with the amplifier, which was giving sharp spikes on the display which were probably unrelated to real gamma-ray signals (essentially what Petrasso *et al.* were commenting upon).

On 27 April he sent a memorandum to Pons and other central figures in the university to the effect that he believed that there was no evidence for a 2.2 MeV gamma-ray arising from neutrons being captured by hydrogen in the water bath, noting that the signal peak is at 2.5 MeV and suggesting that this may be due to some other 'unknown' neutron interaction.

This still leaves unanswered how the original 2.5 MeV peak (p. 284) was moved to 2.2 MeV (p. 285) in the Fleischmann and Pons paper. Neither Hoffman nor Hawkins, who had measured gamma-spectra, knew the answer when I asked them.

I asked Fleischmann about this mobile peak in February 1990. A year after the measurements, in which he had not been directly involved—though they were in his paper—he too was unable to give full details and said that I should ask Pons. However, he suggested that changing from a 'linear to quadratic interpolation' could shift the peak. I asked that if changing the interpolation (a theoretical not a measured change) could shift the peak by 0.3 MeV, how could they locate its position to the three decimal places needed to identify it as neutron capture, and he agreed that one could not.

Unfortunately I have not received any further response.

I was not alone with these worries. Richard Garwin, organiser of the Erice meeting, had commented on the importance of seeing the full data in April 1989 (page 151). Later (16 August 1989) in his capacity as a member of the official US Department of Energy panel investigating test-tube fusion he requested from Fleischmann and Pons information 'that only you can provide'. He noted that the 2.2 MeV peak published in April had become 2.5 MeV in June and remarked, 'In our understanding of results from laboratories all over the world, it is important to us to be aware of problems and sources of error that might affect these results. Therefore I would be most grateful to receive a full explanation of this discrepancy.' Garwin told me a year later that he never received a reply.

There is a further small error in the energy calibration that arose due to Hoffman's experience with germanium detectors (compare the quality in Figure 13 on page 280). This led him to identify the sharp peaks numbers 3 and 8 (page 283) as signals. Number 3 is indeed a genuine peak arising from potassium and was correctly assigned to 1.46 MeV. But *number 8*, which was identified as the signal from thallium at 2.6 MeV, is actually one of those instrumental effects and not a real signal.

So *two* mistakes had occurred. First, the peak that they identified first at 2.5 and then moved to 2.2 had nothing to do with Mount Fusion, and second, having misidentified peak *number 8* as the marker for thallium at 2.6 MeV instead of peak *number 6*, the energy scale was all wrong. A correct scale would have placed the pimple number 7 at 2.8 MeV, even further away from where Mount Fusion should be.

# 16
## CREDIBILITY

On 15 June Harwell Lab, with whom Fleischmann had close contacts, announced that a team of six electrochemists and four nuclear physicists had performed over 100 experiments with as many as thirty test-tube fusion cells and had seen nothing. Harwell had spent the equivalent of half a million dollars and called on scores of specialists from the laboratory's 4000 strong workforce. They made an in-depth fine-mesh survey; they had varied the size, shape and thickness of the electrodes, had used palladium that had eight different metallurgical histories thereby covering the 'cast versus extruded' debate. They had varied the currents and cell designs, run experiments for between a day and six weeks, and analysed the electrodes and electrolyte for traces of fusion products such as tritium and helium, and had included analyses of the lithium content, namely the $^6Li$ and $^7Li$ abundances in case lithium had been in some way involved in nuclear transmutations.

This was the world's most comprehensive hunt for the evidence, set against which many other efforts appear amateurish. The thoroughness of their investigation and its unequivocal results effectively killed test-tube fusion in Britain. Even in the USA the end of the affair seemed to have arrived. Whereas the media in the USA had uniformly referred to the pair during the positive days as Pons and Fleischmann, with the American named first, now suddenly they became headlined in one major paper as Fleischmann and Pons as the tide turned.

But far from giving in, Pons and Fleischmann seemingly carried on as if nothing untoward had happened, much to most people's consternation. Everyone but them was out of step, it appeared. Yet they offered no further solid evidence to support their position; belief or disbelief was a matter of faith more than evidence.

Normally after writing a paper about your research you give talks about it at several institutions. In getting to the results of your research paper you may have had to bring many strands together, and on some of these you may be less confident than others. Although these strands are soundly tied, there are always other scientists with more knowledge about specific areas of the full tapestry, and from their questions and comments the wise speaker will deepen his or her own knowledge. Not infrequently questions from the audience show up deficiencies in the speaker's thesis. Although journals are the archives of research, public acceptance or rejection of that work is more often a result of the performances in front of critical audiences. If flaws are exposed in a talk at institution A, the news spreads out among the experts over days or weeks, reaching those at Institution B where the scientist in question is due to speak at a later date. In the case of test-tube fusion, as Fleischmann, Pons and Jones gave their talks, local experts probed and highlighted the weak points, clarified the uncertainties and collectively helped to push the colossus of knowledge forward. The level of interest was so high that the news spread daily at the speed of light on the computer network.

Many scientists who communicated regularly on the electronic mail computer network also had access to electronic bulletin boards. Just as people used to post notices on the board at their work place for all to read, so now this can be done internationally. You send in your news, on the electronic mail link, to the bulletin board. Every day the latest digest was automatically collected and disseminated, uncensored, for all to read, like a newspaper run co-operatively. The scientists were both the journalists and the readers.

One of these computer mail services went under the codename *sci.phys.fusion* (main board science, subgroup physics, special edition fusion). This was set up in the first days after the March press conference as a medium for sharing technical information among people who intended to design their own test-tube fusion experiments. It was an invaluable source of intelligence as talks given by the main protagonists were reported, the details sent in by scientists at the meetings.

In the normal discourse of science one would expect that a coherent picture would emerge from this, but instead things became increasingly blurred rather than coming into focus. Whereas the electronic bulletin board had begun life as a serious venture to enable research discussions among professionals, its uncensored character led to it being swamped with unsubstantiated gossip and postings from students who had computer access but who were frequently unaware of basic principles of some of the science. Rumours propagated faster than the readers could trace their sources and put order into the stories.

Contributors posted substantive criticisms both of the scientific quality and of the ethics of the two chemists, which in their turn stimulated much heat from supporters and opponents.

The criticisms centred on the announcement of the discovery to the press, disavowing knowledge of similar research elsewhere, initially failing to acknowledge the graduate student responsible for much of the work and supplying his name as an addendum only after being criticised, providing a paper with overstatements, incomplete graphs and ambiguous data and withdrawing publication from *Nature* when referees requested more details, and seemingly being more interested in patent rights than in substantiating their claims.

The paper that Fleischmann and Pons had written on 11 March looked rushed to the eyes of experienced scientists and was described as a 'preliminary' note. The whole basis of the chemists' claims rested on a brief table of numbers for 'excess heat' but nowhere did they report what the errors on those numbers were. Nor was it always made clear that the claims of large excess powers, such as 4 watts output for a 1 watt input, were not present in the paper as measured data, but that some of the 'data' in the paper were projected based on controversial theoretical assumptions. Many other details remained obscure.

Normally in scientific seminars, speakers will show representative samples of their 'raw data'—the actual measurements, the details of when they were taken, how they were analysed and selected before appearing in the summary form in the published paper. These are the periods of test: when experts in the audience will question and criticise the method and opinions will form on the soundness, or lack of it, of the experiment. What the scientific community expected from the talks by the chemists following their press conference was sight of graphs, real data on the currents and voltages, and full details of the cells. Yet weeks or months later many scientists were still complaining of the lack of hard evidence. The talks made claims that cells were now producing more heat than before, that scientific papers describing everything would be produced 'in the summer', 'in the fall', 'by the end of the year'. And questions about details of the experiment sometimes received different answers on different occasions. Far from clarifying details, the various talks—reported on the electronic bulletin board—raised inconsistencies and generated more confusion. However, among the noise there was much significant material and as an instant guide to the feelings of the scientific community, the electronic bulletin board was a leader.

The bulletin board showed that the chemists had supporters as well as critics. One posting proffered a defence against the rushed nature of the paper and the criticism of subverting peer review by arguing that, because Jones' work was providing hot competition, Pons and

Fleischmann had to move fast, and that it was this that excused them. 'P and F were not ready to publish and had said that they would have preferred another eighteen months of experience. But they had no choice. After trying unsuccessfully to get Jones to put off announcing their results, they agreed to prepare a manuscript quickly and submit it simultaneously with Jones manuscript on 24 March to *Nature*. That was 6 March.'

This defence—namely that they were innocent parties pushed into premature publication—was frequently cited by their supporters, but opponents asserted that Fleischmann and Pons had not completed *their* work in the scientifically accepted sense, having not done all the control experiments that were required (in particular those with light water replacing the heavy water) and hence had no credible evidence for fusion *at that time*.

Fleischmann and Pons argued that they had made control experiments with *platinum* electrodes, and also had shown that the heat balanced in the case of an experiment with a palladium sheet, that had been run at low current densities. These controls relate only to the question of heat balance; they do not deal with the central claim—to have evidence for fusion: there was no statement as to whether or not a definitive control had been made, namely whether the heavy water had been replaced by light water. The question of whether or not the chemists had done a controlled experiment with ordinary water was pivotal for many scientists: the electronic bulletin board summarised the feelings of the wider community. The equivocal answers given by the chemists to enquiries about this raised doubts about the conduct of the experiment, and these doubts became public as they were posted on the electronic mail network. Months later I disentangled the probable course of events, which illustrate belief and desire vying with scientific judgement.

## The light and heavy water controversy

A great advantage of the electronic bulletin board was the speed with which information about talks was conveyed, and titbits from local newspapers became available for international consumption. As an example, a scientist from Salt Lake City posted to the electronic bulletin board a report from the *Salt Lake Tribune* of 6 April which read 'B. Stanley Pons said Wednesday (5 April) that he has tried the experiment with ordinary water and it produced no significant heat. This could be evidence that the heating process (seen with heavy water) is indeed nuclear and not chemical.' Here once more we see the importance of 'nuclear' in the claims. The *Tribune* reinforced the importance of the water control and its significance by quoting Utah physicist Michael Salamon 'who said of Dr Pons' experiment ''That's certainly the first thing you have to do''.'

At this stage everything seems fine; they have done the necessary test and it has validated their claims that the heavy water is undergoing fusion. But now the story began to become confusing.

First, Pons' statement of 5 April notwithstanding, tapes of his talk at Bloomington, Indiana, on 4 April show that when asked if he had made a control experiment with ordinary water he said:

'We have not made controlled additions of water. I would imagine that there would be some killing of the process because . . . the probability of fusing protons is much harder than deuterium.'

The words say that they have not made hydrogen–deuterium mixtures, but it was given in response to a question on hydrogen (light water); the impression that it left with members of the audience that I discussed this with was that *no* light water control had been made.

This was the same impression that Fleischmann had given at Harwell in Britain on 28 March but reports of this internal meeting had not been communicated to the network, and at CERN on 31 March.

People at his Harwell seminar on 28 March certainly felt that *no* plain water control had been done. The official report of that presentation was made by Ron Bullough, Chief Scientist to the United Kingdom Atomic Energy Authority, where he says that it was 'surprising and unfortunate that no control experiments appear to have been done.' This summary was even communicated up to British Government who were naturally wanting to be kept informed of what appeared to be the most significant breakthrough in energy generation this century.

News of the reported omission raised some eyebrows in Britain, as I recall from my own experience at the Rutherford Laboratory, adjacent to Harwell, where we heard the report of the Harwell and CERN seminars and registered disbelief that such a basic test had not been done before going public. Sir Denys Wilkinson, a leading nuclear physicist and Vice Chancellor of the University of Sussex, was visiting me *en route* to Erice where a special meeting on test-tube fusion had been hurriedly arranged for 12 April. The confusion over the plain water control worried him to such an extent that after Fleischmann concluded his presentation, Wilkinson stood up and asked the crucial question (as described on page 151) 'What happens when you use ordinary water?'

Fleischmann answered that he was 'not prepared to answer'. As the wording here is ambiguous, I asked Wilkinson if he thought it meant 'unwilling' or 'do not have the information available'. He said that he too was confused by it, but from the intonation interpreted it as meaning 'unwilling'.[1] Tapes of the meeting (page 151) certainly confirmed this impression as Koonin pressed further and Fleischmann said 'I have said quite specifically that I am not prepared to answer that question at this moment . . . You can read into that what you like.'

Richard Garwin, who wrote a resumé of the meeting for *Nature*,[2] concluded from it that 'It became clear that no experiments had been done with ordinary water as a control.' However, this need not be the only conclusion; some wondered if patents were at risk.

Although Fleischmann was unwilling to answer the question, Pons had no such qualms when later that same day, 12 April, in the USA he made his messianic appearance before 7000 eager chemists and test-tube converts in Dallas. Before the speech he gave a press conference in which he appeared to say that he *had* made a control experiment with ordinary water (good news if true) and that it had shown *heat* (disastrous news as this implies heat both with deuterium and with ordinary hydrogenic water; thus nothing to do with fusion at best, and at worse providing evidence that there is some flaw in their experiments).

National Public Radio and the media all duly reported this, much to the confusion of scientists who learned it the next day. After his speech, Pons was asked again the crucial question why he had not reported results done with normal water in place of heavy water; his reply was obscure: 'A baseline reaction run with water is not necessarily a good baseline reaction.'[3]

Confused, the questioner asked Pons to elaborate. Again he spoke in code: 'We do not get the expected baseline experiment. We do not get the total blank experiment we expected.'

This seemed to say that they *had* performed a water control and that they found heat here too.

Deuterium is needed if there is to be any fusion and so light water, $H_2O$, should give no fusion heat. If they were indeed finding heat production with plain water, then their earlier confidence that they had unequivocal proof of fusion was questionable.

By 25 April the *Wall Street Journal* was reporting: 'Mr Pons said he has a plain water experiment producing small unmeasured amounts of heat.' This appears to be referring to an ongoing experiment, an impression reinforced in the University of Utah student newspaper around that time and by conversations with people close to Pons which implied that by April they had an experiment with light water that was showing excess heat.[4]

On 28 April Pons spoke to the scientists at Los Alamos National

Lab and here, once more, there was no mention of plain water controls in his talk; it was again up to questioners to extract this information.

'We tried a blank experiment in *January* [my italics] with ordinary water. We did not get heat in, heat out balance that we expected. We got a small amount of heat—a slightly higher amount of heat out.'

If he had done the experiment, and seen heat, in January then why was there no mention of it in the paper written in March?

Yet by 8 May the story had returned whence it came, with Fleischmann insisting at the Los Angeles convention of the Electrochemical Society that talk of heat from light water was 'ridiculous' and then steadfastly refusing to discuss the experiments.

All possible answers had been given:

(i) The experiment seemed not to have been performed [4 Apr (SP) and reports of 28 and 31 Mar (MF)].
(ii) It had been done but gave no heat [5 Apr (SP) and 8 May (MF)].
(iii) It had been done and gave heat: (a) in January before the press conference [28 Apr (SP)]; (b) in April after the press conference [24 Apr (SP)]; and (c) at unspecified date [SP Apr 10 to C. Martin; Apr 14 to G. Seaborg].
(iv) Refusal to answer [MF 12 Apr].

It was following these revelations that subjective opinion began to polarise. Those who were becoming aware of the conflicting stories began to doubt the entire enterprise, but most were unaware, having only the published paper and remote gossip to go on. Although I had access to the electronic mail reports it was only after going to the various institutions that made videos of the talks and seeing them for myself that the extent of the confusion became apparent.

This range of answers being given on different continents typified the difficulties scientists had of getting accurate, hard information on the details. Without the electronic mail network it would have been impossible to get after the facts. Erice attendees were left with one impression, Dallas delegates another. Had controls with ordinary water been done or not? If they had been performed, then precisely when and with what outcome? There was no clear answer to these questions. If they had not been done by 23 March the two chemists' scientific case was incomplete and should not have published; if they had been done and given the damaging results apparently admitted by Pons three weeks later, then omitting to mention this in the paper was a serious oversight. When the chemists submitted a paper to *Nature* they ran into problems with the referees on various grounds, in particular that the paper showed inadequate data to indicate whether the mysterious heat might not simply be due to electricity passing through the heavy water rather than from an exotic nuclear reaction, and, once again, the criticism about inadequate control experiments using ordinary water.[6]

The crucial time when the story changed into 'heat with light water' was between 5 and 12 April 1989; it returned to 'no heat with light water' sometime during the first week of May.

Two significant events took place in these periods that may have some bearing on these changes.

In their experiments before the press conference there were some that gave no heat,[7] but there was also an experiment with light water that gave a positive result which, at the time, they thought was wrong and terminated it. The first confirmation of their heat generation in experiments using heavy water came from Charles Martin at Texas A & M, but on or about 9 April Martin told Pons that there was heat with light water too. Pons told him 'in no uncertain terms' that he also was seeing heat with light water.[8]

The conclusion from this ought to have been either that something was wrong with the calorimetry, or that the heat is not produced by a nuclear process. However, Fleischmann and Pons believed so much in nuclear fusion, and this was reinforced by what Pons learned from Charles Martin, that once again[9] they argued away contrary evidence, leading to Pons' cryptic 'light water . . . is not a good baseline.' They arrived at a more convoluted explanation built around a piece of news that arrived at about the same time.

Koonin and Nauenberg had written a paper on $dd$ fusion in which they also pointed out that the rate for $pd$ (proton–deuteron) fusion would be faster than the $dd$ fusion rate. Koonin sent a copy of this paper to Pons by electronic mail on 7 April.[10] In Utah the chemists decided that this could cause fusion to occur even in light water: there is about 0.015% natural abundance of deuterium in light water and this, combined with the faster rate for $pd$ fusion, decided them that the heat in light water was also due to fusion! Simons and Walling, in the Utah chemistry department, also wrote a paper at this time[11] with the comment that the Koonin and Nauenberg paper had been given to them by Pons, and with extensive references to the prediction that fusion would also occur in light water, to explain this excess heat that Pons now believed in.

At Erice on 12 April, Fleischmann acknowledged Texas A & M as 'the only experiment I know with a confirmed measurement to date.' So this experiment must have made a big impact on them and, recall, the Texans were seeing their heat with both heavy *and* light water. They had spoken with Pons and gone public during 9–10 April; Fleischmann was in Europe receiving news remotely as he wended his way to the extremes of Sicily where he was questioned about what, unknown to the audience, were fast breaking events.[12] Thus was he 'not prepared to answer.' So there is now a third possible interpretation—not that they had done no experiment, nor that they had patent worries, but that Fleischmann at least was confused.

Pons by contrast showed no such doubts, believing strongly in heat with $H_2O$, as he confirmed on 14 April to Glenn Seaborg, Nobel Laureate in Chemistry and former chair of the US Atomic Energy Commision.

Seaborg had that day personally briefed President Bush in the White House (page 138) about test-tube fusion, and had followed up on this with a phone-call to Pons. Seaborg told him that he had heard a rumour that Pons was seeing heat produced in cells with $H_2O$ and asked if this was so. Seaborg's diary records that 'After some umming and ahing Pons said "yes that is true" and then explained that this fitted in with their ideas . . . he [Pons] believes that the heat is due to *pd* fusion between protons in hydrogen (the $H_2O$) and deuterons in $D_2O$' (there being 1 part in 6000 of plain water that is $D_2O$).

This erroneous belief was also being reinforced by what was currently happening in the Utah chemistry department. Believing that *pd* fusion was occuring in light water with its end product being $^3He$, the local chemists focused on *dd* fusion leading to $^4He$. During the 13 and 14 April experiments had taken place in the chemistry department which seemed to show that there was copious production of $^4He$ occurring in the heavy water cells (page 140). With Texas seeing heat in both heavy and light water cells, with a theoretical paper arguing that *pd* fusion could compete with *dd*,[13] with no one seeing enough neutrons or tritium products, and suddenly the possibility that helium is the nuclear ash and is being detected in quantity, it is easy to understand how excitement and enthusiasm overcame caution and Pons went public the weekend of 16 April.

This heat was then reported in Pons' weekly news conference of 24 April. Asked about the light water experiment, Pons confirmed that it producing heat but did not believe that 'a plain water experiment is a good control experiment'[14]. This was alluded to once more in the question session at Los Alamos on 28 April. However, there were problems arising. The question of heat in $H_2O$ experiments was clearly pivotal and had arisen at a time when their submission to *Nature* was being considered for publication; the referees had raised certain questions about it including 'criticisms about the lack of control experiments using ordinary water instead of heavy water'.[15]

In addition they were preparing to testify before the US House of Representatives Committee on Science Space and Technology on 26

April. At this point Pons withdrew the manuscript from *Nature* claiming that he was too busy to respond to the referees' questions.[16]

In February 1990 I asked Fleischmann if they had made blank experiments with plain water and he stressed that they had made 'any number of control experiments', that the absence of heat for 8 mm diameter rods (in $D_2O$ (sic)) 'verified the calorimetry' and that 'at the time of the paper we had done control experiments on palladium with $H_2O$ under such conditions that there was no excess heat.' I then asked him why he had been 'not prepared to answer' the question put to him at Erice about plain water experiments. He claimed that some of the questions at Erice had been oriented at undermining the patent applications, but then added that he had been unwilling to answer the question because at that stage they had 'not performed the mixtures'.

If *pd* fusion had really been a critical factor in producing heat then it would certainly be the case that $H_2O$ would not be 'a good baseline' experiment. As Harold Furth had pointed out in Congress (page 187) on 26 April, the decisive control would be to mix heavy water into the light water and see if the excess heat rose as the fraction of heavy water increased. Furth's remarks were made in Fleischmann's presence fourteen days after the latter had refused to answer in Erice, so it is not possible to know to what extent this was already in Fleischmann's mind at the time or was *a posteriori* reasoning. But the backing off from the positive reports of heat in light water seem to coincide with Fleischmann's return to the USA and the Congressional hearings on 26 April.

The moment that one seriously thinks about such mixtures one realises that the idea of *pd* fusion generating the heat in the above experiments 'doesn't hold water'.

Nathan Lewis talked to Walling about this during the first week of May (this was before the meeting of the Electrochemical Society on 8 May but after the APS meeting, by which time relations between Pons and Lewis were strained) and pointed out several problems. First, it would take an astonishing coincidence of rate acceleration to bring the *pd* rate just right to compensate exactly for the drop in natural abundance. Even if this coincidence occurred there was another problem. The heavy water samples are not pure, there are traces of light $H_2O$ in them and so the heavy water will *still* show much more *pd* fusion heat than light water because:

(i) heavy water is 'contaminated' with light water more than light water is contaminated with heavy (0.1 to 0.5% as against 0.015%) and so has more chance for *pd* fusion, and
(ii) the heavy water had additional *dd* fusion available as well.

As well as telling Walling about this, Lewis presented these arguments to Chuck Martin of Texas A & M. Martin was regularly talking with Pons and Lewis got this objection to the 'light water fusion' presented to Pons on the phone. So by the time of the Electrochemical Society meeting in Los Angeles on 8 May, the chemists had been told that it was not possible consistently to defend heat with light water as being due to fusion. When the question of light water experiments was raised at the ECS meeting Pons made no clear response, and then Fleischmann cut in and vigorously denied seeing excess heat with light water. There was no further discussion allowed on this aspect of their experiments.

Lewis and Martin were sitting next to one another and were astonished at hearing this denial after the interactions that they had had with Utah during the previous days.[17] It was made such an impact that Martin recorded the fact in his notes at the time.

The footnote to this episode is that in March 1990 Fleischmann and Pons produced a paper in which they claimed to have made two blank experiments with light water *before* the press conference of 23 March 1989. It is not clear whether these had been made before or after the 11 March submission of their first paper. Furthermore there is no mention of any experiments with $H_2O$ that gave excess heat.

Their paper also claimed that it was *Nature* that had generated the antagonism by virtue of its claim that they had not performed control experiments which, Fleischmann and Pons reminded readers, they had, specifically with platinum cathodes and also when they found a null result with $D_2O$ and an 8 mm diameter rod. However, careful reading of *Nature* shows that its editor John Maddox had not made such a general accusation, but had merely said that they had not reported control experiments with *plain water*.


## Back to the future

The full story of the light and heavy water control experiments, and of how beliefs grew and died, influencing the perceptions of the protaganists, will only become known with the passage of time. But the above narrative already shows that the conflicting claims had, at their source, an explanation that may have been more innocent than appeared to many outside observers during the spring of 1989.

The prevarications and contradictions caused much polarisation during April leading up to the APS meeting on 1 May. This was the period when press reports announced replications of test-tube fusion every day while the major laboratories were unable to produce the phenomenon. In particular, the withdrawal of the *Nature* paper when Pons claimed to be too busy to respond to the referees' questions astonished people, as it is a rather surprising response for someone to make if they have data that will make them certain candidates for a Nobel Prize.

Usually in science there is a great pressure to be first, to win the race and gain the honour of discovery. That honour requires acceptance by the community of science which in turn needs refereed publication of all the details necessary for successful replication of the discovery by other scientists. Only then will the claimed discovery be agreed upon and the credits come your way. All research is geared towards eventual publication; gaining funding to support your research is arguably the only venture whose urgency approaches that of getting the results onto paper and staking a claim to priority. And the funding will not be forthcoming until the project is shown to be well founded.

If Pons had done all the checks on his experiment to prove the validity of his claimed discovery—and this would be a singularly great discovery, not just a small footnote in his scientific career—then answering the referee's questions and getting that paper published should have been the most urgent topic on his agenda. That is what other experienced scientists knew would be their priority, and so Pons' claim to be 'too busy' left them incredulous. A posting on the electronic mail network complained that Pons withdrew his manuscript after the editors asked him for changes that he said he was too busy to make, even though he was apparently not too busy to testify before Congress.

This willingness to testify before Congress instead of carrying through their scientific work rankled with many. On the one hand it ran counter to the usual norms, but that of itself doesn't mean that it is wrong. Any scientist has the choice of whether they choose to place higher priority on telling Congress of their work, though they should not be surprised if the scientific community then feels miffed at being put lower in the priority list. Some of the criticism was clearly due to this; it was a psychological reaction to what was perceived as arrogant maverick behaviour. Yet there was also a more serious concern. Pons' reaction fitted in with a growing feeling that the chemists did not really have the goods that they were claiming before the world. Word was leaking out about the light water confusion and that the gamma ray data had the wrong shape to be real; there was also rumour, that eventually was verified to be correct, that the two offerings at 2.2 MeV had originally started out at 2.5 MeV and mysteriously moved. The feeling was that the referees were asking questions about these things, that the chemists did not have the results to satisfy the referees and that 'busy' was an excuse.

300

The confusing answers that the chemists had given to questions about the control experiments with ordinary water, and the negative impact these had made, were in full view again here. According to the *Wall Street Journal* for 27 April, the editor of *Nature*, John Maddox, said that the referees' remarks included 'criticisms about the lack of control experiments using ordinary water instead of heavy water.' As the *Wall Street Journal* also noted, 'the points raised by Nature's reviewers appear to have been the same ones raised by scientists world wide.' These comments from Maddox and the obscure answers from Pons and Fleischmann to their critics told a clear message to the sceptics: there was no proof that the experiment had been satisfactorily completed and the referees' criticisms were unanswerable. Following these remarks Maddox received a letter from Pons' lawyer.

The suspicions that all was not well were reinforced when some of the results that the chemists were claiming were proven to be wrong. The chemists then withdrew some of their claims, but only after others had exposed them (e.g. the detective work by MIT that discredited the gamma-ray peak). This all added to the feeling that the work was suspect or rushed. It was these niggling suspicions that the chemists weren't being scientifically scrupulous that formed the hidden agenda in the Baltimore American Physical Society meeting on 1 May. Few wished to articulate their feelings until they were quite sure that there was nothing in the scientific claims: there is a trait in human nature that excuses the actions of winners after the event. But with the news that several experiments seemingly refuted the chemists, so grew the confidence of the physicists to come out and attack what many perceived to be unprofessional conduct.

Regular followers of the day to day developments had picked up the inconsistencies in the reports on the plain versus heavy water experiments, and many knew that their own experiments (which were already more sophisticated than the original Fleischmann and Pons work) disagreed with the claims of the chemists, and were even able to show that the chemists had exhibited false 'data' in the form of the purported photon signal for fusion. This made many scientists nervous about the credibility of the enterprise, and led to the roasting at Baltimore. The resulting wrath was widely reported following the Baltimore Convention, and its tone led many to accuse the physics community of being unprofessional. Part of it was good-humoured, a response to the chemists' day of rejoicing at the expense of the physicists at the Dallas meeting (12 April). But part of the criticism was very serious, and had the background build-up been better known then the concern about honest science and the headline remark from Koonin that we were suffering from the 'incompetence and perhaps delusion of Messrs Fleischmann and Pons,' might have been better understood.

# 17
## 'IT'S NOT FUSION'

For those who had read the correspondence between MIT and the Utah chemists in *Nature* on 29 June, there was little doubt left that some, at least, of the claims for fusions were flawed. On the electronic mail networks through the spring and into the summer there was still an ongoing debate under the heading 'It's Not Fusion'. Although no one had replicated the dramatic claims of heat and fusion products that had emanated from the University of Utah, there were, nonetheless, interesting claims being made in several quarters that small bursts of neutrons were being seen, perhaps due to 'micro-hot' fusion. The idea here was that the stresses induced in the palladium caused it to crack. The intense electric fields in the crack may accelerate isolated deuterons and give them enough momentum to crash into one another and fuse. At best these were at levels similar to those reported in Jones' experiments, and the phenomenon was not really new, having been discussed in the Soviet Union in the 1970s. It was an interesting sidelight to the test-tube fusion quest, but not really relevant to the main question: if Fleischmann and Pons were correct in their measurements of the heat—and there were sporadic reports of small amounts of heat similar to theirs, in particular coming from Hutchinson's group at Oak Ridge whose cell was the first that I had seen, back in May—then was it because there were still errors unaccounted for in the measurements, was it due to some unidentified chemical effect, or was it due to a nuclear process, in particular fusion of deuterium as they had originally suggested? Time will tell whether either of the first two is the explanation, but science can already eliminate deuterium fusion as the culprit.

To see this, lets first recall some of the essential features in physical and chemical reactions.

The production of energy in any reaction is accompanied by change. When a fire burns, carbon and oxygen are consumed and carbon dioxide is produced; chemical change has taken place, the basic elements, carbon and oxygen, survive but linked in new combinations.

In a nuclear reaction the elements themselves transmute—the neutrons and protons in their atomic nuclei survive but combined in new ways. Detecting the heat energy showed that *something* had happened; detecting the *products* would show if it was chemical or nuclear in origin.

Fleischmann and Pons claimed to measure heat when they loaded deuterium into palladium. If these elements were combining to form a new compound, palladium deuteride *PdD*, this would be a chemical reaction. Fleischmann and Pons argued that the amount of heat was too great to be due to this alone, and asserted that a nuclear reaction must be at work. Their whole motivation had been driven by their belief that deuterium would undergo nuclear fusion, so if the heat was indeed due to this, the deuterium would have disappeared and new elements and nuclear fragments formed. The proof of fusion required finding those new entities, helium, neutrons, tritium and gamma rays, *in amounts commensurate with the amount of heat generated.*

'Follow the neutrons,' was the advice given to Tom Wilkie, the science correspondent of the London *Independent* newspaper, when he contacted the fusion experts at Culham's JET (Joint European Torus) laboratory in Oxfordshire. Wilkie had been a professional physicist familiar with the nuclear field before entering his present career as one of Britain's leading science reporters. He understood much of what was going on from his own direct experiences but with all the confusion and rumour, Wilkie told me, it was useful to have 'a solid rock to hang on to. ''Whatever else goes on don't forget that if there is really fusion then there must be neutrons. Where are the neutrons?'' '

If test-tube fusion between two deuterons had occurred (*dd* fusion) then tritium and a proton or $^3He$ and a neutron should have been produced. For every watt of heat generated some $10^{12}$ neutrons and tritium nuclei would be created. Beams of deuterium can be made in nuclear accelerators and smashed into one another and the outcome studied. These experiments show that the rate for production of these two alternatives are roughly equal, and this equality is preserved independent of the energy that the incident deuterons have. (At least, this is known to be true for energies greater than a few keV.)

Jones' group at Brigham Young University saw neutrons, and that was one of the reasons why their results were taken seriously by the scientific community. The amount of neutrons being produced in his reaction was interesting for science but would not solve the world's energy problems. Fleischmann and Pons claimed big heat, and it was that that excited the media. Had that heat been due to fusion as understood by the accumulated wisdom of 50 years' research into nuclear processes, then they should have been fried by the vast

303

quantities of neutrons pouring out. Yet, as Jones said, the amount of neutrons that he was measuring relative to what the chemists should have been seeing was as a dollar bill compared to the US national debt.

The size of this discrepancy was not appreciated by many of the general public who had the impression that the mismatch was maybe a factor two, five or ten, and that some better understanding would solve this. But in fact it is not so simple. The shortfall is a factor of a trillion, like the blink of an eye compared to a thousand years. Put another way: the neutrons that Jones saw would account for less than one-trillionth of a watt and give temperature rises less than the momentary fluctuations in the air on a still day. Either the heat claimed by Fleischmann and Pons wasn't there (i.e. was an error), or was energy, now liberated, that had been put into the material during its preparation, such as stress and strain energy or electrical energy stored during the many hours passing current during the loading of the electrode with deuterium. Fleischmann and Pons always claimed that the amount of heat was too big to be this and hence they appealed to nuclear fusion as the source; but if it is, then where are the fusion products, the conclusive evidence? Neutrons are too few in number to carry off more than a trifling amount of the heat, so something else must be happening. The obvious excuse was to say that the fusion must be producing tritium much in excess of the neutrons, and that this is how the heat is generated.

Naively one expected the tritium production rate to be similar to that of the neutrons. Originally Fleischmann and Pons had claimed to see both of these products in similar quantities, and used this to support their 'evidence' for fusion. But as the neutrons were produced a billion times too slowly then so must be the tritium; rather than help the Fleischmann and Pons claims, this made them even more incredible.

A perverted form of argument then appeared in some quarters along the following lines. 'Suppose that Fleischmann and Pons are right in believing that there is fusion taking place, even though they have no evidence for it (their neutrons and tritium measurements account for nugatory amounts of the heat). There are many experiments which have conclusively shown that the production of neutrons can account for no more than a trillionth of the heat. So maybe the tritium production rate is much *larger* than the neutron rate; the fact that they are known to be the same when deuterons collide at keV energies— hot fusion—need not imply that this be true under the conditions of test-tube fusion where the collisions take place at eV energies, a thousand times smaller.'

Supporters of this way out even made an argument why this could be so. They first pointed out that the fusion is inhibited by the electrical repulsion between the two nuclei. So it is easier for two deuterons to bump together if they are oriented such that their neutrons are close rather than their protons. In the case of *dd* collisions this will shake a proton loose (leaving tritium) more easily than a neutron. The tritium production will be much larger than the neutron production, as required.

Theoretical estimates suggested that this could cause a preference for the tritium route by a factor of two or three at most, whereas billions would be required to explain test-tube fusion this way. 'So much for theory; maybe the environments of the palladium makes it more complicated,' but exactly how this miracle should happen wasn't explained. Another point that was conveniently ignored was the lesson to be learned from the fact that there are no data on the ratio of neutron to tritium production down to the low energies appropriate to test-tube fusion: the reason that we have no data is because the event rate is too small to measure and so the fusion all but vanishes. So how will you ever explain watts of heat this way?

Notwithstanding these arguments some test-tube fusion advocates became excited by claims that tritium was being sighted in India and by some scientists at Texas A & M University. Tritium was widely regarded as the smoking gun for *dd* fusion; as tritium doesn't occur 'naturally' the presence of tritium in the 'fusion cell' must have come from fusion even though there was not enough of it to equate with watts of heat. However, this appears to be another example of Rutherford's criticism that 'observation of an element has been confused with its creation.'

Tritium can indeed be created by *dd* fusion; it is also produced when neutrons hit $^6Li$ (lithium) whose nuclei each contain three protons and three neutrons. The impacting neutron splits the lithium into helium-4 (two protons and two neutrons) and tritium (one proton and two neutrons). Cosmic rays hitting atoms in the atmosphere can produce tritium which falls in rainwater, ending up in the rivers and oceans. This alone gives roughly one triton to every $10^{18}$ protons or ten thousand tritium atoms per gram of water.

Tritium is produced in quantity in reactors by bombarding lithium-6 with neutrons. This tritium is mostly used in thermonuclear weapons. In the 1950s, explosions of such bombs in the atmosphere spread tritium around the globe and this also rains down, increasing the cosmic-induced tritium 50-fold on average. Tritium decays at a rate of about five per cent per annum, so the remnants of 1950s weapons tests are depleting and the intensities can vary considerably from place to place (the purest tritium free water is probably that from deep ice samples in Antarctica). It is thus important to know how much tritium is in your water before you begin the electrolysis.

When using heavy water, as is the case in the test-tube fusion experiments, it is even more important to know the tritium content. The normal manufacture of heavy water enriches the tritium content and the act of electrolysis enriches it even more. The reason is that the lighter hydrogen electrolyses off leaving enriched deuterium in the remaining liquid, and in turn the electrolysis of deuterium oxide (heavy water) liberates deuterium, leaving the heavier tritium in the liquid. So at the start of electrolysis you have a little tritium among a lot of deuterium; the deuterium boils off and the tritium stays behind. You now top up the liquid with heavy water which itself contains a little tritium. The percentage of tritium in the replenished liquid will be higher than at the start. The more top-ups, so the greater will be the tritium content.

Careful monitoring during the course of an experiment can keep account of this inbuilt tritium growth. What is more difficult to know is how much tritium was in the electrodes to begin with and escaped into the liquid or was liberated among the gases during electrolysis. This could be a particular problem for experiments using nickel anodes. Nickel isn't mined as an element but is found in ores, and there have been suggestions that smelting to extract the nickel can contaminate it with tritium. Some metals liberate hydrogen (and tritium) gases easily, others such as steels tend to retain it. So nickel can have significant amounts of the gas hidden within. During the passage of electrical current in electrolysis, this tritium can pass into the liquid, gas off at the cathode or even enter the palladium. Indeed, as electrolysis is the way you separate tritium from deuterium—by condensing it at the cathode—sighting it there in small quantities isn't very convincing proof for fusion.

Memos from groups at Oak Ridge working on test-tube fusion and sent to the lab directorate at the end of August said that tritium was seen at levels expected for electrolytic enrichment. Similar memos were sent to Peter Bond by the Brookhaven Laboratory experimenters on 1 September.

These data on the lack of tritium were either unknown or wished away by test-tube fusion supporters. Instead they took courage in other reports that tritium had been seen at levels far above neutrons and above expected contamination (for example, see Chapter 14).

But those who believed that the lack of neutrons could be hidden by blaming the heat on the tritium production, and that the tritium seen in these experiments was from fusion, were ignoring some basic nuclear physics.

Tritium produced in *dd* fusion,

$$d+d=t+p$$

emerges with 1 MeV energy. The behaviour of tritium at this energy is well known; people have studied it a lot because of interest in *td* fusion. *td* fusion is a rich neutron source

$$t+d={}^4He+n$$

and the rate at which *td* collisions produce neutrons is well known at and around the 1 MeV energy region. Now, tritium being produced in an environment containing lots of deuterium, such as a test-tube fusion cell, would hit some of the deuterium and produce neutrons, the actual amounts depending on how dense the deuterium is. So even if the initial *dd* collision produced only tritium and no neutrons directly, there would still be neutrons produced by knock-on interactions of that tritium. You can't hide the neutrons away totally.

So, how many neutrons must there be?

In the test-tube fusion cell the deuterium will be in the heavy water or in the cathode, in this latter case forming palladium deuteride. Collisions between 1 MeV tritium and palladium deuteride will produce one neutron for every ten thousand incident tritium nuclei. The heavy water is an even more important target. Here the deuterium atoms are more densely packed (remember that contrary to popular impressions the deuterium is closest packed in the original heavy water) and the incident tritium produces neutrons ten times more efficiently: one neutron per ten thousand tritium nuclei.

The end result of this is that every watt of power in *dd* fusion yields a thousand billion neutrons each second ($10^{12}$ neutrons/sec) via $d+d+n+{}^3He$; even if the direct neutrons are suppressed somehow and all the fusion goes into tritium ($d+d=t+p$) there must be a flux of $10^8$ neutrons/sec from the subsequent *td* collisions. These neutrons themselves hit protons in the water surrounding the electrolysis cell giving more deuterium and a characteristic 2.2 MeV gamma-ray, or may instead hit deuterium giving tritium and a gamma-ray of energy 500 keV (0.5 MeV). This results in one hundred million ($10^7$) gamma-rays per second with energies in the 500 keV energy region. These are easy to detect. Failure to see them will eliminate tritium production from *dd* fusion back in the cell. You can try arguing one path away but consequences keep cropping up; you can run but you can't hide.

The reports of tritium coming from India, where a series of experiments at the Bhabha Atomic Research Centre kept seeing the stuff, made no comment on the absence of neutrons or on the 500 keV photons. Wherever the tritium was coming from it was not from *dd* fusion.

Faced with this, another miracle was needed.

If this was fusion then it was fusion without radiation; the heat was the only clue. A likely possibility was that two deuterium nuclei combined to form helium with no neutrons, protons or tritium fragments. The mass of the helium is less that the combined masses of the two deuterons, which results in 24 MeV of energy being released as a gamma ray. In the normal hot fusion this process is some millions of times less likely than the tritium and neutron production routes that we have met so far. Could this really be enhanced millions of times relative to those modes and also manage to hide the 24 MeV photon, which would be easy to detect and of which there was no sight in any test-tube fusion experiments?

The simplest conclusion was that fusion is not happening. The only possibility in an almost hopeless situation was for supporters to propose that new physics was involved due to the presence of the palladium crystal lattice, enhancing the rate and taking up the 24 MeV of energy, sharing it among the whole set of palladium atoms which would start vibrating and release the energy as heat without radiation. Many of the people thought of this possibility as the only chance for 'radiationless' fusion, and a few wrote papers claiming priority for the idea. In Europe Mick Lomer, Director at Culham, thought of this within hours of learning of test-tube fusion and sent a memorandum to theorists at his lab and at Harwell asking 'Who will work out a rate for me?' Walling at the University of Utah also thought of this and, though he was unable to compute a rate to support the idea, it was taken up by Pons and advertised as an explanation, in part because he then believed that his group had evidence for helium being produced in the cells. As described earlier, these claims were flawed.

But there were other tests possible to see if this particular form of fusion was happening, and these all came up negative too.

If two deuterons fuse to make helium-4 the excess comes off and there must be high energy radiation, gamma-rays or electrons and positrons. The fusion supporters wanted to hide these in the palladium lattice by some unspecified mechanism, and this lack of specificity made it hard to refute them. But here is the problem. It takes only a few electron volts of energy to eject electrons from the periphery of palladium atoms. If fusion occurs inside the palladium, 24 *million* electron volts of energy is released per fusion, and this energy has to be shared among these atoms without even one millionth of it knocking an electron out of them. *Some* radiation would emerge, either electrons ejected from atoms or X-rays as the atoms are disturbed, but none was seen.

In the US Congress on 26 April Fleischmann and Pons said that they were entering into negotiations with Los Alamos National Lab who would look for signs of fusion products in their cells, but 'foot-dragging by the University' aborted this collaboration. Then in May, MIT and other institutions offered to test the electrodes for helium, but this was declined by Utah who said that 'other arrangements' had been made. Finally, it was announced that the Bartel Institute in Washington State would analyse some electrodes; the results were eagerly awaited and were still unannounced at the first anniversary of the original press conference. I asked Martin Fleischmann about this and he claimed that the experiment was 'ambiguous', as an electrode was analysed from a batch of material that had not given a lot of heat. He and Pons had been in Europe at the time the tests were made and had been under the impression that all electrodes had been from the same heat-producing stock. 'Nonetheless, if you are optimistic you could say that there is an accumulation, but that experiment has to be done again,' he said.

A year and a half after these original claims there has still been no evidence offered to support helium production.

## *The nuclear detective*

Fleischmann and Pons had been claiming fusion partly on the basis of a gamma peak at 2.2 MeV which did not exist. The full details of the way that this data had been obtained and the questioning of how the peak had travelled from 2.5 MeV in March, to 2.2 MeV in the paper as support for fusion, and now back to 2.5 MeV again only emerged in the correspondence in *Nature* between the Utah group and the MIT team, which was published in June. The first hints of this emerged at the American Physical Society meeting in Baltimore on 4 May, and with the media beginning to turn against them, they had to come up with some explanations by the meeting of chemists in Los Angeles on 8 May. Under questioning that day Fleischmann admitted that the gamma peak was wrong. So was there any evidence for fusion, any sign of the fusion products—neutrons, tritium, helium and gamma rays?

There were considerable doubts about Fleischmann and Pons' claims to have detected fusion products, and the reports at the Baltimore APS meeting showed that no one else was seeing significant amounts. If there was to be any credible proof that the heat was due to fusion, then fusion products had to be produced in amounts commensurate with the heat. Fleischmann and Pons insisted that they produced heat; in these circumstances the natural process of science would be that radiation monitors would be brought to bear on such cells. This is how collaborations develop and knowledge improves: bring nuclear experts and their instruments into the laboratory where heat-producing cells are operating.

The University of Utah thus arranged for a team of physicists including Mike Salamon, Ed Wrenn and Haven Bergeson to make radiation measurements in Pons' laboratory.

Forcing deuterium deep inside palladium metal where unknown reactions take place, leaving submicroscopic traces buried among trillions of atoms in the crystalline lattice of palladium and then finding them sounds an impossible task, worse than tracing one criminal in a huge city in the pitch dark when all you have is a crude identikit picture. However, it is not like that. It is more like looking for such a person if they were clothed with flashing lights that announced 'Here I am' every time anyone came near. Inside the palladium there are so many electrically charged particles around, such as electrons and atomic nuclei, that they emit gamma-rays and X-rays when even small traces of fusion products pass by. The fusion products have in effect, to pick their way through a dense forest of trillions upon trillions of atomic sensors. The resulting gamma-rays and X-rays reveal not just their presence but also show what they are.

The physicists' plan was to look for gamma rays—the flashes of light—by using a 20 cm by 10 cm block of sodium iodide crystals, a substance that will send electrical signals to a recorder when gamma rays strike. Salamon and Wrenn worked together in setting up the detectors, placing the sodium iodide underneath the table that supported Pons' cells so as to avoid interfering with normal lab activity by Pons' group. Any gamma rays produced in the cell would have to pass through the cell, a water tank surrounding it and the table top before entering the detector, so in a separate laboratory they built a replica of the whole set-up in order to test the detector and calibrate it. They put some standard gamma emitters on the top of the table in exactly the same position as cell 'number 2-1' in Pons' original configuration, in a water-filled tank identical to that used by Pons, with the detector in its exact relative position. The scientists then compared what the standard sources were actually putting out with what the detector on the floor recorded. In this way they worked out how much the table and apparatus affected the ability to record gamma-rays, and knowing this they could, in Pons' lab, compute the true activity in the cell from the number of gamma-rays that his detector picked up.

In early May, when Salamon started his investigation, Fleischmann and Pons were still claiming to see these gamma-rays and were citing them as evidence for fusion. Petrasso and colleagues at MIT had just presented their criticisms at the American Physical Society Meeting and a preliminary version of their paper had been sent to Fleischmann and Pons. Some people were already prepared to write off the neutrons and conclude that either there was no fusion or that some other 'aneutronic' fusion was responsible for the heat. Salamon at last was ready to make a definitive test in the lab where Pons had claimed to have heat-producing cells.

Salamon measured gamma-rays for 831 hours, five weeks from 9 May until 16 June, and gathered a mountain range of some prominent peaks. He saw the clearly identified signal of potassium at 1.46 MeV, correctly identified the signal from thallium at 2.6 MeV and in the middle two small hills from bismuth, one at 1.76 and the other at 2.20 MeV.[1] Everything about the detector was working properly and the line of peaks showed that he was in the right place and knew exactly where to look. But at 2.22 MeV there was nothing; Mount Fusion was not to be seen.

Simply put, there was no *dd* fusion producing neutrons. Stated more precisely, if there was any such fusion taking place it could account for no more than 10 picawatts, $10^{-11}$ or ten trillionths of a watt.

The next channel to test was *dd* producing tritium and protons. This had been a leading candidate for the test-tube fusion process both because of reports from Bockris of Texas A & M University and also from several groups in India, all of whom were claiming to find significant amounts of tritium after completing their experiments. A second reason why many people suspected that this channel might be the culprit was because the reaction products—the protons—do not get out of the palladium electrode (the range of protons at 3 MeV energy, as in this case, is only 30 microns). So this seemingly could avoid the paradox of watts of fusion power with no observable particle emissions.

Or so many people thought. In fact this reaction produces a strong and distinctive gamma-ray signal as its products, tritium or protons, fly out into the surrounding palladium and unused deuterium.

The tritium will eject neutrons from the deuterium. Salamon's team had already ruled out neutrons from the direct *dd* fusion by the absence of the 2.22 MeV gamma-ray; this equally well eliminated neutrons ejected by tritium collisions and thus ruled out the tritium–proton channel.

There is a double check on this channel, since not only can the tritium bump into its surroundings and give off gamma-rays but the protons do so as well. The key here is that a 3 MeV proton passing near to palladium atoms disturbs the nuclei of those atoms. The palladium nuclei then emit gamma-rays with energies of 0.37, 0.43 and 0.51 MeV (and also a very weak line at 0.56 MeV), very different from the 2.2 MeV gamma-ray but equally distinctive as a coded message.

A colleague, Walter Schier of Lowell University in Massachusetts, made measurements for them, exposing some palladium to a beam of protons whose energy had been tuned to 3 MeV in a particle accelerator. The gamma-ray peaks shone out like beacons. Knowing from this that if there were any 3 MeV protons around in the palladium he could detect their gamma-ray signature, Salamon looked through his data to see if there were any sign of these gamma-rays being produced in Pons' lab.

There was none; there were no 3 MeV protons in the palladium and so, by implication, there was no tritium−proton production from *dd* fusion, thus confirming what he had deduced by the previous method. If there was any such fusion too feeble to measure, then this measurement alone limited it to be less than ten milliwatts. But the absence of neutrons ejected from the deuterium, the other test that they had used, placed an even more restrictive limit of less than a microwatt, one-millionth of a watt.

The possibilities for fusion are fast disappearing here. When two deuterium nuclei collide they can shake a neutron loose (forming the $^3He + neutron$ channel) or shake a proton loose (as in the tritium − proton route. All that remains is that either the two deuterons bounce off each other, which gives nothing, or else they coalesce to form a nucleus of $^4He$ and radiate the mass mismatch in the form of a gamma-ray of 24 MeV energy. In hot fusion this route is a million times smaller than the previous two routes. If the heat measured by Fleischmann and Pons is due to this, then this channel would have to *dominate* by over a million—an amazing turn-around, a relative change of a million-million. The final possibility was that deuterium had fused with protons that were plentiful in the cell. In this case the products are $^3He$ and a gamma-ray of 5.5 MeV energy. And here again Salamon saw nothing; the maximum power outputs through these channels would have been picawatts.

So what could it be?

Walling and Simons at Utah had in the first days after the 23 March press conference invented a scheme to avoid radiation, a scheme seized on by Pons and advertised widely. The idea was that the gamma-ray produced in $d + d = {}^4He + \gamma$ never escaped the atom, but instead ejected electrons from the periphery.

But here again there would still be tell-tale gammas produced. The initial gamma may have disappeared, effectively amplifying the electromagnetic field of the atom which knocks out an electron, but the sudden acceleration of the charged electron radiates light—it's known as 'bremsstrahlung' radiation—which Salamon's apparatus would be able to detect. And the result—nothing; at most 10 billionths of a watt.

The conclusions of the paper written by Salamon's team are unusual in that they portray Pons' reaction to these results. During five weeks in spring, a critical period when the world is being told of ever-better heat production in Pons' lab, Salamon's team is monitoring these cells for radiation and concluding that 'if a heat excess were to have occurred during our period of observation . . . no known fusion process significantly contributed to that excess.'

Pons had been reported to have claimed that he had a cell that had boiled, and the media had suggested that fusion teapots would be the outcome. During May there was indeed a period of two hours when the cell did just that. This could be the crucial evidence, the ghost of fusion caught at it while under observation. They checked their data for this critical period but there was nothing untoward happening, none of the spectral features of fusion. That would seem to clinch it: no fusion, but Pons retorted that they should 'not reference these events as being due to release of excess thermal energy' since this boiling might very well have a conventional explanation. Salamon wryly notes: 'Unfortunately we have not received any numerical data on excess heat production during the 831 hours of our monitoring, so we are not able to correlate the absence of nuclear signatures with the presence of anomalous heat.'

That Pons would not give Salamon any numerical data is a strange way to go about science if you have nothing to hide. However, he did claim that there was a two-hour segment in which 'there was excessive thermal release from the cell' but that, in an amazing and apparently unfortunate coincidence, this happened during a period when all of Salamon's detectors were out of commission following a lightning storm. 'Your computer and detector were not under power at that time since they had not been reset from a power failure which had occurred in the lab', Pons told him.

A colleague, Kurt Drexler, suggested a way of making the detectors speak from the grave, even when all power had been lost.

The trick has to do with the sodium in the sodium iodide detector. The team had been using it to detect gamma rays emitted by the neutrons after they stopped in the water. But some neutrons pass through the water and hit the sodium in the detector directly. In so doing they can convert it into radioactive sodium-24, sodium with one more neutron than the normal form (sodium-23). Now, sodium-24 has a half-life of fifteen hours and will be emitting detectable radiation days after it has been activated. If there had been fusion neutrons produced by Pons' cell during the thunderstorm when the power was off, they would have activated the sodium which would have been playing the message for days thereafter, to be heard as soon as the power came back on. Pons apparently was not aware of this nuclear spy.

They checked the records, and here again there was no sign of fusion having taken place. This limited the power as at most a microwatt, one millionth of a watt.

On 16 August in answer to a question at the EPRI Conference at the University of Utah, Pons had shown data indicating a low level excess heat output occurring over an extended period of time. Kevin Wolf, a nuclear physicist from Texas, asked if this had occurred during the time when the Salamon team had their detector in operation, and Pons replied that it had. Salamon concluded that 'If this is the case, the excess does not originate from known nuclear processes.'

There is a final twist to this tale. Pons claimed that Salamon's data were taken over a cell which was not producing heat, and so have no relevance to the question of whether *heat-producing* cells emit fusion radiation.[2] So what we need are measurements of a gamma spectrum taken over a cell that Pons agrees is a heat producer. The answer is in Figure 14. This is a spectrum that Fleischmann, Pons, Hoffman and Hawkins produced in *Nature* on 29 June in an attempts to deflect criticism by the MIT group. In their own words they admit: 'During the period of the measurement it was generating excess heat at the rate of 1.7 to 1.8 watts.' *There is no gamma peak at 2.22 MeV. There is no fusion signal.* As Petrasso and collaborators summed up: 'This imposes an upper bound' of a nanowatt.


## The beginning of the end

The experiment had been made in May and June 1989. Pons was made aware of the results at that time and asked to comment on the draft paper, written in July. He made no request that it not be published and it was duly submitted to *Nature*.

The paper in its original form was rather long and needed shortening, and also Rich Petrasso went over the original draft in fine detail and made several suggestions which helped to sharpen some of the discussion. This refereeing process, both the formal peer review by *Nature* and the personal one, where you ask respected colleagues for their opinions, and rewriting of the paper, mailing back and forth and correcting proofs, all took time and the paper was eventually published in *Nature* in March 1990, at about the same time as the National Cold Fusion Institute was hosting the 'First Annual Cold Fusion Conference' (or, as *Nature* referred to it, The First 'Annual' Cold Fusion Conference).

In 'normal science' the appearance of such a paper, on top of all the negative results on nuclear products from many laboratories throughout the year, would be the final straw, reducing enthusiasm for a 'Second' Annual CF Conference, and helping to bring the story to a close. And for the *scientific* story it will probably be viewed eventually, from the longer perspective of history, as a watershed in the refutation of the claimed nuclear phenomenon. However, in the summer of 1990, it has brought to a head some of the less savoury *non-scientific* aspects of the episode, and raised questions about conflicts of interest with regard to some authorities in the University of Utah, their relations with the National Cold Fusion Institute and their attempts to preserve the illusion that Fleischmann and Pons have demonstrated watts of power from cold nuclear fusion.

On 3 April 1990 Gary Triggs, who is Pons' lawyer, wrote a letter to Salamon in which he called on him and his scientific co-researchers to retract the paper they had published in *Nature*. The letter advised Salamon that any damages suffered 'by my clients proximately caused by any act or omission on the part of yourself or any other co-author on the subject of the paper will not be tolerated. I have been instructed by my clients to take such legal action as is deemed appropriate to protect their interests in this matter.'

News of this was distributed around the scientific community via Douglas Morrison's 'Cold Fusion News' on 27 May. He summarised what were the feelings of many with his headline, 'Lawyers threaten legal action: Academic freedom menaced', and also, 'Stop press: Financial scandal breaks—this might be the end.'

Morrison commented, 'The letter talks of "my clients"—who are they?' The *Tribune* provided the answers: '[Triggs] has received payments of more than $50,000 from the University of Utah for legal work related to Cold Fusion.'

'Triggs represents Stan and Martin and is advising them on a number of issues, including patent prosecution issues', reportedly said Greg Williams, of Van Cott, Bagley, Cornwall and McCarthey (the state's attorneys). 'He is not representing the University.'

The news article continued:

'Mr Triggs was paid out of funds in the Office of Technology Transfer, the University department that assists in turning useful scientific research into profitable ventures. The Office is one of the departments under the supervision of [University Vice-President for Research, James] Brophy.'

' "How they [pay for] it is their business" said Mr Triggs. "I simply send the bills" . . . [he] said he had not billed anyone for the Salamon letter. "I think that is between me and my clients, but I think it's a matter that affects the entire [test-tube fusion] program".'

Salamon's reaction to the legal letter was unequivocal. He noted that Triggs had sent several letters to people threatening them with legal action, and then commented 'These people include several of my colleagues and myself at the University of Utah (even a University of Utah undergraduate received such a letter). I am extremely disturbed, in fact disgusted, that the University has apparently been financially supporting such detestable activity, activity which is antithetical to the spirit of free academic enquiry.'

'What is particularly obscene about this is that my colleagues and I at the University of Utah are being threatened with legal action for honest scientific work done at the behest of the University of Utah, and yet the administration has so far refused to provide us with any legal counsel whatever.'

This new issue of why the University was supporting some employees and not others was raised with the University.

'Dr Brophy said that Mr Triggs was "certainly" not working for the University when he wrote the letter, even though he was involved in a legal matter between faculty members.'

The paper in *Nature* and the lawyer's letter each appeared in March/April 1990 coincident with the Cold Fusion Conference at the National Cold Fusion Institute. The profile of test-tube fusion was beginning to sag until it was raised by the imminent meeting, and the institute's finances were not as healthy as many had hoped. The 4.5 million dollars, given from the State of Utah, was meant as seed money with the hope that industry would contribute more millions, but this apparently had not developed as well as had been hoped.

These financial concerns were alleviated when, at the first anniversary, 23 March 1990, it was announced that an anonymous donor had contributed half a million dollars.

# 18
## THE FIRST ANNIVERSARY

After a year we were left with small amounts of heat at best, nothing that suggested a solution to the world's energy needs, and no commensurate amounts of radiation. Faced with this non-evidence for test-tube fusion most of the thousands who had attended the conferences and watched their computer mail avidly in the early days had by now returned to the pursuits which they had interrupted twelve months before. The attendees at 'The First Annual Cold Fusion Conference' held in Salt Lake City at the end of March 1990 were for the most part the true believers, with a handful of sceptical nuclear physicists asking critical questions.

The media regarded this as rather supportive of the phenomenon: 'Utah keeps embers of cold fusion aglow' said *New Scientist* and 'Sparks still flying over cold fusion', reported the *Financial Times'* Clive Cookson who, twelve months earlier, had started the excitement with his scoop in advance of the press conference. Cookson summarised the year, noting that initially everything had seemed to be straightforward and that with hundreds of laboratories around the world working on it, it had been almost certain that they would 'establish whether the claim was valid in a few months.' He noted that a year later the 'truth' remained elusive, and that there was a widespread impression that the entire affair had been a 'ghastly scientific mistake', this view being particularly strong in the UK. Moreover the impression in the UK was that test-tube fusion was being pursued only by a small band of true believers.

'That is definitely a wrong impression,' was the quote from David Worledge of the Californian Electric Power Research Institute. Martin Fleischmann was reported to have said 'on balance, the results were a massive confirmation of the generation of excess heat—and a confirmation that neutrons are generated at low rates.' However John Huizenga, co-chair of the DOE commission whose report on test-tube fusion came out in November 1989, commented to co-panellist Steven Koonin that nothing much had changed, and that the conference

317

merely reinforced the lack of enthusiasm for the phenomenon that the DOE report had noted.

Those who chose to believe in test-tube fusion had to accommodate two assumptions against all known evidence. First, that fusion took place a factor of 55 orders of magnitude, $10^{55}$, faster than ever seen before, and that the energy was taken up by the palladium metal as heat and not by energetic particles such as neutrons, protons and tritium. No satisfactory theory could accommodate these requirements but, as always, the defence was that experiment must be the ultimate arbiter. And Fleischmann in the concluding talks at the conference seemed to have found that very proof in the pioneering papers by Rutherford and his contemporaries in the 1930s. To a standing ovation from the partisan audience he claimed that the physicists had failed to study their own literature, that the proof of test-tube fusion was there already and that they had overlooked it. This was a dream—to have the nuclear physicists destroyed by their own heroes. What was it all about?

I had been to talk with Fleischmann about six weeks earlier in England, before he left for some months in Utah. At the end of our meeting he showed me copies of some old papers, saying 'these are gold-dust.'

The first was a letter to *Nature*, dated 17 March 1934, from Rutherford, Oliphant and Harteck. This was the first study of $d+d$ interactions, and its main feature was that it was the first sighting of tritium, which was formed as a product in the $d+d \rightarrow t+p$ channel. In those early days the names were different from those familiar today: instead of deuterium they referred to 'diplogen', and the nucleus of the atom was called 'diplon' instead of deuteron.

When I was able to get to a library I got hold of the old journal and found the famous paper. It was unclear what sustenance there was for test-tube fusion devotees. There was a vague suggestion that the $t+p$ channel was slightly larger than $^3He+$ neutron, but the startling feature when I read the paper was the fact that the deuteron beam had an energy of 20 keV, which is an energy similar to that in a tokamak, or, in temperature terms, hotter than the centre of the Sun— nothing to do with test-tube fusion.

The second paper that had excited him was one by Philip Dee which Rutherford had transmitted to the Royal Society for publication in their proceedings. Fleischmann had shown me this in his house and regarded it as suggestive of test-tube fusion in solids from the early era.

Dee had been using a cloud chamber to record the results of $d+d$ reactions. The cloud chamber reveals the passage of moving electrically charged particles by forming trails of liquid drops much like the vapour trails in the sky which reveal the flight of a jet plane. The conservation of momentum implies that the two particles that emerge from the $d+d$ reaction—the triton and proton—will fly off in opposite directions, that is, if you ignore the momentum of the incident deuteron. If the incident deuteron is moving fast, bringing in a lot of energy and momentum, the triton and proton will both be thrown forwards and the angle between them will be less than 180°.

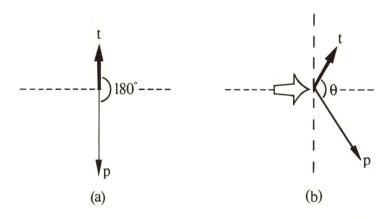

(a)                    (b)

Figure 18. $d+d \to t+p$ (a) at rest, (b) in motion. In (b) the angle $\theta$ between the tracks is less than 180°.

The experiment used beams with 160 keV energy and for this the angle between the outgoing particles should be 162°. And most of them were, but there was a handful of cases where the particles seemed to go off in opposite directions—at 180°, as if the incident deuteron had little or no energy. Dee noted that these back-to-back tracks were the result of nuclear transmutations 'effected by *slower diplons* which have lost energy by (prior) collisions in the target'.[1] So seemingly here was evidence that low energy 'cold' fusion might have taken place. What is more, the target deuterium was in a chemical compound impregnated into mica, so it was in a solid. Had there indeed been cold solid-state fusion observed over half a century earlier, soon after the discovery of deuterium, and had it been overlooked all these years?

Unfortunately not. First of all, the photographs do not show that the tracks are *exactly* back-to-back, and so you cannot eliminate the possibility that the incident 'diplon' had even one keV of energy which is already similar, in temperature terms, to the centre of the Sun. But even apart from this there is another much sharper reason why these pictures from the past offer no succour to aficionados of test-tube fusion. The fact that the images show clear proton and triton tracks emerging from the mica with their full energy shows that the *usual* fusion has occurred—the products have not been hidden from view. Yet the whole thesis of test-tube fusion involves making theories to explain the failure to see any such products. There are *no* energetic tritium or protons seen in the experiments studying test-tube fusion in deuterated palladium.

After Fleischmann had advertised these papers to me in February I read them and discussed the one by Dee, in particular, extensively with Rich Petrasso. We then learned that Fleischmann was alluding to these papers in talks, such as the one at Los Alamos early in March 1990.

I learned of this from Bill Johnson, a nuclear scientist who had led most of the cogent discussions on *Bitnet*, who offered his personal impressions of this talk. A year's passage had reduced interest in the phenomenon; Pons had talked to a packed house in 1989 in the largest lecture room on the site, whereas Fleischmann's 'witty, articulate' talk only partially filled a more modest venue. The talk in large measure covered the same ground as Pons' earlier one, with only a few pieces of news. Throughout the year attempts to obtain new data and details of the experiment had not succeeded, and Johnson described the occasion thus.

'A certain liturgical atmosphere has developed regarding the Utah group's experimental technique: some of the experimental set-up is described as though by ritual, ritual complaints and suggestions are raised by the audience and they are ritually answered 'that's in the paper' or 'we're working on it'. I have been to Catholic masses that were less predictable.' Johnson also commented that Fleischmann had cited the early papers as supportive of test-tube fusion, though without convincing the audience.

Suspecting that this might be a warm-up for Fleischmann's presentation at the forthcoming 'First Annual Cold Fusion Conference', due to be held within two weeks, and being unable to attend it myself, I urged Petrasso to go and ask some probing questions about the confusing and contradictory claims that were being made. Initially he had not intended to go, but at the last minute changed his mind when he learned of some belligerent statements about him quoted in the Salt Lake City media, allegedly made by Pons, to the effect that the nation's physicists were so 'violent' about getting information that they were 'taking display information off television screens and writing technical papers against us' and which specified 'one angry researcher' who published in *Nature* basing his case on such dubious material.[2]

After all these months there was no suggestion in this statement that the data in question could be wrong, or had been withdrawn.

320

Petrasso was concerned that the conference might present too much of a one-sided account to the media if sceptics were not present, and decided to attend.

The university authorities were evidently pleased that Nobel Laureate Julian Schwinger had accepted an invitation to speak. While big names are always good for publicity and for getting political support, the scientific value depends on content and Schwinger's ideas, like all the others, 'had to be imprecise just to cope with the vast differences in data.' So reported the *Salt Lake Tribune*.

The talks bore some of the hallmarks of religious belief rather than of scientific endeavour, in that negative aspects in data were wished away. Thus 'vast differences in data' was taken as a stimulus for considering 'imprecise theories' rather than as an indication that all that was being seen were different random fluctuations or matters unrelated to one another, and certainly with no relevance to the much advertised watts of heat.

The director of the National Cold Fusion Institute, Dr Will, also excused away failure to verify the phenomenon, acknowledging that few results had been produced by the institute outside of the Fleischmann and Pons work, but he claimed that that was because the institute was still young, having opened only in August 1989. However, as the *Salt Lake Tribune* noted, 'the institute's level of funding exceeds anything available to those elsewhere who have claimed larger successes and they have only been working a year.'

There was considerable nervousness in some quarters about how things were developing. The paper by Mike Salamon and the group from the University of Utah physics department, in which they had limited the radiation and fusion pathways in Pons' lab, had just appeared in *Nature*. This paper had been submitted in mid-1989 and, after being refereed and modified, it had finally gone to the printer, coincidentally appearing just prior to the conference. Pons' lawyer was threatening action, claiming that the data were incorrect, that the paper misrepresented the situation in Pons' lab and that its appearance in *Nature* at that time was designed to undermine the conference.

*Nature* itself carried an editorial with headline 'Farewell (not fond) to cold fusion' containing a hypothetical footnote: 'The search for what was called "cold fusion" continued for several years, much as the search for the Philosopher's Stone persisted in the face of repeated failure and enlarging common sense, in this case sustained by cash from the State of Utah and grant-making agencies that should have known better.'

With such accusations in the air any criticism of test-tube fusion at this time touched on sensitive nerves. During a press conference

Will, the Cold Fusion Institute director, said of Petrasso's questions: 'One wonders if he is as critical of his own research. The gentleman sees science as black and white.' Hugo Rossi, Dean of the College of Science at the University of Utah and organiser of the conference, then reminded everyone that the conference existed to stimulate discussion, that presentations of observations and conclusions are not scientific unless they can hold up to the most thorough examination and critique, and that he welcomed Petrasso's contributions.[3] Will then interjected that he wished that 'the comments by Dr Petrasso would be constructive.'

Whether Petrasso had been constructive or destructive rather depended upon your point of view. As an example, Dr Srivasanan had reported how his group in Bombay had found tritium in their experiments. Petrasso asked if the group had tested for the presence of tritium on their electrodes *before* the experiment started so as to ensure that they did not have a contaminated cell to begin with. 'I was given a long lecture, but all I wanted was a simple yes or no,' Petrasso told me. 'I interjected and he answered—"no"—they had not done this check. Also it seems they saw this phenomenon only once and have not reproduced it.' If you want to believe in this as supportive of test-tube fusion, then Petrasso's question was destructive; if important issues about the conduct of experiments are to be brought into the open then Petrasso was being constructive.

(It was therefore particularly ironic that on 7 and 8 June the *Wall Street Journal* and *New York Times* reported that 'Contamination at Three Labs Casts Doubts on Results Pointing to Cold (test-tube) Fusion' and 'Scientists says Cold Fusion Tests may have had some Impure Rods.' Texas A & M, where a group involving J. O'M. Bockris had been claiming to produce large quantities of tritium in its cells and giving comfort to believers, was shown to have been using palladium rods that were contaminated with tritium before the experiment began.)

And so at last the final talk of the conference was given. Martin Fleischmann gave a witty summary including jibes at the physicists who are always ready to invent new particles to explain weird phenomena, and alluding to their predilection for giving them Greek sounding names ending in -*on*, he proposed the dubi*on* and the Morris*on* (an allusion to Douglas Morrison who had authored the electronic mail newsletter on test-tube fusion throughout the year). Then he came to his *pièce de résistance*. With *Nature* currently so antagonistic to test-tube fusion, it was interesting that it was in *Nature* that Oliphant, Harteck and Rutherford had first presented their discovery of $d + d \rightarrow$ proton + triton which Fleischmann had previously cited to me as relevant to test-tube fusion. Rutherford had then submitted Dee's paper to the *Proceedings of the Royal Society*, and Fleischmann now presented

322

it to the audience at the cold fusion conference. He commented that physicists had not done their homework well, and were omitting to read their own literature which contained evidence for test-tube fusion of $d+d$ in metals, courtesy of Dee in 1934.

It was a brilliant piece of oratory and the audience gave him a standing ovation. Douglas Morrison was near the front and all around him were standing and applauding, so he could not see what was happening in the rest of the hall. Rich Petrasso was in the middle. After all the other talks there had been questions but the audience had taken over like at a political convention and there would be no chance for questions here. So Petrasso walked down the aisle, through the rapturous applause, and asked Fleischmann if he could make a comment. Fleischmann agreed.

Test-tube fusion theories discussed throughout the meeting by John Hegelstein, Giuliano Preparata and Julian Schwinger were suggesting that the non-appearance of fusion products was because the energy from fusion was diverted into the lattice of palladium atoms, hiding it from external view other than as heat.[4] This had been a central theme in the theoretical discussions at the meeting. Petrasso made a mild joke to pick up the style of Fleischmann's presentation remarking that in addition to the Morrison there was another relevant particle—the 'skepticon'. Then he made his crucial remark:

'I too am a great admirer of British science and I have with me here the very papers to which you referred. Dee's work had nothing to do with test-tube fusion for the reason that (in Dee's photographs) the triton and proton products come out and escape *with their full energy*.'

This implied that Dee's work was irrelevant to those theories and, therefore, offered no support to the general attempts to give credibility to solid state fusion at room temperature without energetic nuclear radiation.

To which Fleischmann agreed that Petrasso had a valid point.

In the audience Morrison and others were aware of Petrasso's light-hearted opening remark, but the audience was now leaving and many, including the media, missed Petrasso's crucial intervention. The First Anniversary Conference was over.


## Scandal

This is almost the end of the science story, save for a judgemental summing up in the next chapter which evaluates what had been learned from this episode. The confusion of that year was like a supernova, bursting unexpectedly into view and then fading from the eyes of all except a minority of supporters, many of whom were at the anniversary meeting. It had appeared at first to be a matter for science but, increasingly it became clear that other forces were at work. The scientists at the centre of the action became entrapped; patent attorneys and eventually legal issues seemed to play a central role.

Within a month of the first press conference the state authorities were discussing putting money into a 'National Cold Fusion Institute' based on the belief that the fusion phenomenon was real. By the end of the year the statements about science coming from Salt Lake City seemed ever more out of touch with everyday reality.

The relationship between the National Cold Fusion Institute and the University of Utah also was coming under scrutiny.

The National Cold Fusion Institute was funded with 4.5 million dollars from the State of Utah, which was meant to be seed money. These funds were fast running out in the spring of 1990 and there were as yet no convincing and published results to show. The appearance of Salamon's paper, with its negative results, was not helpful to those who hoped to present a positive picture of test-tube fusion. Industry had given little money to the NCFI and at the Supervisory meeting on 23 March 1990 it was said that an anonymous person had given a donation. However, following some investigative journalism by a local reporter and Gary Taubes, a science writer, it transpired that the mysterious donor was none other than the University of Utah itself, and the money, half a million dollars, was in a foundation deriving its income from rent on office space in the University Research Park (where NCFI is located) and from proceeds from university-owned patents.

The story should have appeared in the *Salt Lake Tribune* on Thursday 31 May, but as Dr Brophy, the Vice-President for Research, did not return the Tribune's phone calls, they gave him another day and the news broke on 1 June on the Channel 2 News at 10. Dr Brophy said that 'anonymous donation' was 'inappropriate terminology' and that they would not use that terminology any more. The article in the *Tribune* revealed that the money was referred to in the Institute's quarterly report as 'external funding', and a 3 May press release called it a 'recent anonymous gift of $500,000' which had been used to bring in three visiting scientists from Malaysia, Poland and South Dakota.

The revelation prompted disdain from members of the panel that oversees the State's five million dollar investment. Panel Chairman Raymond Hixson reportedly said 'I'm a little upset about it because it was inferred that it was a private, anonymous donor and therefore we could consider it as an effort by the University to secure outside funding. This, to me, is not quite the same thing. I'm going to have to look into it. At this stage, I can tell you I'm not very happy about it.'

Faculty members also were not happy. College of Science Dean Hugo Rossi and 22 faculty members representing all of the college's departments issued a statement: 'After learning that an unnamed donation, described as "external funding" in the last quarterly report of the Cold Fusion Institute, was in fact, from the University Research Foundation, I called a meeting to discuss the implications of this apparent deception.'

'The perception that desperate means are now being used to continue support of claims unsubstantiated by peer review of the data is unavoidable. It was the sense of the group that a complete and objective financial audit and scientific review must precede any further state or university funding of this project, and in a separate document we are asking the chair of the board of directors of the institute as well as the university, to convene such review panels in consultation with the faculty of the University of Utah.'

Douglas Morrison, in his cold fusion newsletter, reported that 'Dr Rossi, who was at one time the interim director of the fusion institute, said he wanted to make it clear that the statement came from faculty in all departments in his college, including the Chemistry Department, to put to rest the notion that there is a battle between physicists and chemists over how the fusion situation has been handled.'

According to the *Tribune*, the University President, Chase Petersen said, 'There was no intention to mislead anyone,' and continued, 'we assign hundreds of large and small amounts of money to different parts of the university during the year, and we rarely discuss the source of the particular fund, nor does anyone generally care.'

The Electric Power Research Institute, which was about to sign a contract for $160,000 to support research at the NCFI, decided to wait until the dust settled.

On 4 June a letter was on its way to Salamon from attorney Gary Triggs, who had earlier put pressure on Salamon's team to withdraw their paper. This second letter apologised for any 'misconception' that the first letter had caused, and said that there was no intent 'to limit in any way the lawful exercise of academic freedom' and that Fleischmann and Pons now intended to settle the dispute in 'the court of science'. Even so, it repeated the earlier complaint that the publication of the paper in coincidence with the March test-tube fusion conference seemed to be 'another in a long series of negative acts by *Nature* to damage cold [test-tube] fusion in general and [Fleischmann and Pons] in particular.'

The same day pressure on Petersen was growing as the faculty senate voted to ask the State of Utah's governing board for higher education to examine his competence. On 11 June Petersen announced that he would retire during the next academic year.

The last remaining straw that the few surviving test-tube fusion supporters clutched to also began to disappear. On 6 June the *New York Times* and *Wall Street Journal* announced that the tritium, on which the aficionados had been pinning their hopes, was entering the experiment as a contaminant. J.O'M. Bockris of Texas A & M University had frequently advertised his sightings of large amounts of tritium to support the reality of test-tube fusion, and his testimony twelve months earlier had helped to gain the funding of the NCFI.

In *Science* on 15 June the story plumbed new depths with suggestions that the possibility could not be excluded that someone had deliberately been introducing tritium into the experiment in the Texas laboratory. This development raises important issues with regards to the policing of scientific research, an area which scientists and administrators have continued to be rather cavalier about—but that is another story. The administrative scandals and the increasing air of unreality surrounding the episode brought to an end, in June 1990, the most bizarre 500 days in the history of modern science. The fallout, however, will continue for years.

# 19
## ASSESSMENT

In the spring of 1989 Jo Redish had assured people at the Baltimore APS meeting that 'the standard workings of science' would find answers to the questions: Is there useful heat produced during electrolysis of heavy water in the presence of palladium?' and, if the answer was yes, then 'Is nuclear fusion the process responsible?' A year later much had transpired that was not in the 'standard' working of science. These untoward events included Pons' lawyer sending a letter to a group of scientists whose results did not confirm test-tube fusion, allegations in the media that Pons was accusing fellow scientists of stealing ideas or suppressing data that would confirm test-tube fusion, an accusation made by Fleischmann that a scientist investigating test-tube fusion was involved with a group for whom an arrest warrant had been issued in Utah, open verbal attacks on the editor of a leading scientific journal, incorporation in the paper of data that had been measured by scientists who were not initially acknowledged, those scientists whose names were advertised as authors being sometimes unaware of what the paper contained or where parts of the data came from, and mysteriously mobile data measured at one value, moved elsewhere and then moved back again later without a common agreement among the signatories on how this was justified. The standard workings of science had been submerged by 'the malign influence of extraneous considerations.'[1]

To understand some of the reasons for these singularly unscientific events one should first realise that many of them were the actions of individuals under intense pressure month after month, and who from time to time reacted irrationally while in the glare of media attention. Fleischmann and Pons believed strongly in their results and had become convinced that Jones had pirated aspects of their work; having worked in secret they were cut off from the day to day sceptical questioning by fellow scientists that is an essential part of the usual scientific process, and so were unaware of flaws in the nuclear physics aspects especially. They made a premature and highly publicised announcement which received early confirmations, and this must have encouraged them that their gamble to go public had succeeded.

But then refutations appeared from several influential laboratories and institutions which also received headline treatment in the media. The two chemists and the University of Utah were by then public figures, featured in magazines around the globe, their faces instantly recognised, their privacy gone. Public and corporate expectations had been raised and continued to grow due to further highly publicised claims from the University of Utah during April, even while it was becoming clear that the claims of fusion were flawed.

The two chemists had not expected such an intensive response to their press conference, had not anticipated society's craving for the instant fix that radiation-free fusion promised, and were ill prepared to deal with the ensuing demands. Fleischmann, who was more experienced and was away from the storm centre during the weeks immediately following the first announcement, was better situated. Pons, by contrast, was in Utah where almost everyone recognised him as a hero, believed that test-tube fusion would bring fortunes to the state and for whom every positive piece of news was like sweet music. When the first doubts were raised—about heat with light water or the absence of nuclear products—events cruelly conspired to deceive: *pd* fusion could cause the former and *dd* fusion producing helium (which was apparently found in mid-April) gave the latter. Thus the phantom encouraged him onwards when caution was needed. The local climate and these events in his own department helped sustain the belief that everyone else was out of step.

The media were regularly assured that confirmatory experiments were about to be announced, but instead refutations were increasingly evident. Those close to Fleischmann and Pons chose to believe that there was some conspiracy, that people with positive results were being prevented from publishing them; those in the community finding negative results became irritated by the unfulfilled promises that positive results would be presented to back up the claims being aired to the media. Regrettably the pressures encouraged increasingly belligerent responses and suspicion.

Fleischmann and Pons' problems arose initially because they went public too soon with immature results.

So *why* did they publish? If you have results worth publishing in which you are confident, then you publish. If you have not performed all the necessary controls, or have not completed parts of the experiment sufficiently carefully, then you should not publish. There is no absolute—one is never truly finished—but there is a judgement to be made: 'can I defend this against my colleagues' questions?' After all, knowledge is not what is printed in the journals but what is shared among people's minds. If your colleagues do not incorporate your claims into their mental world view then there is little point in writing them down. It is one's peers who decide on the worth of your efforts, and the reaction of the majority of scientists to the Fleischmann and Pons paper was uniformly damning.

Fleischmann has said that the paper 'had to be published . . . though I would have preferred to have backed it up with some further work.' The key is that 'we had written a number of patents by that stage and the view of the university was that we should announce this by a press conference. It was really the patents that were driving this.'[2] And the rush was because of their belief that Jones had pirated the ideas and was about to go public himself.[3]

Initially all went well as confirmations from around the world were announced within three weeks, much sooner than the negative results. For many outside the scientific community, and even within it, these early positive announcements created belief that the astonishing claims might even be correct. The funding of the National Cold Fusion Institute and the reactions in Washington were much stimulated by them. However, in general it is wiser to wait until it is clear whether or not a consensus is appearing; a few early confirmations prove nothing. Indeed the above sequence of events is rather natural in 'pathological science', as Douglas Morrison pointed out in his electronic newsletters.[4]

Interviewed on a BBC *Horizon* television programme Morrison noted the following.

'When a dramatic result such as room temperature fusion is announced, people all over the world start to work on it. Many of these experiments are done very carefully and see nothing; the rest may be less expert and by chance and natural fluctuations a few will see what might be positive signals. Now the psychology enters. These possible positive signals are felt by the scientist concerned to be a confirmation of a previously announced result, and so they don't need to worry about making lots of detailed cross checks: they go ahead, announcing their confirmation by a press conference.

'Meanwhile there are many people doing careful experiments who are finding no signs of the claimed phenomenon. Imagine how you would react if you were in this position. You have a dilemma. If you announce that you are seeing nothing you are saying that some colleagues are wrong. This is not easy, after all maybe it is you that are wrong rather than the original group. So you go over and over your data, double checking everything to see if there is anything that you may have missed. This all takes time and so, early on, it is confirmations that are in the lead, in the news, and in the public perception.

'Then one or two groups have the courage to publish negative results, which gives others the courage to publish and then there are lots of people who have been waiting in the wings with their negative results.'

In the present episode the deluge occurred on 1 May at the American Physical Society meeting in Baltimore, and the media interest by then was so high that it made headlines. With seven days between this and the Electrochemical Society convention in Los Angeles, the media saw test-tube fusion as full of the ingredients of controversy that make good copy. It became a regular feature, especially in the leading US newspapers, and was still commanding space more than a year later.


## The media record

One of the most singular features of this episode is the way that it has been played out in the full glare of media publicity. The news first broke in the *Financial Times* and the *Wall Street Journal*, the leading *financial and business* daily papers in Europe and North America respectively. The media were used at the time as a means of propagating a positive impression of test-tube fusion, but can now be used as an archive to compare claims and promises with what actually occurred, and their daily reports, presented initially in a frozen timeframe, can be combined into a moving record to see if the story moves forward smoothly or if inconsistencies emerge.

The most extensive documented records of day to day activity are in the publications that originate in and around Salt Lake City, such as the *Salt Lake Tribune* and *Deseret News* and at the local television stations. Initially much of their reports were very positive towards test-tube fusion, and presented a picture of an external scientific community that was antagonistic to the local effort. However, by 1990, as it became increasingly clear that the original claims had been excessive, and as controversy grew around the administration of some of the affair, an air of open scepticism began to show.

The most positive reports were in an *ad hoc* monthly newsletter *Fusion Facts*, which claims to provide 'factual reports on cold fusion Developments' and emanates from the University of Utah Research Park (but in small print inside the newsletter notes that it is not affiliated with the University of Utah). It includes regular articles extolling the benefits that solid-state fusion will have for the automotive industry, for reducing environmental pollution and for corporate America in general. A casual reader, having access only to this as an information source, would have little reason to doubt that solid state fusion was not well on the way to industrial exploitation, and that there was any question about its general acceptance as a proven phenomenon.

If the Utah media were accurately reporting Fleischmann and Pons' perception of events then it is apparent that their view of what was going on is rather different from that of many other scientists. In the *Deseret News* Fleischmann reported that the scientific community had been lax in trying to meet with them, and had not made enough efforts to find out what had been done. Pons claimed that 'of the tens of thousands of people who got interested in the science, no scientist ever asked to see our original data.'

But these claims do not fit with the version that has been portrayed within these pages, and with statements that have emanated from Utah earlier in the year. Pons and Fleischmann repeatedly stated that they had provided sufficient details to duplicate the experiment. 'They just don't read the paper,' said Pons in an interview. Similarly University of Utah officials said '. . . their first paper gives all the necessary clues' (*Salt Lake Tribune*, 25 May and Fleischmann explicitly stated that they had revealed all details (*Erice* 12 April). If this is true then any failures to replicate cannot easily be blamed on laxness in seeking explanations. Indeed, some groups were trying hard to discuss their work with Pons and seeking details of his work.

For example, in the days immediately following the 23 March 1989 press conference, Caltech had set up cells and sought detailed characterisation of Fleischmann and Pons' experiment in order to duplicate it as precisely as possible.[5] A group at MIT also sought detailed information but were reduced to using indirect means, obtaining some original data and discussing them with the people who had actually made the measurements (Hoffman and Hawkins). From this they realised that the data were flawed but the $\gamma$-ray data from Fleischmann and Pons were shown in print only in late June 1989, and Garwin at Erice (12 April) had specifically remarked that it would be 'nice to see the whole spectrum.' Harwell had also gained detailed early advice from Fleischmann; when they pointed out that the $\gamma$-ray peak (as evidence for neutrons) shown during his 28 March talk could not be correct, he said he would have to contact Pons. When I asked about these data in 1990 I was referred to Pons. Salamon was told by Pons that he (Pons) was aware the peak was wrong—but there has never been any display of original data emanating from Pons that show where the $\gamma$-ray peak in the chemists' paper comes from.[6]

When some critical questions were asked of Fleischmann or Pons, clear answers were avoided in a variety of ways. Questions about light water controls received answers which led to no clear conclusion (if the statements by Pons and by Fleischmann in April and May are taken together, Chapter 16); as an extreme they were even refused ('I am not prepared to answer that question,' (page 151)).

By 1990 it was clear that the scientific community uniformly did not confirm Fleischmann and Pons' original claims for nuclear fusion. The pair's tactic at the 'First Annual Cold Fusion Conference' in March 1990 was to concentrate on the heat and electrochemistry, avoiding nuclear physics or suggesting that hitherto unknown and secret nuclear processes were at work.

As a result of experiments performed in other institutions, and flaws exposed in Fleischmann and Pons own experiments, the chemists had withdrawn most of their own 'evidence' for nuclear fusion, and retreated to the heat as their only signature. Yet Pons was reported to be critically accusing parts of the community for having spent millions of dollars doing 'poor science' who then 'hung it up and took the blame out on us'.[7]

This attempt to wish away their failures and put the blame elsewhere was symptomatic of a year during which the University had been happy for the media to present a positive image of solid based research, with exciting new developments, which then turned sour as the problems began to show up. Rather than accept that the problems were with the claims of Fleischmann and Pons, *a posteriori* reasoning was used over and over to 'explain' why other people were failing. As this position became increasingly difficult to defend so did the excuses become more extreme and pejorative, culminating in bizarre accusations against some scientists and parts of the scientific media.

When the scientific journal *Nature* criticised their methods, it was the journal that was deemed to be at fault by the chemists. In March 1990 Fleischmann and Pons asserted[8] that *Nature* had been responsible for much of the antagonism due to the journal's 'allegations that we had not carried out blank experiments before the publication of our preliminary note'; Fleischmann and Pons then pointed out that they *had* performed controls, specifically 'one blank experiment (sheet electrode at low current density) which gave an exact . . . balance.'

In evaluating Fleischmann and Pons' criticism of *Nature* it is instructive to read the specific 'allegation' that appeared in the journal.

On 27 April, in the issue that Fleischmann and Pons were accusing, *Nature* had criticised the lack of 'rudimentary controls of running electrolytic cells with *ordinary water rather than heavy water*' (my italics). And again referring specifically to the lack of *water* control it described it as a 'glaring lapse from accepted practice'. This, and only this, lack of control was criticised in that article. Yet, as we have seen, (pages 113, 147–8), Fleischmann and Pons had both given the impression the no $H_2O$ experiments had been made, no unequivocal 'yes' was given when asked even though by April 1990 their updated paper claimed differently.[9]

The scientific media, in the form of *Nature*, that spoke out against some of the claims emanating from the Fleischmann and Pons supporters was attacked in its turn. Meanwhile the general media was being used to propagate publicity for the chemists' claims that were less than fully substantiated.

Among the original stories fed to the media, to substantiate the claims of a well established phenomenon, was that Fleischmann and Pons had been experimenting with their electrolytic cells for some 5½ years. University of Utah officials repeatedly stated that Fleischmann and Pons had been perfecting their technique for that period of time (*see*, for example, the *Salt Lake Tribune* of 2 May and the *New York Times* of 3 May), and used this 'fact' to explain away the difficulty that many scientists were reporting in duplicating their work. The claim was that as Pons and Fleischmann had spent years on the procedure, how could someone effectively claim that the technique didn't work after only a couple of months of effort? But almost from the beginning, inconsistencies with this view were noticeable, clues that the work had not fully occupied that time, that the positive work had been at the last moment and that the programme was in its infancy.

Immediately after their 23 March announcement there were hints that the experiments were incomplete: scientists were questioning the lack of adequate neutron measurements data, and the lack of information on controls. The 'preliminary note', published in the *Journal of Electroanalytical Chemistry*, didn't help much either; the scanty amount of data (which were not amplified by a complete paper throughout the entire year) only added to many scientists' growing frustration and to a feeling that the preliminary note contained essentially all of the work, not merely a summary. In an interview, Brophy stated that Pons and Fleischmann had signed an agreement with Johnson Matthey 'a few months ago after they started getting positive results' (*Salt Lake Tribune*, 13 May). The idea and preliminary work may have taken place five and a half years before the press conference, but it was just one of many activities, as Pons' curriculum vitae showed, and only came into the front line late in 1988 as 'positive results' appeared.

The pace of discoveries after 23 March 1989 also raised doubts about the advertised history.

In their original announcement, Pons and Fleischmann had claimed that they had achieved a test-tube fusion reaction which was generating four times as much heat as was input, for a period of 100

333

hours. By 18 April, this figure had reached 8 to 1 (*Salt Lake Tribune*, 18, 19 and 20 April). Bursts of heat lasting as long as two days, and producing up to 50 times the input energy were reported in early May (*Salt Lake Tribune*, 9 and 10 May), and by mid-May, Pons and Fleischmann had observed bursts producing 100 times the input energy (*Salt Lake Tribune*, 21 May). The duration of a sustained reaction had been stretched to 800 hours by 13 May (*Salt Lake Tribune*, 13 May).[10] On the surface these press reports presented a picture of exciting progress encouraging belief that the phenomenon of solid state 'test-tube fusion' would be taken on board by the community of science. But many experienced scientists felt that if Fleischmann and Pons had been studying the phenomenon taking place in their cells for 5½ years, then they would have accumulated enough observations to have known what to expect. And yet, new observations were being made on a weekly basis![11]

The measurements of nuclear products, of neutrons and tritium, were vaguely reported and sparse. Neutrons were reported from one rod, tritium had been measured with another, and no attempt had been made to measure heat and products from a matrix of rods to determine how the 'unknown nuclear reactions' depended on volume, or whether or not they were in proportion to the heat—even though in absolute amounts they were billions of times too small.

The absence of neutrons and tritium in quantity made the nuclear thesis founder. Then Pons announced that $^4He$ was being seen in amounts that explained the heat. Given the small amount of observed tritium and neutrons, it was clear that the heat could only be from *dd* fusion if one rewrote the nuclear physics texts, and that the only remaining pathway was $^4He$ production. The fact that this route is orders of magnitude less that the neutron and tritium production in high energy fusion was no problem for believers, who were prepared to enhance the fusion processes by 55 orders of magnitude anyway. During five and a half years Fleischmann and Pons apparently had not bothered to look for $^4He$, yet within a few weeks *after* the press conference preliminary measurements had been made and were suddenly being presented as the key to the entire phenomenon.

They had no good evidence for nuclear products; their claims for fusion were based primarily on the heat. The *measured* amounts were rather modest compared to the *theoretical extrapolations* that made the headlines. Their main belief in the heat was based on prolonged bursts of power that totalled megajoules, seemingly in excess of what chemical processes would liberate. Fleischmann said at Erice[12] that he believed that such bursts would liberate 50 megajoules in 1000 hours, but they had 'not had time to do such an experiment'; yet they had found time to pursue 29 other joint projects and numerous independent ones.

Limited though their evidence was, Fleischmann and Pons were certain that they were correct. They knew of Jones' neutron data—the phenomenon was therefore there, and only the details remained to be defined. Early on they stated that they were looking forward to their peer's scrutiny (*Salt Lake Tribune*, 24 March), and that they encouraged verification attempts (*Salt Lake Tribune*, 28 March). On 27 March, Pons told *Tribune* reporters that he had heard that Los Alamos National Lab 'had repeated his experiments with success' (*Salt Lake Tribune*, 28 March). This was later denied (*Salt Lake Tribune*, 19 April). One article stated: 'So far there have been no confirmed reports that it has been duplicated, but the researchers still say they aren't surprised. Dr Pons has said the experiment is more complicated that the press reports made it out to be, and only recently have the technical details reached the scientific community.' (*Salt Lake Tribune*, 9 April). On 10 April, Texas A & M University announced that they had duplicated the results. 'A bustling, near breathless B. Stanley Pons could hardly contain his . . . excitement. "I'm getting happier by the minute . . . I always knew this would occur in time, without doubt". ' (*Salt Lake Tribune*, 11 April). Clearly, Pons and Fleischmann had expected confirmation. They were anticipating it. This state of mind is inconsistent with the notion that they had intentionally withheld essential information from the scientific community, and Fleischmann explicitly stated that they had not[13]; conversely it tends to undermine claims that failures to replicate the phenomenon were due to 'incorrect procedures'.

As more and more failures were reported, Pons and Fleischmann began to struggle to explain why. The 'extruded versus cast' palladium rod issue was such a case of speculation after the event: Pons claimed that only one in 15 or 16 cast rods were faulty, while 50–70% of extruded rods failed (*Wall Street Journal*, 25 April). In this article they openly acknowledged that at the time they wrote their paper they did not appreciate the cast versus extruded palladium issue. However, these figures alone couldn't explain the widespread failures at other major laboratories. Fleischmann, who was a long-time Harwell consultant, had advised the Harwell lab in its efforts at duplicating Pons and Fleischmann's results. A spokesman for the lab stated: 'Our understanding from Mr Fleischmann . . . is that we have the conditions dead right.' (*Wall Street Journal*, 27 April) When the Harwell lab gave up the effort after spending the equivalent of $500,000 and making 127 varieties of experiment, University of Utah officials were mystified. 'It's disappointing that they have not been able to do the experiment properly,' said James Brophy in an interview (*Salt Lake Tribune*, 16 June).

For a few days in mid-April 1989 Pons believed that he had identified the nuclear products, that there was $^4He$ produced commensurate with the heat. Had this been true the phenomenon would have been verified and theorists would have to rewrite the textbooks. Unfortunately the $^4He$ measurements in the gases were flawed. However the new choice was that the palladium rods were stuffed full of $^4He$—'that would be the ball game', as Brophy put it (*Salt Lake Tribune*, 21 May)—and so attention turned to analysing the rods. The University of Utah people were confident that it would be there, and the media were teased with promises that collaborations with unnamed laboratories who were analysing the electrodes would reveal this soon.

Although the analysis should not have taken very long, no results came out. The DOE panel were told 'soon'. At a meeting in October 1989, sponsored by the National Science Foundation and Electric Power Research Institute, it was said that the result had been made meaningless by mishandling of the rods. In February Fleischmann said to me that the results were 'ambiguous' as, apparently, rods that were not heat producers had been analysed for helium. At the First Annual Cold Fusion Conference in March 1990 talks by D. Thomson of Johnson Matthey and Nate Hoffman of Rockwell gave the impression that no helium had been found.[14]

So here again we see a fluidity in the evidence and confident assertions from Pons being undermined by subsequent events. All of this can only be consistent with the idea that Pons and Fleischmann had only a few months' worth of experience with whatever phenomenon was taking place in their cells, and that positive statements were being made based on incomplete data which were disproved—with less publicity—as more extensive tests were made.

As I researched this history it became clear that this was indeed the case, yet it was also very obvious, even though at the time of the 23 March 1989 press conference their experiments were rather limited, that they were convinced that they had evidence for nuclear fusion: they said that any other explanation was 'inconceivable'. Most other scientists did not agree with that, and during the year the evidence has been reduced rather than added to. How did this strong conviction arise and push them beyond the brink such that contrary evidence would be ignored or even changed in order to help substantiate the claims?

## The origins of belief

An important piece in the forming of their belief was the role of Jones, even though, as it subsequently transpired, his data had little relation to those of the chemists and have not yet been universally accepted as being significantly above background. We began with a question: How did an idea that could have crystallised anywhere, at any time in half a century, be delayed until the late 1980s and then emerge simultaneously at two points only a few miles apart? There is clearly more to this than meets the eye and it is in answering this that much of what subsequently transpired begins to make sense.

A facile answer would be that one group had the idea and the other, discovering this from the academic gossip column and near location, then took it for their own. I found no evidence for this. A second possibility is that the ideas were not new, and had happened before on several occasions but had died. Eventually these two groups learned of one another and the confidence gained from their mutual interaction brought each to fruition. There is some support for the notion that the ideas had been thought of several times before; in electrochemistry we knew of such attempts at least 60 years ago, and for fusion in the Earth we have the Indian experience, as well as claims from an Arizona geologist to have had the idea in the 1960s.[15] What distinguished the 1980 efforts were two unrelated early circumstances. For the BYU effort there was the serendipitous presence of Jones in Palmer's lab; Palmer had an idea and Jones was a specialist able to put it to the test in a custom-built experiment. For the chemists in their turn there was the mysterious disappearance of the palladium block that set them onto their course. The explanations of how the saga developed really begin with that.

The vaporised palladium block is a media favourite for which there is little evidence. The cynic might say that it encapsulates the whole story in that the claims are at the same time too extreme and also too modest. A four inch hole in a solid concrete floor (as claimed by Pons in the *Wall Street Journal*, 27 March 1989) is too much to be simply a chemical explosion of a few grammes of metal. Yet it is far too little if that metal had undergone a spontaneous nuclear fusion explosion. A micro H-bomb, for that in essence is what an uncontrolled fusion reaction is, should have been more damaging and irradiated the surroundings.

One suspects that if the incident occurred as early as 1985, then subsequent events have caused it to take on grander significance than it actually had nearer the time. If it had really been so dramatic and decisive so early on, then Fleischmann and Pons exhibited a remarkable lack of urgency. Pons' name appears on an astonishing number of papers, more than a hundred during 1985 to 1988, and none is essentially on test-tube fusion, which would have left him little time for serious study of that subject. Not until late 1988 is Hawkins brought in and the DOE grant application made. If the palladium vaporisation event was so singularly dramatic and in 1985, it seems surprising that they did not immediately drop all else, fearful that someone somewhere would have the idea and beat them to the prize, and concentrate on the fusion project instead of their 29 other joint projects between 1985 and 1988.

If the early work had been perfunctory and vaporisation of the palladium block had occurred in 1988 rather than 1985, the progression of events would be easier to understand. Statements made by Fleischmann or Pons in recorded talks that I have studied, or that have been made to me have tended to be rather less specific about this timing than some of the media statements attributed to them, and Pons' notebooks have not (as of July 1990) been made available, so it is hard to document the chronology of some of their early work. Jones' claims for priority are easier to document as there is his paper written with van Sieclen in 1985 to substantiate the origins of his work and notarised logbooks that he has exhibited publicly, even though his published experiments did not take place until 1989.

## The meeting

The two groups were doing rather different science, indeed it was that complementarity that helped fuel the profound emotional reactions. Both teams drew succour from each other. When they first met, each team had a *belief* that they had set out to prove, and also had some marginal evidence supporting that belief. However, each group perceived that the *other* had the crucial missing ingredient. For Fleischmann and Pons, the discovery of Jones proved disastrous; Jones, by contrast, probably gained a lot from it.

The following is one possible interpretation.

Before the end of 1988 Fleischmann and Pons had formed a belief that fusion should be taking place, releasing energy (heat) and nuclear radiation. They saw heat at levels which gave them hope and the palladium block vaporisation added to this. They had made no nuclear radiation tests worth speaking of before they met Jones (they had monitored the lab to ensure that there was no health risk and had thereby *limited* the amount of neutrons—what was good for their health was bad for their thesis). It was the meeting with Jones and his announcement that he was seeing low level neutrons that elevated the project into the front line.

For Fleischmann and Pons this is the key to Pandora's box. 'Voilà! There is indeed fusion. Jones shows it is there, we can do it,' one can imagine them thinking. Jones is also talking a lot about helium outgassing and emissions from Mount Mauna Loa, but they also perceive him to be interested in fusion as a source of the Earth's *heat*. They fear that if Jones so much as *mentions* heat their own patent dream could be put at risk. This fear that Jones might scoop the heat measurements forces them to a hasty conclusion on the nuclear measurements, and encourages them into obsessive security.[16]

For Jones the picture is almost a mirror image of the above. His motivation came from Palmer's thoughts about helium-3 outgassing; the idea of the Earth's *heat* being due to fusion was a secondary idea, but one that clearly excited him. He is a nuclear physicist and does not measure heat; he is seeking the neutrons and sees *hints* of them at a low level which may be real or may be background from cosmic rays or the surroundings. He needs a better detector in order to measure the energy spectrum of the neutrons before he can be sure, and then Fleischmann and Pons turn up, whose submission to the DOE announces that they *detect heat*. As Jones' neutrons were the key for Fleischmann and Pons, so is their heat the key for Jones.

Not until the last day of 1988 is Jones' energy-discriminating neutron spectrometer ready to make what he hopes will be the definitive measurements. In the early weeks of 1989 he accumulates enough of a signal to confirm to his satisfaction that (a) there *are* neutrons, and (b) the hints of neutrons already seen during the previous years were real. Thus although the data published by the BYU team were measured only in 1989, Jones can support the claim that he saw hints of neutrons long before.

So although both groups stress that they had been independently working on the problems for several years, when they first learned of each other *neither* had the 'definitive' data that they needed; for both groups these were not fully in place until 1989. Both groups were believers, and each had circumstantial evidence whose status would only become clear after subsequent events.

We can all think of examples in our own experience where a later event confirms earlier suspicions. In experimental particle physics this is very common: data are accumulated over many months and the first hints of a small enhancement in the number of events at a particular point on a histogram may signal the discovery of a hitherto unknown particle. The proof requires more data, and as these arrive the peak on the histogram grows definite; the new particle is indeed there—we were right all along. 'At which point did you make the discovery?' is then a question with no simple answer. Rarely does it matter. For Fleischmann Pons and Jones it helped to fuel a needless controversy.

Fleischmann and Pons believed that someone in Jones' collaboration had decided to push forward with solid state fusion research motivated by the knowledge of their work. Whether such opportunism occurred wittingly or unwittingly, and to what extent, may never be fully known, but in the intense passion of wishing to win what appeared to be a glittering prize it was easy for Fleischmann and Pons to imagine that they had been outmanoeuvred unethically: through the autumn until Christmas 1988 when Jones and Pons were communicating, Jones' group had no *definitive* neutron data; by 2 February 1989 they had marginal data that they were going to announce to the American Physical Society and had sent out an abstract into the public arena.

Hawkins and Fleischmann with Pons were by then trying to gather their own data with increasing desperation. Mistakes were made, which is hardly surprising in the circumstances, but the pressure also affected their judgement; the spectres of Jones, Rafelski and the Brigham Young University group increasingly appeared as demons.

Convinced that the nuclear phenomena were real, and fearing that they might lose all to their rivals, Fleischmann and Pons let the world know what they had done, even though a scientifically acceptable confirmation was not yet to hand. At that stage there was no reason to suppose that it would not be forthcoming, and they carefully labelled their paper a 'preliminary' note.

If there was indeed substance to their suspicions then it is hard to know how else they could have proceeded. Their preliminary note was dated 11 March; Jones' abstract was dated 2 February—so claiming priority for Fleischmann and Pons through these written announcements alone might be moot.[17] One view is that Jones' abstract was matched by the chemists' *preliminary* note and that the joint papers to *Nature* would establish the phenomenon 'officially'. Jones *et al.* saw it differently; for them the 'preliminary' note was a paper and as such violated the agreement to joint submission.

Fleischmann and Pons were so sure that subsequent experiments would prove them to be right that the press conference—held *after* the preliminary note was written—seemed a justifiable risk. What they did not anticipate was the intense reaction that their claims would arouse in the media and throughout science, and that serious flaws with the nuclear physics data would so quickly become apparent.

The effect of the meeting can be gauged by assessing what would likely have transpired had they worked independently through to a conclusion. Jones gained much whereas Fleischmann and Pons suffered.

Had they not met, Jones would have talked about his preliminary results at the APS meeting and published them. This might have interested a few specialists, some geologists may have been enthused by it, but the evidence for neutrons emitted by fusion was so borderline that it would hardly have generated much excitement in the physics community, let alone have thrust Jones into the limelight. As it turned out in reality, this will probably be the most cited paper that Jones and collaborators produce in their entire careers.

The latter remark may well apply to Fleischmann, Pons and Hawkins too. Certainly their paper was one of the most widely quoted in all of science in 1989—a point that 'Fusion Facts' was quick to seize on and draw comfort from, as if citation and accuracy go together. However, when one considers the wide criticisms of its quality, the number of errors, the withdrawal of much of its nuclear data and the rather general agreement that some of its methods were not scientifically sound, one sees that the uncritical appeal to a citation index as judgement of quality can be flawed. Given the increasing usage of such citation indices in decision making, sociologists of science and policy makers may wish to include this particular case in future studies.


## Heat

Had the chemists not met Jones, they would have continued studying the calorimetry, the heat, for another two to three years. They would have quantified its production characteristics and, with more careful investigation of nuclear radiation in consultation with experts in the university or with Harwell in collaboration, would probably have established that the heat had nothing to do with nuclear physics. Instead they might have discovered an interesting effect in the electrochemistry of palladium that they would announce to some acclaim, stimulating research into the solid state physics and chemistry of the extended electrolysis of heavy water with palladium cathodes. Though the hydrogen–palladium system is well studied by metallurgists and others, electrolysis over periods of months is a relatively unexplored area.

The poor quality of the nuclear measurements and the unsatisfactory way that the two chemists dealt with serious criticism has led many people to write off the whole experiment as an example of pathological science. And indeed it may turn out to be, but while I am certain that

341

there is no evidence for sustained fusion, nor has there ever been, the possibility that there is an interesting heat generation phenomenon may still be open. There are some careful experiments which report heat produced in bursts and which cannot be casually ignored. The questions of whether such bursts are produced within the cell, or by some external random influence, and whether they have any application, remain to be settled.

A pattern may be emerging from the worldwide data. There are some qualitative similarities present in experiments that produce heat and some of those that do not claim an effect. The differences appear to be quantitative and depend, in particular, on the group's assessment of intrinsic errors in the measurements. as follows.

Experiments often show a heat deficit in the early hours which then moves towards a positive excess. Whether or not you regard this excess as significant depends on how large you assess the intrinsic errors to be. The net effects tend to be of the order of a few percent 'excess' at best; I put 'excess' in quotes because the measured heat is usually *less* than the input but it is then assumed that power has been expended in electrolysing the heavy water and 'could be recovered'. This assumes that none of the heat has already been recovered by recombination and so counted twice. As Lewis put it at Baltimore: 'The cell is not a heat engine, it is a refrigerator; but a refrigerator that did not run as cold as expected.' Concerns like this caused Harwell to assess the intrinsic measurement errors of the experiments to be of the order of ten per cent, and so there was no measured heat imbalance for their cells. Fleischmann and Pons, Hutchinson at Oak Ridge and some others claim that they can run their experiments to an accuracy of better than one per cent; for them ten per cent heat excesses are 'real' and so they claim to see interesting heat bursts lasting several hours. However, some cells give nothing, and even those with bursts give nothing much of the time.

Whether there is an effect depends on how large the true errors are. This is an issue of theory and data analysis which the electrochemical community must come to an agreement upon, and a consensus can be expected in due course. Qualitatively there do seem to be bursts of heat whose origin has to be explained. The question is whether they are 'interesting', in the sense of being a genuine effect originating within the cell and instigated by the electrolysis of *heavy* water, or uninteresting, in the sense of being the result of incomplete bookkeeping in the heat ledger.

Even if they turn out to be the former, there is no reason to believe that one can extend the bursts. The wishful thinking that one can do so has to face the restriction that energy is conserved—you cannot get out more than is there. You have a finite bank account. If you draw out 90 per cent of your savings and spend them you can live like a millionaire for a few hours but not for longer, at least, not unless you *are* a millionaire. Can deuterated palladium live like a millionaire? It can, but only if it uses its nuclear account. However, there is no evidence that electrical currents can fund it.

## Fusion

*There is not, nor has there ever been, any evidence for nuclear radiation produced in the cell commensurate with the levels of heat claimed.*

There is no valid evidence for the production of helium, gamma rays, or neutrons, and the few places that claim to see tritium (though in amounts that are nugatory on the heat scale) tend to be in environments where it is easy to have tritium contamination. There is a 'Catch 22' situation: people who are experts with neutrons or tritium tend to work in places where there are plenty of sources for the material. Guarding against contamination when looking for low levels is easier when you are not in such an environment. The onus is on those experiments to demonstrate that the tritium is not from contamination, especially since the absence of neutrons shows that the tritium is not being produced energetically (i.e. as in nuclear reactions involving deuterium—recall that the energetic tritium will knock neutrons out from the local environment and into the detectors, yet none are seen). There have been rather persuasive indications that some of the most extreme claims for tritium were seriously flawed.[18]

The whole hype and the intense media interest came as a result of the flawed claims to have seen nuclear products. As the editor of *Nature* reminded readers: 'Even the journalists present [at the Utah press conference on 23 March 1989] would not have believed them were it not for their claims to have detected neutrons and gamma-rays, which claims were quickly shown to be insubstantial.'

In normal science, the withdrawal of evidence for the products of *dd* fusion would convince everyone that there was no such fusion. And for most scientists it did, though Pons and Fleischmann chose to insist that failure to see the products showed not that *they* were wrong, but, instead, that 60 years of nuclear physics must be wrong.

Recall how the nuclear story developed. Initially they claimed to see both neutrons and tritium in similar quantities. Soon some of the neutron evidence had been shown to be wrong, and the rest was being widely criticised. The tritium evidence was also widely criticised.

They were already aware that even had their tritium and neutron

343

data been valid, it would still not have accounted for more than a trifling amount of the 'excess heat' that they were claiming. This was the problem that anyone who believed in fusion had to explain: where is the heat coming from? As even these trifling amounts of nuclear products became suspect the message that it was not nuclear should have become clear. But already there was great pressure on the central parties in the affair to be proven to be right; there *had* to be fusion, that was the keyword. If it was not the conventional *dd* fusion, and this fact was already rudely apparent, then the saving of face demanded that there be some nuclear mechanism at work.

There are known routes whereby light nuclei, such as deuterons and lithium, can interact without producing neutrons and tritium. Some people seized on these, such as *dd* producing helium-4, or lithium-6 and deuterium fusing to make helium, and appealed to a miracle that would somehow switch off all the other possibilities as well as elevating these special pathways by a staggering 50 or more orders of magnitude (note, that is $10^{50}$ not merely 50). A quick experiment seemed to show helium being produced in the required amounts and Pons announced this on television, even though some control experiments that would have shown that the helium was from the air and not from fusion, as pointed out by Lewis, had not been done. The desperate need to find some evidence for fusion was recalled by in an editorial in *Nature*.

'The discovery of $^4He$ in the gases coming from a cold (test-tube) fusion cell was briefly trumpeted as a demonstration that deuterons . . . were indeed fusing but by a reaction that produced $^4He$ and a gamma-ray rather than tritium and proton or $^3He$ and a neutron as conventional nuclear physics would have it. Inconveniently, however, materials scientists pointed out that if helium had been formed in the palladium electrode it would have stayed there . . . not have evolved in the gases. But because Pons and Fleischmann had given out the suggestion in their usual teasing and informal way, it could be retracted without much pain on their part.'

With each failure to see fusion products commensurate with the heat claims, new ways of modifying the standard nuclear physics were proposed, but each in its turn ran into difficulties and was eliminated. At the year's end the favourite was that the fusion products all stay hidden within the palladium, only the heat being left as testimony. To support this Fleischmann suggested at the First Annual Fusion Conference that *dd* fusion occurred in metals at very low energy as evidenced by historical experiments performed by Philip Dee, with Rutherford's knowledge, in the 1930s. Not only did those images from Dee's experiment not prove *dd* fusion at *thermal* energies but the fusion products emerged clearly from the target, in violation of the 'secret fusion' advocated by the test-tube fusion supporters.

It is such manifest inconsistencies, self-contradictions and prevarications that have left much of the scientific community sceptical, at best, of Pons and Fleischmann's claims for nuclear fusion.

And when Salamon and a group of physicists, who had been invited into Pons' laboratory to measure radiation from cells there, reported that they saw no significant gamma radiation at all, a variety of contradictory suggestions were made:

(i) Salamon's team was looking at cells that were not heat producers; (ii) his team suppressed evidence of radiation[19] (which is inconsistent if indeed these were dead cells); and
(iii) one cell did produce heat in a burst but Salamon's detector was temporarily out of action.

Salamon's team then makes a clever experiment seeking *residual* radioactivity in their sodium iodide detector and, finding none, can eliminate radiation even during this heat burst.

In the twelve months from March 1989, Fleischmann and Pons had travelled far. Their problems had begun within days of the press conference and, if there were a single moment which signalled the impending doom, it was when someone at Fleischmann's talk in Harwell on 28 March pointed out that the gamma-ray peak, supposed evidence for neutrons from fusion, was wrongly placed. This brings us to another disturbing element of the episode: the 'evidence' for neutron capture revealed by the gamma ray at 2.5 MeV, the 'mobile peak' that went from 2.5 to 2.2 MeV by means that have never been satisfactorily explained.


## Unstable elements

It is a well known trait of human psychology that people can become so committed to a preconceived belief in something that contrary information is ignored or reinterpreted to fit with the 'facts'. In society one occasionally learns of cases where innocent people have been convicted because there was evidence that was ignored or 'explained away' because it did not fit with the preconceived notion.[20] The test-tube fusion episode contains examples of this.

Fleischmann and Pons believed in test-tube fusion and events conspired to reinforce that belief. The arrival of Jones added urgency, and also helped to confirm their belief.

345

They thought they had proof when Hoffman and Hawkins presented the gamma-ray measurements to them, and so they went public. When the proof was shown to be flawed—the day that the Harwell scientists told Fleischmann that the gamma peak at 2.5 MeV was at the wrong place to be evidence for neutron capture—they did not draw the obvious conclusion and withdraw, but decided that something must have been wrong with the calibration.

The data were flawed through admissible error which is regrettable but not uncommon, and is one of the reasons why peer review is so important: if scientific discoveries are to be the basis for industrial development, then society must retain confidence that it is being presented with goods that have had some quality control. The problem for Fleischmann and Pons was that this error only became public after the press conference, by which time the two chemists were public figures, exposed in a way that they had not foreseen, and under intense pressures. Their manuscript was about to be published in the *Journal of Electroanalytical Chemistry* and a version was also under consideration by *Nature* containing a potentially disastrous error which, if it appeared in public, would destroy a substantial piece of their evidence for fusion.

Fleischmann was absent in Europe, and Pons in Utah had to deal with this emergency involving data that had come primarily from Hoffman and Hawkins. If the dream of every scientist is that they will make a revolutionary discovery, then their nightmare is that the day after going public and becoming famous an unforseen flaw will destroy them. For very few does the dream come true; for Pons at the end of March the nightmare too was threatening, though he did have five days in the sun.

When it first became clear that their evidence was flawed they should either (i) have withdrawn their claim immediately or (ii) calibrated the detector against standard reference sources to see if the peak had been mispositioned, hopefully finding that its true position was indeed at 2.2 MeV.

Neither of these was done. Had the second option been pursued it would have shown that the peak was indeed not at 2.2 MeV, a fact subsequently admitted by its reappearance at 2.5 MeV in their 29 June *Nature* article (without comment about the original 2.2 MeV positioning). The result would have been that the excitement over test-tube fusion would probably have died within a few days.

Translating the peak to 2.2 MeV without making adequate checks, and in failing to mention in the paper how the energy scale calibration had been made, was a serious error of judgement; though with their belief so strong that their test-tubes were indeed producing fusion and its associated radiation, one may understand their predicament.

The peak was presented in the electrochemical journal as if demonstrably measured; there were no statements in the published paper to suggest any doubt, and no hint that the peak was mobile; it was apparently well researched evidence supporting their claims for *fusion*. Fleischmann, who was in Europe about to give a major speech at CERN, did not have access to the original data sets of Hoffman and Hawkins from which this isolated peak had been extracted, and so it would be difficult, at that time, to do more than take on trust that the relocation to 2.2 MeV could be substantiated.

Repositioning the peak at 2.2 MeV in place of 2.5 MeV was not enough due to the presence of an (immovable) nearby peak at 2.615 MeV (*see* pages 284, 285). Previously this had been immediately to the right of the signal peak when the latter had been at 2.5 MeV, but now that the signal peak had been translated to 2.2 MeV the intervening energy scale no longer fitted, and so this too had to be changed. To justify this a 'quadratic interpolation' was invoked in place of 'linear'. Nuclear physicists to whom sodium iodide detectors are standard equipment uniformly agreed that 'If it's not linear you throw it away'; Knoll's textbook[21] states this paradigm, namely that sodium iodide's 'extraordinary success' stems from its 'excellent linearity'. Unfortunately I have been unable to ascertain from any of the authors how this quadratic interpolation was arrived at.

Having investigated the origins of the peak further, it was decided that moving its position had been the wrong choice and an entirely new peak was offered in an erratum. This erratum also included Marvin Hawkins' name in the author list.

However, this peak had nothing to do with gamma rays. There is no sign of any such structure in the spectrum that appeared in *Nature* on 29 June, carrying the names of Fleischmann, Pons, Hawkins and also Hoffman, nor was there any mention of any signals at 2.2 MeV. The original peak at 2.5 MeV was restored as the purported signal, the first time that all but a few scientists had seen it. How the 2.5 MeV peak became translated to 2.2 MeV, and why it was subsequently replaced by another peak, also at 2.2 MeV but with the wrong shape to be a genuine gamma ray, is still a mystery.

The scientific community were largely unaware of all this but experimental physicists, at least, realised that the displayed gamma-ray peak could not be real; its shape gave it away, as the first paper from Petrasso's group showed.[22] Although it was clearly not real there was no statement possible then as to what it was. The use of the television spectrum was a bait by MIT; whether it came from Pons' lab or not, its effect was to make the chemists reveal their full spectrum for the first time and thereby enable MIT to expose the fallacies.

The community had to extract the details from Fleischmann and

347

Pons—the range of gamma peaks only appeared after Petrasso had raised the matter, the question of whether they had made control experiments with light water and with what result received a range of confusing and apparently inconsistent answers, the errors on the 'excess heat' were not quoted. As these details emerged, so were several of the claims shown to be insubstantial. Fleischmann and Pons explicitly disowned portions of their data only after others showed them to be flawed.

It was MIT's exposure of the gamma-ray problems and Caltech's demonstration that some of the extreme claims for power production were based on theory and not on direct measurement that began to undermine confidence in test-tube fusion, and it was Harwell's extensive experiments, following early advice from Fleischmann, that terminally convinced most people that test-tube fusion was a chimera.

In Britain the latter effectively killed off the subject in both scientific and public interest. In the USA, with the DOE enquiry ongoing and the protagonists in Utah seeking to maintain interest, the saga continued. In Utah, the university authorities were eager for the scientific status of test-tube fusion to be clarified and so Salamon, Wrenn, Bergeson and collaborators were allowed into Pons' lab to make measurements of radiation.

During five weeks in May and June 1989 they monitored the available cells and found no gamma rays, limiting specific nuclear channels to factors of a thousand to a million below that required for generating watts of heat. During this period dramatic claims were being made by Pons that heat bursts had occurred in at least one cell producing significant excesses over and above the input energy: unfortunately Salamon *et al.* did not access radiation from such cells during this period.

It is particularly sad that this group of scientists should be accused in a letter from C. Gary Triggs, a lawyer and personal friend of Pons, that they had selected data and predesigned the experiment to yield negative results. Offers of support for their legal defence poured in to the headquarters of the American Physical Society. The resulting media attention had a welcome outcome in that Triggs wrote a further letter apologising for 'misconceptions', and Fleischmann and Pons said that the issue would be left to 'the court of science'. But the episode can have done little to enhance the image of science and has raised anew issues about scientific freedom.

In parallel with this the National Cold Fusion Institute, and the University of Utah's role in its funding, were coming under increasing scrutiny. In turn the competence of the university president, Chase Petersen, who had played such a central role in propagating test-tube fusion, was questioned and he announced his resignation effective

July 1991. And accompanying these developments worries about the possibility of fraud in a corroborating experiment in Texas were being raised. The rapid sequence of regrettable events has brought the perception of test-tube fusion far from the brilliant heights that Utah hoped for in March 1989.

The year had begun with a story in business journals, the *Financial Times* and the *Wall Street Journal*, announcing 'Breakthrough in Quest for Fusion Energy'. Those wonderful days in late March 1989 were uplifting, a period that scientists and many others will long remember. But a little over a year later the *Wall Street Journal* was announcing 'Cold Fusion Scientists' Lawyer tells Skeptic to Retract or Face Suit' as news of the threat against Salamon and his colleagues broke.

This is not the way that science should proceed, and I hope that readers who have persisted this far will realise that this is not normal science.

The seeds for the disaster were sown when the first errors in measurement were pointed out to Fleischmann and Pons: by attacking when they should have retreated they set themselves on a dangerous course. They should have withdrawn their nuclear 'evidence' and insisted on the heat, which may have some scientific substance (though non-nuclear), and although this would have created embarrassment it would have left them better placed in the longer term. Instead, their strong belief that they had more evidence than in fact they had, led them to present data that had been obtained more by enthusiasm than by careful science, and to get into ever more convoluted explanations as the basis for test-tube fusion was questioned. Some of this may have been understandable early on, given the pressures at the time, but many months have now passed and the evidence for a nuclear phenomenon has been depleted rather than added to. There comes a point where one has to accept the message of the data—that absence of evidence is evidence of absence.

There is pressure on Fleischmann and Pons, and on those who supported and encouraged the adventure, to be justified in claiming fusion and, even though the details have to change, for the essence—*fusion*—to be true. They are the victims of their own excessive claims.

349

# EXCESS HEAT IN CALORIMETRY

If you input electrical power to a cell which is sealed, and no gases evolve, all of the power would go into heating the electrolyte.

When you electrolyse heavy water you evolve fuel—$D_2$ and $O_2$—and electrical power is used up in the process, leaving less to heat the liquid. This is what is happening.

Electric current provides negatively charged electrons, denoted $e^-$, which enter the electrolyte via the cathode. This disturbs the electrical forces in the $D_2O$ molecules and splits the molecules into $D_2$ gas and negatively charged ions, $OD^-$;

$$4D_2O + 4e^- = 2D_2 + 4OD^-$$

The negatively charged ions migrate to the positively charged anode where they return the electrons, forming $D_2O$ and $O_2$;

$$4OD^- = 2D_2O + O_2 + 4e^-$$

To see what is the net result of this electrical charge passing through the electrolyte substitute the second equation into the right hand side of the first one. The amount of electrical charge carried by four electrons has caused the splitting of two heavy water molecules:

$$2D_2O = 2D_2 + O_2$$

This has taken energy which is contained in the deuterium and oxygen gases. Electric current in amps is the flow of one coulomb of charge per second, so the more current that flows, the more evolution of gases there is, and the more power is used in producing them. For an amount of current $I$ amps, the power used in evolving deuterium and oxygen is $1.54 \times I$ watts. So if all the gases evolve, the net power entering the electrolyte and available for heating it will be given by the net power entering the cell *less* the $1.54 \times I$ watts which is exiting to the atmosphere in the gases. The result of this is that the cell will run *cooler* than the power you put in, the latter being $IV$

351

if $V$ is the potential energy difference of the anode and cathode in volts—the 'voltage of the cell'. The power available for heating is thus $I \times (V - 1.54)$ watts, which is known as the 'Joule heat'. An example is that it is possible to operate a cell such that power goes entirely into evolving gas and the cell temperature does not change.

These different quantities, namely total power input ($I \times V$) and Joule heat ($I \times (V - 1.54)$) are important in the interpretation of the numerical results given by Fleischmann et al.

The paper gave a table of 'excess' power in watts for a set of three rods (length 10 cm, diameters 0.1, 0.2 and 0.4 cm) operated at two different current densities (8 and 64 mA/cm$^2$) and a shorter rod (1.25 cm) for each of these diameters and operated at the higher current density of 512 mA/cm$^2$, making a total of nine entries in all.

It also quotes this excess power as a percentage of breakeven (where 'breakeven' of 100% implies that the 'thermal output equals the input'). The breakeven percentage was expressed in three forms :

(a) based on Joule heat, that is $I(V - 1.54)$, call this $A\%$;
(b) based on total energy, that is $IV$, call this $B\%$;
(c) based on a cell of 0.5 volts where energy in the fuel is recovered, call this $C\%$.

From these you can immediately deduce the voltage of the cell since from (a) and (b) you find

$$A(V - 1.54) = BV$$

and from (b) and (c) we have

$$BV = 0.5C$$

Whereas the excess power was being quoted to four significant figures, the percentages had been rounded off to integers, and this prevented sensible analysis in some cases.

As an example of the application of the above, the 0.4 cm dia. $\times 10$ cm rod at 64 mA/cm$^2$ current gave $A = 66\%$, $B = 45\%$, $C = 438\%$. $A$ and $B$ show that the cell voltage is 4.8 volts; hence if one could run a cell at 0.5 volts one would obtain the 438% return at $C$. This appears to be the origin of the claim to have measured 4 watts output for a single watt input. However, it only later became clear that the 0.5 volt data in column $C$ were theoretical projections, not measurements. Lewis at the Baltimore conference on 1 May 1989 even argued that it is thermodynamically impossible to operate a cell at such a voltage.[1]

As an example of the dangers of rounding off to integers there is the 0.1 cm diameter rod at 512 mA/cm$^2$ current. The results were quoted as $A=5\%$, $B=5\%$, $C=81\%$. It was only when Fleischmann and Pons' follow-up paper of March 1990 appeared that it was possible to reconstruct the actual numbers: $A=4.55\%$ (which had been rounded *up* to 5), $B=5.5\%$(which had been rounded *down* to 5). With these values it becomes clear that $C=82\%$ and the cell voltage is essentially 9 volts.

(In this latter paper it becomes clear that in comparing their watts and volts quoted, the inferred currents scale as the surface area of the *sides* of the rods. However, for the short rods of 1.25 cm, the ends contribute 4 to 16% of the total surface area as diameter goes from 1 to 4 mm.)

With a current of 0.2 amp and a voltage of 9 volts we can calculate what the actual heat inputs and outputs are for the cell. This gives a rather different perception of what is going on than the excess percentages give at first sight.

Total power input $(IV)=1.8$ W
Power into electrolysis $(1.54I)=0.31$ W
Input power as defined by F and P $(I(V-1.54))=1.8-0.31=1.49$ W
Input power measured by temperature rise $=1.57$ W
'Excess' defined by F and P $=1.57-1.49=0.08$ W
Fraction of total input power $=0.08/1.8=4.4\%$ 'excess'

However, return to the top line and start again:

Total input power $=1.8$ W
Input power measured by temperature rise $=1.57$ W
*Deficiency*$=1.57-1.8=-0.23$ W
'Expected' deficiency if all gases evolved $=-0.31$ W

So we see that the cell is actually *consuming* energy (net $-12.6\%$). However if all of the $D_2O$ has been converted into gases, which have not recombined, the 'expected' power loss would be $-0.31/1.8=-17.2\%$ and so relative to the *expected* loss we have a net 4.6% gain; it is in this sense that Fleischmann *et al.* claim 'excess' power.

This highlights one of the concerns: the magnitudes of the 'excess' are less or comparable to the amounts used up in evolving the gases. If gases are recombining, returning heat to the cell and heating it up without being accounted for, the 'excess' could disappear.

[1] G. M. Miskelly *et al.*, *Science* 246 (1989) p.793.

# FUSION DOES *NOT* GIVE THE EARTH'S HEAT!

Can fusion account for the $^3He$ production *and* the Earth's heat? The answer is that it may account for the former but only for at most ten parts per million of the heat. The original motivation of Palmer was to explain the $^3He$ production; the idea of heat production played a prominent role in the BYU–Arizona preprint but in the published version the emphasis had changed back to the $^3He$ production.

The following pieces of arithmetic are interesting in that they highlight significant *differences* from the motivation of Fleischmann and Pons. First, they deal entirely with fusion in sea-water—essentially *plain* water ($H_2O$ with one part in 6000 $D_2O$) *not* heavy water. Secondly they highlight that the *heat* of the Earth owes little if anything to fusion.[1]

*Helium-4 and helium-3 production*

Every second, averaged over the Earth's surface, 40 billion atoms of $^4He$ gas pour out per square metre. In the same time only some hundreds of thousands of atoms of $^3He$ emerge.[2] Small though this number is, it is nonetheless large compared with what would be expected from radioactivity.

All of the $^4He$ is produced by radioactive decay processes, alpha decays of thorium and uranium, and the energy released in these processes, supplemented by decays of potassium, tidal friction and primordial heat, can account for all of the Earth's heat.

The sums are as follows.

---

[1] Numbers can be obtrained from B. Mamyrin and I. Tolstcklin, *Helium Isotopes in Nature* (Elsevier, 1984), and L. Mchargue, P. Damon and H. P. Dart III, *Cold Nuclear Fusion, Helium Isotopes and Terrestrial Heat Production* (Department of Geoscience, University of Arizona, 1989).

[2] More than 50 000; less than 800 000.

354

The total heat output from the Earth is $4 \times 10^{13}$ watts and the surface area is around $5 \times 10^{14}$ m$^2$ so, as an order of magnitude, the output is about 80mW/m$^2$

The radioactive decays that liberate $^4He$ release between 5 and 6 MeV, or $10^{-12}$ joules (watts $\times$ sec) apiece. With $4 \times 10^{10}$ of these decays per square metre every second this alpha radioactivity alone accounts for 40 mW/m$^2$, namely half of the heat. Considering that there are other radionuclides liberating energy together with the other sources mentioned above, it is clear that there can be only a small amount of heat unexplained at most.

To geologists and geophysicists this is rather well accepted. If there is any puzzle it is the origin of the $^3He$. The radioactive decays of the elements that account for the $^4He$ and the heat give less than 1% of the $^3He$. A fusion rate of $10^{-23}$ fusions/deuteron in every second—at the limit where Jones *et al.* claim a signal—could account for the $^3He$ production arising from *pd* fusion in sea-water.

The key numbers are as follows.

There are $1.4 \times 10^{24}$ grams of water in the mantle reservoir.

One mole (18 g) of water has $6 \times 10^{23}$ atoms, in each of which there are two free hydrogen atoms (whose nuclei are protons). Hence there are some $10^{47}$ such protons in the waters. Now 1 part in 6000 of water is $D_2O$, so this yields around $10^{43}$ deuterons in the oceans in total, or $10^{28}$ 'beneath' each square metre of surface.

A rate of $10^{-23}$ *pd* fusions, $p + d = {}^3He + \gamma$(5.4 MeV), per second would therefore generate 100 000 atoms of helium—3 per square metre each second.

## $^3He$, fusion and heat

Suppose that these $^3He$ atoms originate in *pd* fusion.

Each individual fusion produces a single $^3He$ and liberates 5.4 MeV or about $10^{-12}$ joules (watts x seconds). So if there are $10^5$ such occurrences in each square centimetre per second this will give a net heat outflow of $10^{-7}$ watts/m$^2$ or one ten thousandth of a mW/m$^2$. So $^3He$ production by fusion at a rate of $10^{-23}$ accounts for the $^3He$ outgassing in a steady state in line with Palmer's idea, but would give only about one-millionth of the heat.

This is satisfactory because the Earth's heat is accounted for from well known mechanisms already; there is no need for fusion to explain it. So it is possible that, if Jones' experiments do reveal a *dd* fusion rate of $10^{-23}$, they could offer encouragement for $^3He$ production. However, it is also clear that this nuclear fusion has little to do with the Earth's heat source and that Fleischmann and Pons need have had no worries on this score.[3]

So, at the end, the Brigham Young collaboration and the University of Utah work were rather far apart. Fleischmann, Pons and Hawkins measure heat but see no nuclear products: Jones' team detect products which may be relevant for the production of $^3He$ in the Earth but have nothing to say about heat.

[3] Note that Jones *et al.* in their paper, *Nature 338* (27 April 1989), p.737, concentrate on helium production and make no mention of terrestrial heat.

# NOTES

CHAPTER 1
1. S. Koonin; APS meeting Baltimore, 1 May 1989.
2. J. Maddox, editor of *Nature*; G. Triggs' (lawyer for S. Pons) letter to M. Salamon, March 1990.
3. R. Feynman, in *What Do You Care What Other People Think?* R. P. Feynman and R. Leighton (Unwin Hyman, 1989).

CHAPTER 2
1. For the original story read F. Paneth and K. Peters, 'on the transformation of hydrogen into helium', *Berichteder Deutscher Chemischer Gesellschaft*, vol.59, 2039, (1926) and *in die Naturwissenschaften* vol.14, 956 (1926). Repetition of the experiments in the Baker Laboratory at Cornell University showed that they underestimated the effects of some of the errors (in *Berichted. Dt. Chem. Ges.* vol.60B, 808 (1927) co-authored with P. Guenther). Finally Paneth wrote a paper in *Nature* vol. 119, 706 (1927) in which he notes that glass does not give off helium in the presence of oxygen or in a vacuum but will do so when heated in a hydrogen atmosphere.
2. Published by Cambridge University Press.
3. Muon catalysed fusion, a form of 'cold fusion', is described in Chapter 4.
4. Test-tube fusion is but one type of attempt to achieve cold fusion, hence the insistence on 'test-tube fusion' in this book to distinguish it from the more general field of cold fusion.

CHAPTER 3
1. J. Pollock and S. Barraclough, *Proc. Roy. Soc. New South Wales* vol.39, p.131, 1905.
2. A cyclotron is a form of particle accelerator, a descendent of Cockroft's original accelerator.
3. A control experiment is one that should give a null or other explicitly predictable result if the physical processes at work have been correctly identified. Typically one repeats the original experiment with one crucial feature changed, in this case the deuterium replaced by hydrogen.
4. Kurchatov's talk in 1956 referred to neutrons produced in *linear* discharges and the Harwell people already knew that that was not the best route to pursue. (B. Pease to FC)
5. This is why scintillation counters would not suffice.
6. A problem with the accounting is that only a modest proportion of the fusion reactions occurring in JET and similar machines are 'thermonuclear'; the precise percentage is still being investigated.
7. No one in the fusion community seriously anticipates that *dd* fusion will generate a useful energy source; the *dt* reaction has a better reaction rate (cross-section)

and yields more energy per fusion, however it will make the machines radioactive. Hence to date everyone is studying *dd* to learn better about the general behaviour of the plasmas under these extreme conditions.

8. *New Scientist*, 29 October 1988, p.27.
9. *New Scientist*, 26 November 1988, p.19.
10. *R and D in Nuclear Power*, published by Her Majesty's Stationery Office, London; *New Scientist*, 11 February 1989.
11. *Nature* 336, p.3, 3 November 1988.
12. *Science*, April 1989.
13. Hunter's turbulent time in office ended with his resignation in November 1989; *Nature* 342, p.212.

CHAPTER 4

1. C. Frank, *Nature* 160, p.525, 18 October 1947.
2. C. Frank in *Bristol University Newsletter*, 29 June 1989.
3. A. D. Sakharov, *Passive Mesons*, Internal report of the Lebedev Institute 1948; communication from L. Ponomarev to FC. A modern review is that of L. Ponomarev, to be published in *Contemporary Physics*, 1990.
4. S. S. Gershtein, Lebedev institute thesis 1957; 'Nuclear reactions in hydrogen associated with muons'(in Russian).
5. As recounted by J. D. Jackson to FC, 6 July 1990.
6. L. Alvarez, Nobel address, 1968. The discovery paper was in *Physical Review* vol.105, p.1127, 1957.
7. *New York Times*, 29 December 1956.
8. JDJ to FC, July 1990.
9. J. D. Jackson, *Physical Review* 106, p.330, 1957.
10. V. P. Dzhelepov *et al. Soviet Physics JETP* 23, 820, 1966.
11. E. A. Vesmen, *JETP Letters* vol.5, 91, 1967.
12. Ponomarev's article in *Contemporary Physics* gives more details about this and other items.
13. LP to FC, April 1990.
14. This paper, published in *Journal of Physics*, G, 12, 213, 1986 suggests that piezonuclear fusion in the liquid metallic hydrogen core of *Jupiter* may account for that large planet's excess heat. It makes no statement about the Earth's heat (*see also* appendix).
15. The material in this section is based on S. Jones' personal memoranda; conversations with SJ and/or Paul Palmer in November 1989 and with SJ on various occasions in 1990.
16. Paul Palmer and Steven Jones to FC, November 1989.
17. Note the original idea comes from the observation of *helium-3* and then there follows the idea of *heat*. Contrast the work of Fleischmann and Pons, later, which sees *heat* but no copious nuclear products.
18. The Earth's heat alone would take two years to melt a sheet of ice 1 cm thick. Natural radioactive decays of uranium and other abundant elements account for at least 80 per cent of this heat, possibly even more.
19. S. Jones' personal memoranda; SJ to FC October 1989.
20. Quoted in *Business Week*, 10 April 1989.
21. In December 1988 the American Physical Society had invited Jones to speak about muon catalysed fusion at their May 1989 meeting. His abstract, mailed early in February, was the first hint that the organisers received that a new development would be discussed (J. Redish to FC).

CHAPTER 5
1. Conversation between MF and FC, 8 February 1990.
2. In Fleischmann's opinion the failure of some groups to see heat production in cold fusion cells is because they did not scale their apparatus suitably.
3. *New York Times*, 9 May 1989.
4. A fact which some supporters of solid state fusion take to heart.
5. H. B. Mark Jr and B. S. Pons, 'An in situ spectrophotometric method for observing the infra-red spectra of species at the electrode surface during electrolysis', *Analytical Chemistry* vol.38, p.119, 1966.
6. SP presented to US Congress 26 April 1989.
7. Fleischmann at Erice on 12 April 1989 made an insightful remark that the media missed: he said that they pursued the work for five years 'on and off'.
8. $H^+$ ions diffuse into the palladium until the load is about one hydrogen per palladium atom. The palladium atoms form a regular crystalline structure known as 'face-centred cubic' and the hydrogen ions take up sites among the palladium atoms. As this happens the palladium atoms are displaced slightly as the whole metal takes up the strain, expanding several per cent. It takes several hours to charge up the palladium fully, after which hydrogen gas begins to bubble off.
9. Chemical potential is essentially defined to be the change in energy per added particle if the volume and entropy (order-disorder in the system) stay constant.
10. It was clear to them very early that the number of neutrons was not commensurate with the heat. This created interest in lithium being a source and helium being the product. However, this overlooks that lithium-6 interacting with a deuteron will also produce neutrons in association with beryllium-7. In addition the presence of lithium-7 produces neutrons, and furthermore there is no evidence for helium either.
11. The explosive power from *dd* fusion in a typical palladium rod, if it all goes, is akin to that from over one tonne of TNT.
12. There are conflicting versions of when this event occurred. The Utah authorities have propagated the version that this took place early in the story, around 1985. I have adopted this in the main text but, as will become apparent, it raises some unanswered questions as to what filled in the intervening years. I asked Martin Fleischmann when this event took place and he alluded merely that it was 'some time ago'. In Erice on 12 April 1989 he said that they had started off with a variety of experiments which 'eventually finished up using massive pieces of palladium such as this cube . . . (and then) we used lower currents and palladium sheets'.

Their talks uniformly emphasise the singular impact that the explosion made on them; it appears in bold face in their paper yet a version of their DOE application written in 1988 makes no mention of it and describes only the palladium sheet; there is well documented evidence that they had a concentrated research programme since 1985 devoted to other projects and there has been persistent anecdotal suggestion on the electronic bulletin board that it occurred in September or October 1988 just prior to the student, Marvin Hawkins, being brought in to run the experiments with the rods.

I welcome input here from informed readers in order to help complete the record.
13. Which rather undermines some suggestions that Jones pirated aspects of the proposal.
14. B. S. Pons, bibliography up to 23 August 1988, presented to US House of Representatives Committee on Science and Technology, 26 April 1989.
15. Among these it is instructive to note his collaboration with Fleischmann. They had not jointly published until 1985 when they produced three papers. This grew to five in 1986; five in 1987 and sixteen in 1988. None of these bears crucially on the research into solid state fusion. Their early works concentrated on

359

microelectrodes and by 1988 their interests extended to mathematical modelling for systems of conducting polymers and studies of liquid and gas phase chromatography. This suggests a time-consuming programme (not to mention Pons' other papers) of which the fusion research was merely a part.

16. M. Hawkins to FC March 1990 and 22 June 1990. M. Fleischmann to FC February 1990: 'All of the experiments in the paper—I had either done those myself or working with Marvin Hawkins. Stan had done something, but at that stage, principally, most of those experiments had been done by me or/and Hawkins.'

CHAPTER 6

1. SJ to FC, November 1989.
2. *Science* 244, 28.
3. *The Scientist*, 1 May 1989.
4. I regularly review similar applications and in turn my own applications are reviewed anonymously by five or six scientists. Although there is no official 'policing', the DOE funding officers have a rather clear picture of what is going on and where; DOE officials annually visit the sites where they are supporting scientists and so have a continuing awareness of how the research programmes are progressing. In particular it would be quite natural for the DOE to encourage a collaboration that would improve efficiency or even to mount some level of enquiry if one or other of two funded groups claimed that their work was being pirated. The latter did not happen in the present case as Fleischmann and Pons eventually chose to accept funding from sources other than the DOE.
5. SJ letters to friends; SJ to FC March 1990.
6. MF to FC February 1990; M. Hawkins to FC April and June 1990.
7. Based on conversations with MF February 1990; MH April and June 1990; analysis of tapes and scripts of talks and papers by MF, SP and/or MH.
8. Plus an insignificant hint that the *background* radiation in the lab rose slightly.
9. Material based on conversations with R. Bullough and D. Williams December 1989 and February 1990; with MF February 1990; RB memoranda. The timetable and course of events incorporates material from S. Jones memoranda and conversations with him in December 1989; conversations with M. Hawkins and R. Hoffman and also reports in *Science* vol.244, p.28, p.420; *Science News* 135, p.196; *The Scientist* 1 May 1989 and *Business Week* 8 May.
10. At Erice on 12 April 1989 Fleischmann said that this was based on known properties of that type of detector and took account of the reduced efficiency due to the relative positioning of the detector and cell. They did *not* make a calibration of its efficiency by means of a standard neutron source of known intensity.
11. S. Jones memoranda on cold fusion history.
12. Hoffman was used to using high resolution solid state germanium detectors where the peaks are orders of magnitude *narrower* than with sodium-iodide that he had to use on this occasion. This led to some misidentification of the $^{208}Tl$ peak.
13. However the 2.5 MeV peak has nothing to do with neutrons, nor does the energy of the gamma-ray have anything to tell about the initial energy of the incident neutron. Any neutrons slowed down by collisions in the waterbath surrounding the cell are captured and emit a gamma ray whose energy is 2224 keV (about 2.2 MeV). The neutron combines with a proton in the water to form a deuteron whose mass is 2225 keV less than the combined masses of a free neutron and proton. It is this mass excess which transforms into energy ($E = mc^2$), 1keV going into kinetic energy of the deuteron and the remaining 2224 keV being emitted as a $\gamma$-ray that is the signal. But Pons did not learn this until too late, after the press conference; see Chapters 9 and 15 for more on this.
14. *Nature*, 29 June 1989.

15. CC to FC August 1989; MF to FC February 1990.
16. SJ to FC December 1989.
17. The paper sent by Fleischmann and Pons the next day was also received on 24 March, but as it was later withdrawn by them there was no dispute about priority for dates here.
18. S. Jones memoranda; reports in *Business Week* 8 May, *The Scientist* 1 May, KSL TV video tapes of news conference.
19. HB to FC, February and April 1990 vskip 0.1in

CHAPTER 7
1. R. Bullough to FC, December 1989.
2. FC conversations with Harwell scientists.
3. Reported in *The Scientist* 1 May 1989. S. Jones' notes record that this was at the suggestion of the DOE.
4. *See also* the DOE report in Chapter 14.
5. MF to FC February 1990.
6. Yet in 1990 Fleischmann and Pons wrote a paper in which they claimed to have made experiments with light water before 23 March 1989.
7. R. Bullough, Chief Scientist at Harwell, memo of meeting on 28 March 1989.
8. Report by R. Bullough, Chief Scientist at the UKAEA, March 1989.

CHAPTER 8
1. In January 1989; according to Pons' response to a question in Indiana on 4 April 1989, page 295. Two experiments with light water made before 23 March 1989 balanced and were cited in a paper by Fleischmann and Pons in 1990. It is not clear whether these experiments had been made before 11 March 1989 when their first paper was written.
2. Palladium heated to 850°C is porous to hydrogen but not to other gases and hence makes a good filter.
3. *Twistor*, published by William Morrow, New York 1989.
4. Seaborg won the prize for chemistry in 1951, the same year that the physics prize was shared by Cockroft.
5. This panel was chaired by physicist Norman Ramsey and chemist John Huizenga.
6. GS diary, GS to FC 22 June 1990.
7. To detect helium as a purported nuclear product, it would be crucial to melt the palladium rod. Results of such a test have been promised throughout 1989 but, a year later, still have not been disclosed.

CHAPTER 9
1. CM to FC 25 and 30 May 1989.
2. *Science* 244, p.284.
3. *Science* 244, p. 285.
4. The peak with 1000 events that Fleischmann now alluded to was eventually exhibited in the errata to their paper, and published in *Journal of Electroanalytical Chemistry*, vol. 263, p.187 – 188. However there was not simply a change of scale, for it also had a different shape (*see* Chapter 15), a fact that Fleischmann seemed unaware of when at Erice.
5. When the question of triatomic molecules was first raised there was an impression that they were negatively charged. As Eden and Lei's mass spectrometer could not detect negatively charged ions this seemed to be no problem. It was when they learned on the day after the press conference that the triatomics were *positively* charged that the problem became central.

361

6. The equation $E = mc^2$ relates mass and energy. The mass difference between a deuteron and the combined masses of a free proton and neutron is the 'binding energy' of the deuteron amounting to 2.224MeV, hence the release of this energy, carried off by the γ-ray, when the neutron is captured by a proton in the water bath.

7. Potassium-40 which is 1 part in 10 000 of all potassium.

8. COMPTON EDGE AND ESCAPE PEAKS

The gamma rays captured by the sodium iodide (*NaI*) detector deposit their energies which are measured and plotted as a spectrum. Suppose that we had a gamma ray of 2224 keV energy as is generated in the capture of a neutron by a proton. The naive expectation would be that this would give a single spike, S in the figure, with a slight spread in energy (about five per cent due to the intrinsic limit on the ability of the detector to resolve energy). In practice the signal is more complicated because the gamma ray may interact with electrons in the atoms of the *NaI* crystals and deposit only a part of its energy, the remainder then escaping from the crystal. This gives a characteristic autograph: instead of a single peak S there is a broad shoulder (C to CE) and two additional bumps (I and II). These are known as the Compton edge (CE) and the single and double escape peaks (I and II).

*The Compton Edge*

The gamma ray may scatter from an electron in the crystal (known as Compton scattering after the physicist Arthur Compton). The amount of energy given to the electron depends on the angle through which the gamma is deflected; the bigger the angle so the larger is the energy given to the electron. When the deflection is very small the electron gains almost no energy; the electron gains its maximum energy when the gamma ray is turned back through 180° at which point the electron gains 1990 keV energy (if the original gamma energy is 2224 keV).

An electron with 1990 keV energy or less will travel no more than about 3 mm before depositing its energy in the sodium iodide crystal; but in contrast the *scattered* gamma-ray may escape from the crystal and deposit nothing. Hence the energy spectrum due to these *electrons* gives a broad shoulder all the way from zero energy up to the maximum—the 'Compton edge' (CE).

*The Escape Peaks*

The electromagnetic fields within the crystal may cause the gamma to convert into an electron and a positron (the positively charged 'antiparticle' of the electron with identical energy $E = mc^2$ of 511 keV). This positron almost immediately annihilates with one of the electrons in the material, producing a *pair* of gamma rays, each carrying 511 keV (figs I,II). If both of these gamma rays are captured in the *NaI*, depositing their energies in addition to the electron's, then all of the 2224 keV energy is recorded and contributes to the spike S. But if either (I) or both (II) of the gamma rays escapes from the crystal then 511 or 1022 keV will be lost respectively. Thus there will be two peaks; the one due to single escape (I) occurring 511 keV below the spike S (hence at 1713) and the double escape (II) at 1202 keV.

Therefore what would have been a pristine spike S ends up in practice as a spike S, a Compton edge CE and two 'escape peaks' on a broad background, the essential features illustrated in the cartoon opposite. Compare this with the actual MIT spectrum (Figure 11) which shows the spike, edge and first escape peak. Now compare with the peak displayed by Fleischmann, Pons and Hawkins (Figure 11) which has no such features and as such cannot be a real gamma-ray induced peak.

362

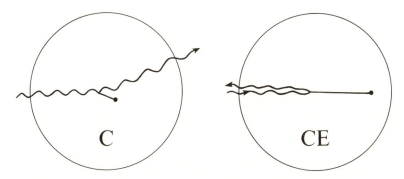

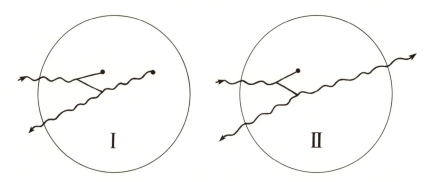

Figure 19. In C and CE the circle denotes the extent of the *NaI* crystal. The wiggly lines denote gamma rays, one entering from the left, scattering and exiting to the right. The solid line represents an electron that has been set in motion and deposits is energy in the crystal.

In I and II one or both gammas escape. The V represents the production of an electron (top line) and positron (bottom line), the latter being annihilated in the material, thereby producing two gamma rays.

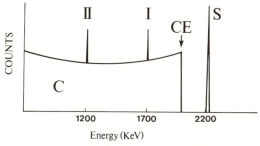

A single spike S at 224 keV is accompanied by other characteristic features at lower energies. Energy is plotted along the bottom axis; the vertical axis represents the intensity of the signal. This is a cartoon and is not an accurate portrayal of the real situation. *See* page 280 for a realistic spectrum and identify these features on it.

CHAPTER 10

1. Only members of the House Committee are allowed to question witnesses and hence questions that scientists present wanted to discuss could only be mentioned rhetorically in their testimony with the hope that a committee member might choose to take it up. One of the participants said to me that this made the session rather insipid as a scientific enquiry.

CHAPTER 11

1. D. Smith in *Caltech Engineering and Science Magazine* (ed. J. Dietrich) summer 1989 p.2–15.
2. My own experiences are similar. I have alluded to some of the problems in Chapter 5; there was also uncertainty as to whether surface area included the ends of the rods, and which figures were reliable given that comparisons of the columns A and B did not always match. I have seen at least three different sets of numbers extracted from these published figures. *See also* appendix, page 349.
3. *New Scientist* 6 May, p.29.

CHAPTER 12

1. Gai's experiment at Yale saw essentially no neutrons coming from the cells and placed limits on them factors ten to a hundred below the levels that Jones claimed to have seen. This suggested to many that Jones' 'signal' was possible due to a background cosmic ray contamination. Later Gai and Jones joined forces but were unable to agree on the conclusions of their mutual experiment.
2. Though at this time Hutchinson and also Scott's experiments began to give suggestions of anomalous behaviour, pages 231, 264.
3. Confirmed by University of Utah spokesman, *New York Times*; 3 May.
4. *New York Times*, 3 May.
5. *See also Science*, 15 June 1990 which has documented these suspicions. If these allegations are true then they are serious not least because testimony on the tritium was a piece of the 'evidence' that helped to convince the Utah legislature to fund the National Cold Fusion Institute with $5 000 000.
6. I asked Fleischmann about this in February 1990 and he was surprised and then suggested that the scale 'must be logarithmic'.
7. *Journal of Electroanalytical Chemistry*, vol.263, pp187–188.
8. Annealed and *not* pre-charged with deuterium. The other two annealed rods had been pre-charged and never gave excess heat.

CHAPTER 13

1. Fleischmann and Pons claimed in a paper in 1990 that they had performed experiments with light water before 23 March 1989. However, *see also* Chapter 16 for the development of this story.
2. Martone communication to ERAB-DOE panel.
3. Granted on 8 May 1949 to G. P. Thomson, a Nobel Laureate in Physics, his idea was not openly published until 1958.
4. *Science* 244, p.144.

CHAPTER 14

1. The diffusion rate of deuterium into palladium is such that a 1 mm diameter rod should be fully loaded in under twelve hours; Fleischmann recommended that one continue for four times as long in order to be safe, hence the times quoted in the text.

2. *Cold Fusion Research*, published November 1989 by US department of Energy, 1000 Independence Ave., Washington DC20585; Report number DOE/S-0073.
3. The technically interested reader may wish to read J. Sherfey and A. Brenner, *Journal of Electrochemical Society*, vol.105, p.655; (1958).
4. Fleischmann and Pons state in their 1990 paper that they took special care to ensure that the electrodes were always submerged.
5. For example, Pons' press conference on 24 April reported in the *Wall Street Journal* (25 April); and the history of plain water experiments described in Chapter 16.
6. M. Oliphant, P. Harteck and Rutherford (*sic*), *Nature*, 17 March 1934, p.413.
7. N. Packham, J. O'M. Bockris *et al. J. Electroanalytical Chem*, 270, 415 (1989).
8. C. Storms and E. Talcott.
9. P. Iyengar, 'Cold (test-tube) Fusion results at the Bhabha Atomic Research Center Experiments', Proceedings of 5th International Conference on Emerging Nuclear Systems, Karlsruhe, 3–6 July 1989.
10. *See also* Petrasso's question at The First Annual Cold Fusion Conference, page 321, which also applies to the Indian experiment.
11. *Science*, 15 June 1990, page 1299; *see also* contrary opinions in *Science*, 3 August 1990, page 463.

CHAPTER 15
1. According to Fleischmann at Erice 12 April, the count rate was on average two per hour background and six per hour signal.
2. MS to FC, April 1990.
3. Contrary to some people's impressions, this need not imply that a sodium iodide detector cannot detect the neutron peak. If there is an intense neutron source, the peak will show up clearly. The bismuth peak is actually a combination of two lines centred at 2117 and 2204 keV.
4. See Petrasso *et al.*, *Nature*, 29 June 1989 in particular references 1,3,5 therein.
5. Testing the detector with a neutron source, preferably before 23 March, would have confirmed if this were the case. Even though they were pressed for time by the presence of the BYU team, this measurement could have been completed within a few hours of obtaining a suitable reference source.
6. More precisely, an upper limit of 400 counts/sec, one-hundredth of the original claim.

CHAPTER 16
1. DW to FC, Asilomar, 12 October 1989.
2. *Nature*, 338 (1989) 616.
3. *Science*, 244 (1989) 285.
4. C. Walling to N. Lewis, May 1989; UU Newspaper, April 1989; C. Martin to FC, May 1990; SP to G. Seaborg, 14 April 1989; GS to FC, June 1990.
5. Notes taken by N. Lewis and C. Martin; to FC, May 1990.
6. *Wall Street Journal*, 27 April 1989.
7. In a report written in 1990 Fleischmann and Pons claim to have performed blank experiments with *LiOH* electrolyte in place of *LiOD*. Presumably these involved $H_2O$ in place of $D_2O$ as questions in seminars revealed that the electrolyte was formed by dissolving lithium in the water, hence $H_2O$ for *LiOH*, though the paper makes no explicit statement to this effect.
8. C. Martin to FC, May 1990.
9. As, for example, reinterpreting the 2.5 MeV $\gamma$ peak, Chapter 15.
10. The BYU-Arizona collaboration (Jones *et al.*) had distributed their own paper at the end of March into early April. Although their experiments had used heavy

water, their motivation had been the possibility that fusion occurs in sea-water, essentially *light* water, by means of *pd* fusion. There were some calculations in their paper suggesting that there was enough deuterium in plain water to generate the Earth's heat. This conclusion is quantitatively wrong (see page 352), but nonetheless may have added to the awareness of the possibility of fusion even in plain water.

11. My faxed copy was dated 18 April.
12. Fleischmann's responses to Garwin about the $\gamma$-peak scale being changed suggest that he did not appreciate that the peak also had been changed, thus providing more hints of their communications problems.
13. Though many orders of magnitude too small to explain the heat—a point which was continuously ignored by aficionados.
14. *Wall Street Journal*, 25 April 1989.
15. *Wall Street Journal*, 27 April 1989.
16. News Conference, 24 April 1989; *Time, Newsweek,* and *Business Week,* 8 May 1989.
17. N. Lewis to FC, 23 May 1990; C. Martin 25 and 30 May 1990.

CHAPTER 17
1. This '2.20' bismuth line is due to two juxtaposed lines at 2.119 and 2.204 MeV.
2. Salamon *et al.* were observing *four* cells nearly continuously, one of which was known to be a blank in that it had platinum not palladium as cathode.

CHAPTER 18
1. Diplon is the old word for deuteron, the nucleus of deuterium.
2. This statement attributed to Pons also asserts that the spectrum was a dummy borrowed for display purposes from another department on campus (reported comments in *Deseret News,* March 1990 and related disavowal of television spectrum in *Nature,* 29 June 1989).

However, there are similarities between the televised spectrum and that taken in his own laboratory, as displayed in *Nature* (vol.339, p.667 29 June 1989) and if the relation between these were better documented then it might be possible to understand better the status of the 'signal peak' claimed at 2.496 MeV by Fleischmann, Pons, Hawkins and Hoffman (FPHH).

If we ignore for a moment the vertical spike (V), we see in the TV, in FPHH and in MIT the characteristic sharp peak from potassium (number 3), a bump due to bismuth at 1.76 MeV, a lower broader bump (which also comes from bismuth around 2.2 MeV), then the hill from thallium (the extreme right end of the MIT spectrum, number 6). (These are compared on pages 168–9.) Notice that for MIT there is nothing prominent to the right of the thallium peak: this is the situation in the normal environment.

But the TV monitor shows a further structure, rising up to a pimple and with a shoulder on its right. The FPHH spectrum also shows such a structure (number 7 in Figure 10, page 165). In both FPHH and the TV spectrum the shape of this structure, its height and its width relative to thallium correspond. It has a shape and relation to the background that differ from genuine $\gamma$-rays as displayed and discussed earlier. It is an interesting coincidence that such a structure is present in the spectrum of Pons *et al.* and in a TV spectrum that is reported not to have originated in his laboratory. If this is the case and we are to assume that fusion cells are not standard lab equipment throughout the university, then the peak number 7 in Pons' spectrum, which he claims is the 'signal peak', can have nothing to do with fusion; its origin may be better understood if the measuring system that was employed in obtaining the TV spectrum is examined and compared with that used in FPHH—whatever features the two experiments have in common will be the likely source of the structures to the right of the point C.

The original TV reproduction in *Nature* shows the apparent 'signal peak' rather more clearly than the image in this book and interested researchers should examine that original. You will notice that the original has been cropped so that nothing is shown beyond the structure that is like the 'signal peak' in contrast to the more extensive image shown here (page 168) which shows more structures to the right hand end of the display, similar to the line drawings of spikes (labelled 8 and 9 in Figure 10) of the FPHH spectrum.

The spike V and the similar spike to the immediate right of the apparent 'signal peak' are cursors—indicators that are displayed on a visual display unit when you wish to highlight a region of interest in which the computer will then perform some numerical accounting. As these cursor marks are displayed on the screen but are not in the computer's memory bank of real data, when you ask the computer to print out a graph of the data it will not plot the cursors. That is why the peak V does not appear in FPHH's graphical plot.

In the original criticisms, the MIT group suggested that the cursor was the source of the structure that Fleischmann Pons and Hawkins had identified as their fusion signal. (The cursor position corresponds to 2.5 MeV which is where the chemists had originally placed their 'signal', a fact of which MIT were aware). However, unknown to MIT at that time, the Utah group had miscalibrated their energy scale such that they believed that the pimply shoulder (number 7) is at 2.496 MeV. A linear scale from the potassium peak (1.46 MeV) to their pimple (2.496 MeV) places the sharp line in the TV picture arising from the cursor at 2.2 MeV!

This is the only sighting I have been able to make of anything that could have been identified as a 2.2 MeV peak. However, FPHH stated in their paper of 29 June in *Nature*: 'We have now confirmed . . . that (this peak) is . . . a screen cursor on the . . . visual display unit' so that removes it as a possible source of their 2.2 MeV peaks displayed in their original papers. However, it is also true that in their full spectrum as exhibited in their 29 June paper there is *no* structure at 2.2 MeV and they do not even mention any such structure. The peak displayed in the erratum to their original paper in *Journal of Electroanalytical Chemistry* has totally disappeared; the peak originally presented at 2.2 MeV has returned to 2.5 MeV.

3. Letter of J. Rossi; to R. Petrasso, 5 April 1990.
4. Schwinger allowed 300 keV of the 1 MeV tritium energy to escape; at this lower energy the tritons need not eject so many neutrons and so the problem of the absence of secondary neutrons could be avoided.

CHAPTER 19

1. *Nature*, 29 March 1990, p.365.
2. MF in BBC TV Horizon/Nova, March 1990.
3. For example, S. Pons in *The Scientist*, 1 May 1989.
4. This refers to a talk given by Irving Langmuir, entitled 'Pathological Science', which was printed in *Physics Today* October 1989. Douglas Morrison has drawn analogies with the test-tube fusion episode in his article 'The Rise and Fall of Cold (test-tube) Fusion, *Physics World*, February 1990.
5. N. Lewis to FC, April 1990
6. F.P.H. *Journal of Electroanalytical Chemistry*, 261 (1989) 301 and 263 (1989) 187.
7. *Deseret News*, March 1990.

8. M. Fleischmann and S. Pons, University of Utah Chemistry Department Report (March 1989).

9. Their insistence that they had performed controls that verified their calorimetry, by having obtained null heat outputs with palladium or platinum electrodes and heavy water, has no direct relevance to the question of whether the heat is due to a *nuclear* process. It is this latter point that the light water control addresses. And it was their insistence that it was a *nuclear* process that had created the interest.

10. Note that Salamon *et al.* took radiation measurements during 9 May to 16 June, but were never told of other cells existing and found no radiation.

11. I am indebted to Patrick J. Smith for collating these news items and for use of his independent assessment that appeared on the electronic bulletin board in 1989.

12. MF 12 April 1989.

13. MF at Erice in answer to a question from M. Broer.

14. D. R. O. Morrison, *Cold Fusion News*, May 1990.

15. Letter from H. P. Dart to Steven Jones, April 1990.

16. MF interview with FC February 1990; and S. Pons in *The Scientist*, 1 May 1989 expressed concern that Jones enlisted Doug Bennison, an electrochemist, after meeting with them.

17. A critical discussion about abstracts and the 'anticipation' of forthcoming results is in Broad and Wade's *Betrayers of the Truth; Fraud and Deceit in Science* (Oxford University Press 1985).

18. G. Taubes, *Science*, 15 June 1990.

19. MF to FC, February 1990.

20. Future historians of science, or present-day believers in the phenomenon of test-tubes fusion, may decide that this present work is an example of a preconceived notion.

21. G. Knoll, *Radiation Detection and Measurement* (Wiley), *see* Chapter 10.

22. *Nature*, 18 May 1989.

# INDEX

airships, helium production 18
aluminium electrode tests 180
Alvarez, Luis 55–6
American Chemistry Society, cold fusion
    session 147, 154
American Physical Society meeting
    aftermath 222–3, 330
    Baltimore 1–2 May 161, 211–18, 299,
      309–10, 327
    Caltech preparation 171, 208–10
    Steven Jones 83, 86, 94
APS *see* American Physical Society
Argentina, fusion 34–7, 38, 43
AT&T Bell Laboratories 149
Atomic Energy Commission, USA 56
atoms, structure 18–19

Baliff, Joe 93
Ballinger, Ronald 171, 173, 188–90
Bangerter, Governor N. H. 11, 172, 253
BARC *see* Bhabha Atomic Research...
Barnes, Charlie 123, 171
    Caltech experiments 192–3, 196, 198,
      202–4, 206, 209
Bartel Institute 309
Beck, Doug 171, 192, 193, 205
Bennison, Doug 99, 108
Bergeson, Haven 103, 310, 348
Bewick, Alan 71, 73
$BF_3$ neutron counter 131–2, 152–5
Bhabha Atomic Research
    Centre 248–50, 307
Blackett, Patrick 52, 53, 55
Bland, Les 159
Bockris, J. O'M. 199, 226, 270, 311,
    322, 326
Bond, Peter 219, 246
    collaboration plea 216
    DOE Washington DC meeting 142–5
    negative results concern 157
    tritium information 306
Bose, S. K., helium supplies 23–4
Brigham Young University 10–11, 17,
    125
    Federal hearings testimony 173,
      180–4

fusion discovery dispute 85, 88, 90–4,
    99–100, 102–3, 337–40
muon catalysed fusion 32
natural helium sources 24–5
neutron detection 160
piezonuclear fusion 58, 60–9, 259
    *see also* Jones, Steven
Bristol University, UK 52
British Nuclear Fuels 50
Broer, Matthijs 149
Brookhaven Laboratory 306
Brophy, James 191
    erratum list query 229
    'four watts for one' statement 124
    fusion discovery dispute 99–100,
      102–4, 209
    fusion research 225, 333, 335–6
    legal action 315–16
    meeting arrangements 21
      March 276–7
    NCFI donation story 324
Brown, Norm 90
Bullough, Ron
    Harwell research 113, 234–5, 244,
      253, 293
    Utah co-operation 89, 92, 106–8, 110
Bush, George 152, 238, 251, 296

Cable News Network Television 163
Callis, Clayton 147
calorimetry
    Caltech experiments 203, 206–7, 214
    experimental methods 258–65
    heat excess 351–3
    Oak Ridge research 231
    test-tube fusion 80, 341–3
    Texas A & M experiments 156–7
Caltech 123, 161
    APS meeting 171, 258
    test-tube fusion research 163,
      192–210, 220, 259, 331
Cambridge Electric Power
      Company 161
Cameron, John 147, 159, 161
carbon electrodes 156

Central Electricity Generating Board (CEGB) 46, 237
CERN *see* European Particle Physics Centre
Chadwick, James, neutron discovery 21
Chatterjee, Dr Shyamadas 24, 26
CIT *see* Compact Ignition Tokomak
cloud chambers, neutron detection 42
Cockcroft, Sir John 34, 37–41
cold fusion *see* muon catalysed fusion; piezonuclear fusion; solid state fusion; test-tube . . .
commercial fusion 27, 29–30, 33
Compact Ignition Tokomak, USA 47–8
Compton edge 279, 280, 361–2
control experiments
  absence comments 128, 130–1, 291
  electrodes 156
  Utah research 147–9, 151, 261–2, 292–301, 331–2
Cook, Martin 24, 25
Cookson, Clive
  *Financial Times* scoop 101–2, 103, 121, 122–3
  news break 23 March 110
  twelve months summary *FT* 317
cosmic rays 52, 53, 61, 75
  background radiation 89–90, 113, 180–2, 197
  natural fusion 24, 25, 305
Craig, Harmon 67
Culham Fusion Laboratory, UK 43–4, 110, 234, 303, 308
Czirr, Bart 63, 65–6

Dahl, Paul 66
de Tar, Carlton 276
Decker, Daniel 180–3
Dee, Philip 318–20, 322–3, 344
Deltmann, Robert 172
*Deseret News* 172, 330–1
deuterium 22
  fusion rates 84, 198, 204
  light water reaction 147, 151
  muon catalysed fusion 56, 57–8
  natural fusion 61–3
  palladium effects 256–8
  test-tube fusion 9, 77–81
  tritium report 133–8
  *see also* deuterons: heavy water
deuterons 22
  fusion processes 27–9, 126, 265–7, 303–8, 311–12, 318–20, 328
  Harwell research 39–40, 42
Dobereiner Cigarette Lighter 138
DOE *see* US Department of Energy
Drexler, Kurt 311
Dzhelepov, V. P. 57, 58

Earth *see* natural fusion
economic considerations
  European research budget 45–6
  US research 46–9
Eden, Van 124, 133–8, 155–6
Electrochemical Society meeting 225–9, 295, 330
electrolysis
  calorimetry heat excess 351–3
  early fusion experiments 20–1, 22–3
  palladium heat effect 341–3
  piezonuclear fusion 64–6
  test-tube fusion 75–80, 84, 86–7, 258–9, 261–2
Ellison, Tim 159
Energia Nucleare e Energie Alternative (ENEA) 240–1
energy production
  calorimetry heat excess 351–3
  chemical reactions 126–7
  early experiments 20, 22–3
  fusion research 5, 8–9
  heat error causes 156–7, 158–9
  modern fusion 26–30
  test-tube fusion estimates 237, 333–4
Erice meeting, Sicily 149–51, 293, 295, 298
errors *see* experimental errors
Eureka programme, Europe 178
Europe
  JET programme 43–5
  research budgets 45–6
  test-tube fusion inquiry 234, 238–9, 250–1, 253
European Particle Physics Centre (CERN) 148, 150
experimental errors 260–3, 342

Fain, Sam 135, 136
Farrel, Ken 220
Fawcett, Ron 97, 99
Federal hearings *see* US House of Representatives Committee...
Fell, David H. 200
Ferguson, Archie 107
Feynman, Richard 13, 192–3
*Financial Times* 101, 121, 123, 317, 330, 349
First Annual Cold Fusion Conference 317, 320–3, 332, 336, 344
Fishburn, Dudley 237
fission energy production 8, 26
  *see also* energy production
Fleischmann, Martin 124–5, 327–9, 347
  background 70–2, 73–4
  calorimetry 258, 260–2, 342, 352–3
  Caltech criticisms 195–6, 204–9
  CERN talk 31 March 148

370

control experiments 147, 149, 151, 293–9
criticisms 218, 220–2, 224, 289–92, 301
Electrochemical Society 225–9
Erice workshop 12 April 149–51
Federal hearings testimony 174, 179, 208, 223, 255
fusion discovery dispute 67–8, 83–104, 111, 275, 336–41
fusion process theory 265, 267–8, 318, 320, 322–3, 344
fusion products absence 302–4, 309, 343–5
Harwell 106–9, 112–18, 128, 147, 148, 235–6
heat excess 317
legal action 315, 325, 348
media record 330–6
neutrons gamma-ray data 149–51, 278, 282, 287–8, 300–1, 309–10, 314, 345–7
palladium preparation question 216
palladium treatment 199–202, 219–20, 256, 258
test-tube fusion 25, 32–3, 43, 75–82, 255, 355
test-tube fusion discovery 7, 9–12, 17–18
tritium production 50
Fogle, Pam 100
Frank, Charles 52–5, 57
Frascati Laboratory, Italy 239–41
Furth, Harold 49, 148, 173, 187–8, 298
*Fusion Facts* 156, 330
fusion product, definition 44

Gai, Moshe 212
Gajewski, Ryszard 64, 85, 90, 99
gamma-rays
  Caltech experiments 195–6
  Compton edge 279, 280, 362
  fusion evidence 128, 267, 307–8, 343–5
  Harwell criticisms 113–15, 148, 149, 235–6
  MIT research 163–70, 227, 366–7
  Utah experiments 95–8, 128, 149–51
  Utah investigation 277–88, 309–14, 331, 346–9
Gandhi, Rajiv 244
Garching Laboratory, Germany 239
Garwin, Richard 150–1, 284, 287–8, 294, 331
Gaudreau, Marcel 162
General Motors experiments 219
Georgia Tech, Atlanta
  cold fusion claim 147
  neutrons report 131–3

retraction 143–4, 152, 154
germanium detector 280, 281, 288
Germany, helium production 18, 23
Gershtein, S. S. 55, 57, 58
Gittus, –, UKAEA 234
Goodman, Charles 158–9

Halsted, Beverley 118
Hansen, Eugene 11
Harrison, Kent 183–4
Harteck, – 267, 318
Harwell Laboratory, UK
  gamma-ray data 276, 282, 331
  research conclusions 223–4, 232, 253, 288, 335
  test-tube fusion research 101, 105–18, 198, 234–9
  Utah co-operation 89, 92, 95, 98–9
  Utah group criticisms 276, 293, 348
  ZETA project 34, 37–44
Hawkins, Marvin 222, 282
  gamma-ray data error 114, 275–6, 331
  heat measurements 86
  neutron gamma-ray spectra 167, 170, 279, 284, 287, 346–7
  test-tube fusion research 82, 103, 314, 338, 340–1, 356
heat
  calorimetry excess 351–3
  natural fusion 354–6
  palladium electrolysis 341–3
  *see also* calorimetry; energy production
heavy water
  calorimetry heat excess 351–3
  control experiments 292–9
  early experiments 22–3
  palladium electrolysis 342
  piezonuclear fusion 65
  test-tube fusion 84
  tritium content 306
Hecker, Sid 191
Hegelstein, John 323
helium 19, 22
  detection claims 202, 226, 227, 265–8
  early fusion experiments 18–23
  fusion evidence 303, 308–9, 312, 343–4
  natural sources 23–6, 61–3, 64, 339, 354–6
  nuclear reactions 27–9
  piezonuclear fusion 64–5
  Utah claims 140–2, 226–7, 334–6
Heydermann, P. 246
Highfield, Roger 111, 123
Hixson, Raymond 324
Hoffman, Bob
  gamma-ray data error 114, 276, 287–8, 331

Utah research   95, 98, 279–82, 314, 346–7
Hoffman, Nate   336
Hooper, Baroness   233, 237
hot fusion   234
  nuclear energy   29–30
  practicality   31–3, 84
  test-tube fusion comparison   84
  see also thermonuclear fusion
Huemmel Island, fusion reactor   36–7
Huggins, Prof. Robert   173, 184–6, 250, 255
Huizenga, John   251, 317
Hunter, Robert   47
Hutchinson, Don   127, 214–16, 230, 263–4, 302, 342
hydrogen   19, 22
  early fusion experiments   18–21
  metallic   77–8
  muonic atoms   52–5
  nuclear reactions   28
  test-tube fusion   76–8
  triatomic molecules issue   155–6

India
  favourable reports   133
  natural helium sources   23–4
  test-tube fusion   25–6, 242–50
  tritium detection   269, 305, 307, 311, 322
Indiana Cyclotron   159–60
Indiana, University of   146–8, 160–1
inertial fusion   32
International Centre for Scientific Culture, Erice   149
Iredale, Peter   107–8, 111, 234
iron electrode tests   180
Italy, cold fusion research   239–41
Iyengar, P.   246–7

Jackson, J. David   56, 57, 127–8
Japanese threat argument   177–8
Jensen, Gary   66
JET see Joint European Torus
Johnson, Bill   320
Joint European Torus   303
  fusion programme   43–5
  test-tube fusion   110, 234–5
Jones, Steven   10–11, 58–60
  APS meeting   213
  Caltech criticisms   203–4
  cold fusion differences   10–11
  cold fusion discovery   17–18, 327, 329, 337–41
  DOE referee   82
  Electrochemical Society   225
  fusion discovery dispute   10–11, 83–8, 90–4, 101–3, 108, 275, 292
  fusion evidence   303–4

muon catalysed fusion   25–6, 32–3, 55
natural fusion   354
neutron detector   160
Oak Ridge research   125, 126
piezonuclear fusion   60, 63–4, 67–8, 116, 236, 255, 267

Kavanagh, Ralph   196
Kellog, Stephen   196, 198
Khabarin, L.   61, 64
Koonin, Steven
  APS meeting   210, 211, 213–15, 301
  Caltech research   171, 192–6, 198–9, 204–6
  control experiments query   151, 294
  DOE panel   317
  Electrochemical Society   227
  fusion rates   130, 296
Kurchatov, Igor V.   38–9

Lamb, Prof. John   99
Landau, –   225
lanthanum electrode tests   180
Lawrence Livermore Laboratory   32, 137–8, 231
Lei, Wu   155–6
Lewis, Nathan   141, 171, 173–4, 344
  APS meeting   211–16, 218, 221, 342, 352
  Caltech research   192, 193–6, 198–203, 206–10
  control experiments   298–9
  Electrochemical Society   227–9
light atoms, structure   28
light water control experiments   147–9, 151, 292–9, 331, 332
  heat error discovery   156–7
  neutron detector flaw   153–4
lithium   28, 305
  Harwell experiments   289
  test-tube fusion   78–9, 344
Liu, Wei, tritium evidence   133–8
Lloyd, Hon. Marilyn   174–5
Lockhardt, Stan   162, 227
Loh, Eugene   277
Lomer, Mick   110, 234, 308
Los Alamos National Laboratory
  collaboration abandoned   190–1
  test-tube fusion research   231, 256, 269–70, 294–5, 297, 335
  Utah collaboration   190–1, 251, 309
Lozowski, Bill   159

Maddox, John   299, 301
Magaziner, Ira   12, 253
  Federal hearings   172–3, 177–9, 191
Maglic, Bogdan   50
magnetic chambers see tokomaks
Mahaffey, James   131–3, 152

Mamyrin, B.  61, 64
Mariscotti, Mario  36
Mark, Harry  72
Martin, Charles (Chuck)
  Caltech co-operation  194, 199, 202,
    207–8
  control experiments  156–299
  Federal hearings  129–31
  heat excess  129, 147, 258, 296
Mason, Prof. Grant  102
Massachusetts Institute of Technology
    see MIT
media reports  330–6
  APS meeting  212, 214–15, 218
  British news scoop  121–3
  cold fusion conference  317, 321
  control experiments  294, 301
  Federal funds hearings  172–91
  Indian research  245–7
  NCFI  324–5
  test-tube fusion  154, 157, 223–5, 251,
    270, 322, 326, 348
  test-tube fusion discovery  9, 10, 118
  Utah press conferences  100–3,
    110–11, 171
  ZETA project, Harwell  41, 42–3
meetings list  222
mesotrons  52, 53
metals, helium presence  61, 64
micro-hot fusion  302
Miley, George  186
military connection  233–4, 241–3, 246
Ministry of Defence, UK  241–2
Miskelly, Gordon  206–7
MIT  221, 301
  APS meeting  218, 227, 310
  cold fusion program  151, 161–71
  Federal hearings  173, 188–90
  Plasma Fusion Center  162
  test-tube fusion  161–71, 283–7
  Utah data  331, 347, 365–6
Morrison, Douglas  322–3
  BBC Horizon TV interview  329
  CERN newsletter  224
  Harwell concern report  224
  legal action report  315
  neutron emission criticism  213
  'pathological science' comment  254–5
  Utah University departments  325
muon catalysed fusion  32–3, 52–8, 125
  fusion rates  204–5
  Idaho experiments  59–60
  natural fusion  25–6, 61
Myerhof, Walter  219


National Cold Fusion Institute  314, 316,
  321, 324–6, 329, 348
National Engineering Laboratory  59–60

natural fusion  24–5, 61–4, 305, 337–9,
  354–6
  Sun  8, 33–4, 84
Nature  12, 275, 282, 286–7, 314–16
  control experiments  291, 298–302, 332
  fusion products  343–4, 346–7
  gamma-ray data  286–7, 302, 309, 314,
    321, 343
Nauenberg, Michael  130, 198–9, 296
NCFI see National Cold Fusion Institute
neutrons  21–2, 126–7
  $BF_3$ counter flaw  152–5
  Caltech research  152–4, 195–8, 204
  cosmic rays  180–2
  detection  39–40, 41–2, 158–9, 230–1
  detection equipment  131–2, 180, 182
  fusion evidence  6, 9, 301–6, 313, 334,
    343–5
  Georgia Tech report  131–3, 143–4
  Harwell criticisms  235–6
  Harwell experiments  107–9, 113–16
  Indian experiments  244–5, 248–50
  Italian experiments  240
  MIT measurements  283–4, 301
  piezonuclear fusion  66–9, 85, 339–40
  test-tube fusion  81, 265–8
  US inquiry  252-3
  Utah experiments  87, 89–92, 95,
    98–100, 276–88
  Utah gamma-ray data  149–51, 331,
    345–8, 366–7
  see also gamma-rays
New York Times, leader article  191
nickel  180, 255–6, 306
Nier, Al  67
nuclear energy  8, 23, 26–7
  fusion processes  27–9
  Indian policy  242–3
  Italian referendum  239–40
  UK research and development  46,
    237


Oak Ridge National Laboratory
  Federal hearings testimony  173, 188
  heat excess  230–1, 263–5, 342
  test-tube fusion  124, 126–7, 214–18,
    220
  tritium detection  306
Oliphant, –  267, 318, 322

palladium  126
  Caltech experiments  198–202
  casting effects claims  219–20, 289, 335
  control experiments  130, 180
  demand  6, 129, 234, 237
  deuterium effects  256–8
  early fusion research  19–21, 22
  electrolysis effect  342
  fusion evidence  312, 323

fusion processes 303, 307, 308
fusion rates 84
heating property 138
hydrogen affinity 9, 19
Indiana Cyclotron experiments 160
piezonuclear fusion 63, 65
test-tube fusion 5, 10, 76–82, 84, 86–7, 337–8
vaporised block event 137–8
Palmer, Paul 24–6, 60–9, 354
cold fusion discovery 17–18, 337
Paneth, Fritz 18–21, 22, 66
Parker, Ronald 48, 167
'pathological science' syndrome 148, 254–5, 329, 341
peer review, scientific work 34, 37
Penner, Reginald 194–6, 206
Perey, Francis 230
Peron, Juan, fusion reactor 34–7
Petek, Misha 264
Peters, Dennis 146
Peters, Kurt 18–21, 22, 66
Petersen, Chase 92–4, 102, 176, 208, 233, 325, 348
Petrasso, Richard 283–4, 286, 310, 314, 320–3, 347
gamma-ray criticism 166–7, 170
Phillips, Lee R. 63–4, 90
piezonuclear fusion 180
APS meeting 213
Brigham Young research 60–9, 83, 236
discovery dispute 83–6, 88, 91–4, 99–102, 338–9
Harwell experiments 116–17
see also solid state fusion
pinch effect, hot fusion 32
pions 52, 53, 54
plasma, hot fusion 32
platinum experiments 156, 292, 299
Ponomarev, Leonid I. 57–8
Pons, Stanley 72–4, 327–9, 349
ACS Dallas meeting 12 April 147–8
BF$_3$ detector 154–5
calorimetry 258, 260–2, 342, 351–3
Caltech criticisms 195–6, 203–9
control experiments 147–8, 292–301
criticisms 216, 218, 220–2, 224, 289–92
Electrochemical Society 225–9
Federal hearings testimony 174, 179, 208, 223, 255
fusion discovery dispute 67–8, 83–103, 106–8, 111, 114, 275, 336–41
fusion products 269–70, 302–4, 309, 343–5
Indiana talk 31 March 146–7
legal action 13, 315, 321, 325, 348

Los Alamos collaboration 251
media record 331–6
MIT criticism 320
neutrons, gamma-rays 160, 276–9, 282–3, 286–8, 300–1, 309–14, 345–8
palladium treatment 199–202, 219–20, 256, 258
radiationless fusion 308
test-tube fusion 25, 32–3, 75–82, 255, 355
test-tube fusion discovery 7, 9–12, 17–18, 43, 255
tritium production 50
Porter, Sir George 101
Powell, Cecil 52, 53
Preparata, Giuliano 323
Princeton Plasma Physics Lab 186, 187–8
Princeton University 32, 47–8
proton–deuteron fusion processes 148
protons
fusion process 8
nuclear structure 19, 21–2

Radiation from Radioactive Substances 21
radiationless fusion 308
Rafelski, Johann 63, 64, 67, 85, 340
Ramsey, Norman 251
Rather, Dan 10
reactors 44
Argentina 34–7
CIT program 47–8
JET programme 44–5
TFTR 47
ZETA programme 37–44
Redish, Edward (Jo) 210, 212, 327
Rees, Prof. Larry 66
Richter, – 35–7, 38, 43
Rochester University 32
Roe, Robert 174–5
Rose, Basil 41–2
Rossi, Hugo 322, 325
Rubbia, Carlo 240
Rutherford, Ernest 18, 21, 267, 270, 305, 318, 322, 344

Sailor, Michael 194–6, 200, 206
Sakharov, Andrei 38, 55
Salamon, Mike 102, 267
control criticism 293
data disputes 313–14
Nature article issue 315–16, 321, 325, 348–9
neutron counter evidence 277–8, 310–11, 344–5
spectra data problem 277–9, 331
Salt Lake Tribune 324, 330–1, 335–6
Saltmarsh, Michael 48–9, 173, 188, 217, 220

Santa Fe meeting  217, 220, 230–1, 258–9, 270
Santangelo, Pat  200–1
Savannah River Reactor  50
Scaramuzzi, Francesco  240–1
Schier, Walter  312
Schloh, Martin  162
Schoettler, John  238
Schwinger, Julian  321, 323
Scott, Charles  231, 263–4
Scraggs, Bud  12
Seaborg, Glenn  138–40, 152, 238, 251, 297
*The Secret of Huemmel Island*  36
Seki, Ryoichi  123, 196
Sieclen, Clinton van  60, 338
Simons, –  296, 312
Sinha, Bikash  24, 25–6, 245–7
Smith, Douglas  193
sodium iodide
  gamma-ray detection  310, 313
  neutrons  277–8, 280–1, 345, 347
solid state fusion  17, 25, 32–3
  *see also* piezonuclear fusion; test-tube fusion
Srivasanan, Dr  322
Srivastara, –  246
Stanford University  184, 259
Steenblick, Rick  132, 154–5
Stolper, Ed  201
strong nuclear force  29
Sun and fusion  8, 31, 33–4, 84
Sununu, John  223, 238
Sweet, Colin  45
Symko, Orest  277

Tanberg, John  20–1, 22–3, 64, 66, 78
Tank, Kurt, fusion reactor  35–7
tantalum electrode tests  180
*Tass*, report 12 April  133
Taubes, Gary  270, 324
Taylor, Craig  277
Taylor, Eric  41–2
Teller, Edward  38, 56
test-tube fusion  5–7, 9–13, 17–18
  APS meeting, Baltimore  211–14
  assessment  327–49
  calorimetry  258–65, 351–2
  Caltech experiments  192–210
  control experiments  130, 292–301
  discovery dispute  9–11, 83–104, 336–41
  Electrochemical Society  225–9
  events April/May 1989  122
  funding  11–12
  fusion process theory  318–23, 344
  fusion products  9, 265–70, 302–9, 343–5
  gamma-ray spectra  276–88, 300–1,

309–14, 345–8, 365–6
  Georgia Tech results  152–4
  Harwell research  105–18, 223–4, 232, 234–9
  international reactions  233–53
  media record  330–6
  Oak Ridge research  214–18, 220, 230–1, 263–5
  Presidential interest  138–40
  Texas A & M research  156–7
  tritium production  49–51
  US Congressional Hearings  48–9
  Utah group criticisms  289–92
  Utah University experiments  74–82
Texas A & M University  194, 202
  cold fusion claim  129–31, 147
  test-tube fusion  220, 225–6
  test-tube fusion research  255–6, 258–9, 269–70, 296–7, 299, 305, 311, 322, 326, 335, 348
  thermometer error  156–7
TFTR *see* Tokomak Fusion Test Reactor
thermometers, earthing error  156–7
thermonuclear fusion  5, 8, 37–45
  *see also* hot fusion
thermonuclear weapons  31, 38
  lithium  79
  metallic hydrogen  77
  test-tube fusion  6, 7, 159, 233–4, 246
  tritium  49–51, 305
Thomson, D.  336
Thomson, G. P.  38–9
Thonemann, Peter  38, 39, 40
Ting, Samuel  148
titanium evidence  180, 240, 245
Tokomak Fusion Test Reactor, USA  47
tokomaks  32
triatomic molecules issue  155
Triggs, C. Gary  315–16, 325, 348
tritium  22
  detection claims  203, 226, 228–9, 235, 265–70
  fusion evidence  302–9, 311–12, 334, 343
  Indian experiments  248, 322
  JET programme  45
  military connection  6, 49–51, 159, 234
  muon catalysed fusion  57–8, 59
  natural fusion  24, 62
  nuclear reactions  27–9
  Seattle evidence  133–8
  Texas A & M experiments  326
  TFTR programme  47
  triatomic hydrogen error  155
  Utah experiments  87, 235
Tufts, Bruce  203

UK Atomic Energy Authority  37, 46, 89, 234

375

United Kingdom
  JET programme  43–5
  research budgets  45–6
  test-tube fusion impact  233, 237–8,
    241–2, 253
  tritium production  50
  ZETA project  37–44
  see also Harwell Laboratory
United States of America
  hot fusion programme  46–9
  test-tube fusion impact  6–7
  test-tube fusion inquiry  208–9, 216,
    223, 238–9, 250–3
  tritium production  49–50
Urey, Harold  22
US Congressional Energy Research and
  Development Committee  48–9
US Department of Energy
  fusion research budget  47, 48
  Oak Ridge Laboratory  214–15, 230–2
  piezonuclear fusion  83, 94
  test-tube fusion  67, 82, 238, 250,
    259–63, 269, 288
  test-tube fusion investigation  139–40,
    142–5, 348
  test-tube fusion review  317–18, 336
  tritium production  49–50
  Utah funding  84–5, 89, 92, 99, 102
US House of Representatives Committee
  on Science and Technology  172–91,
  208–9, 216
USSR
  muon catalysed fusion  24, 25
  thermonuclear fusion  38–9
Utah State grant  11–12, 180, 182
Utah, University of  5, 7, 9, 11, 328
  cold fusion funding  171, 180, 250
  control experiments  292–9
  Federal hearings  172–91, 208–9
  First Annual Cold Fusion Conference
    317, 320–3
  fusion discovery dispute  67, 69,
    83–104, 333
  helium-4 evidence  140–2
  legal action  315–16
  natural fusion theories  25

NCFI links  324–6, 348
neutrons and gamma-rays  276–88,
  309–14
press conference 23 March  121
test-tube fusion  73, 82, 259, 269, 275,
  348
see also test-tube fusion

vaporised palladium block  137–8
Variable Energy Cyclotron Centre, India
  (VECC)  244–8
Vesman, E. A.  57, 58
Viyogi, Yogi  244, 246
Vogelaar, Bruce  198

Wadsworth, M.  268
Walker, –  48
Walker, Robert S.  174
Wall Street Journal  189, 330, 348
  article 24 March  124, 125
  cold fusion article 10 April  131
Walling, Cheves  140, 141, 296, 298–9,
  308, 312
Walton, Ernest  40
Wang, T. R.  198, 202
water see heavy water; light water
Watkins, Admiral  238, 251, 259
White House interest 13 April  138–40
Wilkie, Tom  303
Wilkinson, Sir Denys  128, 151, 293
Will, Dr  321–2
Williams, David  106–13, 116, 128, 224,
  234, 239, 249, 253
Williams, Greg  315
Wilner, Torsten  22–3
Wolf, Kevin  269–70, 314
Wood, Ted  41
Worledge, David  317
Wrenn, Ed  249–50, 310, 348
Wright, Pearce  225

Yudenich, V.  61, 64

Zeldovich, Yu. B.  55
Zero Energy Thermonuclear Assembly
  machine (ZETA)  37–44

376

539.764   Close, F. E.
CLO
          Too hot to handle.

$19.95

| DATE | | | |
|---|---|---|---|
| | | | |
| | | | |
| | | | |
| | | | |
| | | | |
| | | | |
| | | | |
| | | | |
| | | | |
| | | | |
| | | | |
| | | | |
| | | | |

Grant                          ✓

© THE BAKER & TAYLOR CO.